FORESTRY FOR THE NEXT DECADE

Managing Thrust Areas

ABOUT THE EDITOR

A.K. Bhattacharya is an Indian Forest Service Officer of 1983 batch from Madhya Pradesh cadre. He is presently an Associate Professor and Conservator of Forests in Indian Institute of Forest Management, Bhopal, an autonomous organization under the Ministry of Environment and Forests, Government of India.

Dr. Bhattacharya is Post Doctoral in Environmental Sciences and has contributed a large number of articles in national and international journals on a range of topics related to forestry and environment. He also has to his credit a book on *Kandhar Plane Hijack,* based on his own experience as a hostage.

His main contributions to the field of eco-tourism include organizing capacity building workshops and training courses for the local authorities of the various concerned departments, communities and NGOs, extending professional support to the State Government for promotion of eco-tourism, supervising research works on eco-tourism related topics, and preparing Eco-tourism Management Plans for few protected areas.

Forestry for the Next Decade

Managing Thrust Areas

VOLUME – ONE

Edited by
Ajoy Kumar Bhattacharya

CONCEPT PUBLISHING COMPANY, NEW DELHI-110059

Cataloging in Publication Data—DK
Courtesy: D.K. Agencies (P) Ltd. <docinfo@dkagencies.com>

Forestry for the next decade : managing thrust areas / edited by Ajoy Kumar Bhattacharya.
2 v. cm.
Contributed articles; chiefly on Madhya Pradesh, India.
Includes bibliographical references.
Includes index.
ISBN 8180694275 (set)
ISBN 8180694283 (pt. 1)
ISBN 8180694291 (pt. 2)

1. Agroforestry—India—Madhya Pradesh. 2. Agro-forestry—Nepal. 3. Ecotourism—India—Madhya Pradesh. 4. Forest management—India—Madhya Pradesh—Citizen participation. 5. Forest management—Nepal—Citizen participation. 6. Non-timber forest products—India—Madhya Pradesh— Management. 7. Forests and forestry, Cooperative—India—Madhya Pradesh. I. Bhattacharya, Ajoy Kumar, 1955-

DDC 634.909 543 22

ISBN: 81-8069-427-5 (HB) Set
ISBN: 81-8069-428-3 (HB) Part I
ISBN: 81-8069-429-1 (HB) Part II

First Published 2007

Published and Printed by
Ashok Kumar Mittal
Concept Publishing Company,
A/15-16, Commercial Block, Mohan Garden,
New Delhi-110059 (INDIA)

Phones : 25351460, 25351794
Fax : 091-11-25357103
Email : publishing@conceptpub.com

Foreword

Indian forestry is heading towards a transitional phase. The demand and supply gap of forest products is increasing, and as a result now people are more concerned about environment, and social pressures are pushing foresters to adopt a wider approach to forest management. Sustainable development of forests has become the most challenging job for the Indian foresters. It is time to scrutinize past policies and strategies and identify management options to meet the goal of 33 per cent forest cover in the country by 2012, as per the mandate fixed by the Ministry of Environment and Forests, Government of India.

The book in two volumes entitled *"Forestry for the Next Decade: Managing Thrust Areas"* contains articles focused on five major contemporary themes: Agroforestry; Ecotourism; Joint Forest Management; Non Timber Forest Products Management and Public People Private Partnership in Forestry. These are the areas of the present day forestry priorities, which need to be addressed to achieve the goal of the sustainable forest development. It is a significant endeavour to bring aforesaid issues to the forefront and identify their roles to achieve sustainability in forestry. I am sure this attempt will not only help Government officials, academicians, researchers and students involved in forestry research but would also encourage further research and emphasis in these areas.

I compliment Dr. Ajoy Kumar Bhattacharya, IFS, Conservator of Forests of Madhya Pradesh Forest Department for bringing out this book for the future discourse.

(G.K. Prasad)
Director General of Forests and Special Secretary
Government of India, Ministry of Environment and Forests

Preface

The management of our forests is moving towards sector wide approaches and focus is on supporting sustainable livelihoods of rural and tribal population with their active involvement on natural resource conservation and utilisation. Though we have developed many participatory policies and devolved authority to manage forests sustainably, still their roles on active forest management are largely undermined or often ignored. Likewise, most of the recent policies and programmes have placed more emphasis on improving forest cover outside forests. People generally account environmental goods received from the forests and often ignore environmental services and non-tangible benefits. Hence, emerging challenge for the foresters and forestry sector professionals is to look forests beyond its physical boundary. Now, the time has come to adopt more holistic approaches and strategies in our forest conservation and development endeavours.

Sustainable development of forests remains the most challenging task for the Indian foresters. We need to deliberate on "how can we integrate other themes or areas of forests, where lot of our professionals are working for its development, such as tree outside forests, ecotourism, joint forest management, conservation and management of NTFPs, public private partnership etc. With this book in two volumes, I am confident that most of our challenges on managing the forests for next decade can be addressed.

I appreciate the efforts of Dr. A. K. Bhattacharya, IFS, Conservator of Forests and Chief Executive Officer, Madhya Pradesh Ecotourism Development Board for bringing out this book *Forestry for the Next Decade: Managing Thrust Areas* (*in two volumes*). The book successfully integrates above emerging issues of forestry, based on his national and international experiences and supported by a series of his publications in peer reviewed scientific journals. This book is one of its kinds, which focuses on five major contemporary themes: Agroforestry, Ecotourism,

Joint Forest Management, Non-Timber Forest Products Management and Public People Private Partnership in Forestry. These are the areas of the present day forestry priorities, which need to be addressed to achieve the goal of the sustainable development.

I am confident that this attempt will not only help Government officials, academicians, researchers and students involved in forestry research but would also encourage further research and emphasis in these areas. I compliment Dr. Ajoy Kumar Bhattacharya, IFS and all the authors who have co-authored with Dr. Bhattacharya for bringing aforesaid issues and challenges to the forefront to achieve sustainability in forestry.

I strongly feel that this book will be quite useful to analyse past policies and strategies and identify management options to meet the goal of 33 per cent forest cover in the country by 2012, as per the mandate fixed by the Ministry of Environment and Forests, Government of India.

M. Hashim
Principal Chief Conservator of Forests
Government of Madhya Pradesh

Prologue

The forests are renewable resources that sustain and support millions of people. India with 1.8 per cent of global forest area is facing a strong challenge of meeting the needs of 16 per cent world's human and 17 per cent of cattle population. The natural forests of the country are confronting threat of destruction and degradation due to demographic pressure of humans and cattle, insufficient infrastructure, inadequate investments and inappropriate planning and management. It is high time to investigate the future approaches, strategies and management options, to prevent the further degradation and improve the forest cover.

The recent trends in forestry sector indicate the emergence of areas, which address issues related to the sustainable livelihood and promotion of participatory management approach for sustainable forest development. In the last decade, expansion of Agroforestry Programmes, Promotion of Ecotourism, Joint Forest Management (JFM) Programme, Management of Non-Timber Forest Products (NTFP) and Public People Private Partnership (PPPP) for the rehabilitation of degraded forests received attention of our policy makers and planners and emerge as key management options for the coming years. Adequate attention needs to be paid in these thrust areas to achieve the goal of sustainability.

Since ancient time, Indian farmers have practised agroforestry in many forms. In recent years, scope of agroforestry has increased considerably because of being an ecologically sound natural resources management system, which diversifies and sustains production for increased social, economic and environmental benefits. There is increasing evidence that biggest threat to our forests is attributable to ever increasing demand of fuel-wood, fodder and unsustainable agriculture. A time bound and need based agroforestry and farm forestry programmes are very crucial for meeting the sustenance demand of forest produces.

These days, Ecotourism is emerging as an important component of the Indian tourism industry and a tool for sustainable development of natural resources. Ecotourism is a responsible travel to natural areas that conserves the environment and sustains the well-being of the local people. It is considered as sustainable, equitable, community based endeavour for improving the living standards of indigenous host communities. Economic benefits to local community, generation of environment awareness among people and involvement of private investment for biodiversity conservation are some of the advantages of this novel concept.

The Joint Forest Management (JFM) concept is sharing of responsibilities, products, control and joint decision making authority over forest lands covered under JFM based on a formal agreement between Forest Department and local user groups. The main objective of JFM is to provide a visible role to the local communities in planning, management and protection of forests and to give them a share in the benefits from these forests. It recognises the livelihood and sustenance needs of the people through the principle of 'care and share'. JFM in India has emerged as a powerful tool for sustainable forestry. The further strengthening of this programme will prevent degradation of forest resources and even promote rehabilitation.

Sustainable management of NTFP, in particular the bamboo and medicinal plants, is essential in view of their importance in uplifting the socio-economic condition of the rural people and in earning foreign exchange for the country. Sustainable harvest, best management practices, *ex situ* cultivation, marketing linkages, certification are some of the major areas which require adequate management interventions over the next decade for the sustainable development of NTFP.

Partnership with Indian corporate sector with "Green face" to attract private investment for the afforestation of degraded forests/wastelands, will not only help in the rehabilitation of these lands but also in generating employment for the local people and raw material supply to the forest based industries. However, in India, forestry professionals and Government have yet to be convinced for increasing private sector involvement for the promotion of forest cover. The process of public-private partnership, the criteria and the implications have to be meticulously examined and the suitable models acceptable to all stakeholders be worked out.

In the above backdrop, this book contains various articles relevant to Agroforestry Systems, Ecotourism, Joint Forest Management, Non-Timber Forest Products Management and Public People Private Partnership in forestry. The papers are the outcome of the research works carried out by the author over the last one decade. Out of 55 papers included in the book, 23 papers have been published in national and international journals. The book is divided in five sections focusing aforesaid topics separately. I congratulate and compliment the authors and co-authors for their fruitful contribution to the world of forestry.

Bhopal **A.K. Bhattacharya**

Contents

VOLUME ONE

PART I: AGROFORESTRY

PART II: ECOTOURISM

VOLUME TWO

PART III: JOINT FOREST MANAGEMENT

List of Contributors

Arjjumend, Hasrat, Development Consultant, Grassroots (India) Trust, New Delhi.

Bala, Sanjay, Post Graduate Fellow, Indian Institute of Forest Management, Bhopal (M.P.), India.

Banerjee, Suchitra, Professor in Zoology, MLB Girls Autonomous College, Bhopal (M.P.), India.

Basnyat, Bijendra, Research Officer, NARMA Consultancy, Kathmandu, Nepal.

Bisariya, A.K., Conservator of Forests, Ujjain Forest Circle, Ujjain.

Boyle, Shobna, Post Graduate Fellow, Indian Institute of Forest Management, Bhopal (M.P.), India.

Chowdhary, Bharat, Post Graduate Fellow, Indian Institute of Forest Management, Bhopal (M.P.), India.

Dave, Ankur, Programme Officer, Vivada Inland Waterways Limited., Kolkata (West Bengal).

Dubey, Manoj K., Research Fellow, Wood Technology Research Centre, School of Forestry, University of Canterbury, Christchurch, New Zealand.

Dubey, Rajeshwari, Professor in Geography, Sarojini Naidu Autonomous PG College, Bhopal.

Dubey, Yogesh, Assistant Professor, Indian Institute of Forest Management, Bhopal (M.P.), India.

Gangopadhyay, P.B., Principal Chief Conservator of Forests—Wildlife, Madhya Pradesh (India).

Hansda, Regina, Research Associate, Indian Institute of Forest Management, Bhopal (M.P.), India.

Jha, Saurabh, Project Manager, Jungle Lodges & Resorts, Bangalore (Karnataka).

Jha, Vanya, Technical Officer, M.P. Ecotourism Development Board, Bhopal (M.P.), India.

Kalra, Shikha, Post Graduate Fellow, Indian Institute of Forest Management, Bhopal (M.P.), India.

Kujur, Samir Stephan, Programme Co-ordinator, Forest Development Agency, Madhya Pradesh Forest Department, Bhopal.

Kumar, Yogesh, Post Graduate Fellow, Indian Institute of Forest Management, Bhopal (M.P.), India.

Nahar, J. Muhammed, Camp Manager, Jungle Lodges and Resorts, Bangalore.

Pandit, Mamuni, Research Associate, Indian Institute of Forest Management, Bhopal (M.P.), India.

Patra, Krishna, Assistant Professor, Satya Sai Women College, Bhopal (M.P.), India.

Rao, S.N., General Manager (Plantations), ITC, Bhadrachalam (A.P.).

Saksena, Vandana, Research Scholar, Centre for Environmental Sciences, Jiwaji University, Gwalior (M.P.), India.

Sekhar, Kunal, Post Graduate Fellow, Indian Institute of Forest Management, Bhopal (M.P.), India.

Shanker, Thriveen, Project Manager, Jungle Lodges & Resorts, Bangalore, Karnataka, India.

Sharma, Kiran, Professor in Zoology, MVM College, Bhopal (M.P.), India.

Sharma, Ravi, Research Scholar, Forest Research Institute Deemed University, Dehradun (Uttaranchal), India.

Sharma, Rekha, Post Graduate Fellow, Indian Institute of Forest Management, Bhopal (M.P.), India.

Singh, R.K., Director, State Institute for Rural Development, Raipur, Chhattisgarh.

Sinha, V.K., Professor, Indian Institute of Forest Management, Bhopal.

Tiwari, Piyusha, Research Associate, Indian Institute of Forest Management, Bhopal (M.P.), India.

Vasudevan, Samita, Programme Officer, Yuva (NGO), Nagpur.

PART I

AGROFORESTRY

1

A Study of Agroforestry Systems in Madhya Pradesh, India

A.K. Bhattacharya, Bijendra Basnyat, Bharat Chowdhary and *Mamuni Pandit*

Introduction

As is well known, agroforestry is a generic term and the various forms, which it can take, can be classified into various groups based on the way trees are planted and the products that are harvested. The study has revealed that agroforestry in this region is still in its infancy. A farmer knows how to plant a tree but fails to identify the right mix of crops and trees. Awareness about management options too is less. This hasn't deterred innovative farmers but it has led to the widespread acceptance of a small number of tree species which have a proven successful track record.

The agroforestry systems in the study area can be classified as *Agrisilvicultural,* and all the agroforestry in the region is limited to this system. This was found to be the only type of practice currently in existence. Further grouping of the systems with respect to time led to the conclusion that all of the systems were of 'Simultaneous' types. Again in simultaneous systems, practices can be organized according to whether they are mixed or zonal. Largely, spatially zoned practices are being followed in this area with very few examples of spatially mixed practices. If classified on a socio-economic basis, the practices are of subsistence, intermediate as well as commercial type of practices.

The Systems

As mentioned above, trees with crops or Agrisilviculture were found to be far more popular with the farmers than any other practice. An account of the various tree-crop combinations found in the study area has been provided here.

Eucalyptus + Wheat + Soyabean

One of the most popular and common tree-crop combinations is Eucalyptus and wheat. Eucalyptus trees have been found in the form of boundary plantations and are mainly planted for commercial purposes.

Mango + Wheat + Soyabean

Mango, being an evergreen, perennial tree, is preferred a lot. Mango trees are found in either separate patches on croplands or in the form of rows between crop areas. Wheat and soyabean are cultivated in different seasons annually while fruiting in mango trees starts around the fifth year.

Gram + Wheat + Orange + Mango

Horticultural trees too are gaining popularity among farmers as fruiting starts early and returns are good. They are grown generally with wheat and gram.

Wheat + Khejari

Khejri, a small to medium sized thorny tree, is grown with wheat. Farmers are fully convinced that this tree species improves fertility beneath its canopy and has no competition with arable crop for soil moisture and nutrients.

Soyabean + Jamun + Aonla + Neem

Although this combination was not a popular one, it was found to be existing in a small area.

Eucalyptus + Wheat + Teak

This combination was found at some places and was one of the few successful ones.

Many more tree species like *Gmelina*, Shisham, Babul, Mahua, Arjun and Tamarind were seen with crops.

Major function of the woody component is for demarcation of boundaries and commercial wood production. *Agri-horti-silviculture* is more popular with farmhouse owners. *Citrus* sp and *Mangifera indica* are most commonly planted woody components. More technical interventions and inputs are provided during the plantation of woody component and less or no intensive care are taken after it is established. No silivulcultural operations as thinning, pruning are carried out although they plant timber species like Eucalyptus and Teak. Whenever, they prepare fields for the seasonal cultivation, management of woody components are also given due care.

The farmers cultivate wheat and soyabean along with woody component. The farmers are cultivating these crops on the basis of their traditional knowledge and skills. No improved varieties of wheat or soyabean are cultivated. Farmers use only bio-fertilizers and very few are applying the chemical fertilizers. Application of fertilizers and irrigation is almost nil for both of crops. Weeding is generally done for soyabean only. No intensive care is given for both the crops as they are planted.

The agroforestry is the traditional practice over India. Farmers are not calculating financial return from it. Whenever they were asked information related to cost and benefits, common answer to these questions are:

> "No sir, actually I'm just not able to recollect the exact amounts on a yearly basis, but yes, I can certainly tell you what I had to shell out at the time of planting these trees."
>
> "Sir, things are not so easy as you might think. Remembering how much fuel-wood, fruits or fodder these 100 trees produce and that too for the past 8 years is not an easy job. I'm sorry."
>
> "What economics! I'm just bothered about my costs. Hardly do I recognize any major financial gains of planting these trees"

However, attempt has been made to calculate benefit cost ratio (BCR) based on farmers' recall methods on the yield over last five years by different crops and woody component. BCR of different agroforestry model is summarized in Table 1.1 below. The BCR was highest in agri-horticulture system (fruit trees with soyabean) followed by agri-horti-silviculture system (mixture of trees, fruit tree and soyabean). The return from agri-silviculture model was low because farmers are not getting benefits from tree economically though it might have contributed ecologically by preventing erosion or increasing the productivity of soil.

Table 1.1: Cost and Benefit of different agroforestry models after five years of plantation per acre

(*In Rs.*)

Sl. No.	*Models*	*Cost*	*Benefit*	*BCR*
1.	Trees with cereal crops	1720	3140	1.8
2.	Trees with cash crops	2150	3800	1.8
3.	Fruit trees with cereal crops	1950	5200	2.7
4.	Fruit trees with cash crops	2180	6425	2.9

Source: Field Survey.

Some Practical Observations

A farmer who plants trees or other woody perennial plants instead of seasonal crops has a different set of commitments. Trees, however, are not a panacea for all ills. To achieve the full potential, a farmer must be able to carry out all those operations that control how an agroforestry system functions in a timely and skilful way. But in the study area it was found that farmers aren't really aware of any new technological or management systems. This can be one of the reasons why agroforestry is still struggling.

There are other reasons too. Poor land productivity, water shortage and lack of technological know how have been found to be the main reasons which prevent farmers from replacing the traditional ways of growing crops with agroforestry. In other words, richer is the environment, greater will be the choice of landuse practices that can successfully be undertaken.

Farmers associate risk with agroforestry and try to avoid the consequences of making a wrong decision. Farm size is an important issue here because farmers with access to only small areas of land protect their familyies' subsistence needs first and plant food and other preferred crops. Thus agroforestry options have been found to be more readily acceptable as the farm size increases.

Agroforestry, a means of minimizing risk, is actually associated with a lot of risks. Then at the same time, agroforestry is not covered by the extension services, and is never considered as a potential intervention. Moreover farmers are more interested in the tangible economic benefits from agroforestry and care little for the long term ecological benefits derived from incorporating trees on farmlands. Younger farmers are more inclined to plant crops, while tree planting is favoured by the older generation.

Conclusion

There is clearly much more to agroforestry than simply mixing a few trees and crops together and hoping for the best. Different tree-crop and their interaction with crops have to be understood, and this hasn't received any research in the area. There is a critical need to allocate scarce resources in the right way.

Exchange of information and ideas between all the groups involved is essential. We can see that agroforestry is really not a new discovery at all, even for scientists. Much of what we need to know in order to supply a scientific basis for advice on how to practise agroforestry successfully has already been studied to a degree in many other disciplines. Certainly, existing knowledge needs extending and elaborating, but, hopefully, progress can rapidly be achieved.

At the same time, we must temper enthusiasm about what agroforestry might achieve with a considered view of its limitations derived from a sound knowledge of how agroforestry systems function and what may constrain farmers who want to use it. The farmer is always the final arbiter.

2

Diversifying Livelihood Opportunities of Rural Poor Through Forestry

A Study of Multi-Layered Irrigated Plantation in Madhya Pradesh (India)

A. K. Bhattacharya and *Bijendra Basnyat*

Introduction

Realizing the role of local communities in the forest management, the National Forest Policy, 1988[1] envisaged that local communities should be motivated for development and protection of forests. Based on the broad framework and guidelines provided by it, the Joint Forest Management (JFM) Programmes were formulated and implemented in almost all States of India. After formulation of JFM resolution, all the forestry activities centred around the basic preamble of people's participation. Giving due recognition to the above policies and resolutions, Madhya Pradesh Forest Department (MPFD) has implemented different Programmes and projects in partnership with local community institutions for facilitating sustainable management and joint benefit sharing under the JFM. Some of the most promising projects include rehabilitation of degraded forests, sustainable employment through irrigated plantation, rehabilitating degraded Bamboo forests, other forest development activities and so forth. These projects focused on improving people's control over material resources and strengthening economic security through income generations, skills oriented training, new technologies development and marketing along with providing ancillary services as health, literacy programmes, legal educations and aids etc. Most of above projects and schemes related to forestry development are implemented through JFM committees. Field experiences and studies indicate that above approaches have made spectacular impact on forests specially in rehabilitation of degraded forest lands. This led MPFD to design and implement multi-layered irrigated plantation (MLP) scheme for improving the livelihoods of people below poverty line along with rehabilitating the degraded forests with active participation of village forest committees (VFCs). This scheme is currently implemented in 69 villages from 10 districts where 1670 families have been participating. The MLP is under its fourth year of implementation and has been gradually expanding to other degraded forests as well. Though MLP has followed very innovative and participatory approaches, there exists dearth of information on process documentation as well as its preliminary assessment for replicating the scheme in other areas. Since it will be too early to assess the impacts of the project within its short duration of implementation, this paper has critically analyzed its processes, approaches, strengths and weaknesses. Understanding these issues and learning lessons from the project provides valuable insights and experiences for replicating it elsewhere. The paper first provides salient features of the scheme and then assesses the strengths and weaknesses of the project on the basis of review of the programme document, interactions with the beneficiaries' families and forestry officials.

The Process

The Evolution of Concept

According to a report of MPFD, annually about 2 lakh ha of forests have been degrading due to excessive biotic pressure in Madhya Pradesh (MPFD, 2003)[2]. The major constraints and problems in rehabilitating the degraded forests include lack of funds and difficulties in systematic and scientific working on forestry management. While the MPFD is facing problems to rehabilitate the degraded forests, development planners have been looking ways to enhance employment opportunities in the rural and geographically disadvantageous locations as a result of increasing population. Having recognized the need to involve local communities in forestry development activities, MPFD had initiated several approaches (Table 2.1) but these approaches did not significantly contribute to improve people's control over material resources and strengthening economic security (Bhattacharya and Basnyat, 2003)[3].

Table 2.1: Involvement of the People in Forestry Development Activities

Sl. No.	Forest development activity	No of committees/ families involved	Area of plantation (ha)	Amount spent (Rs. crore)
1.	Rehabilitating degraded forests	3000 committees	1,00,000	27.9
2.	Sustainable employment through irrigated plantation	1670 families	1,670	10.0
3.	Rehabilitating degraded bamboo forests	56 committees	4,700	1.41
4.	Other Forest development activities	282 committees	2,050	21.57

Source: Bhattacharya and Basnyat, 2003.

As a result, the MPFD initiated MLP, the purpose of which is to improve productivity of the land and make it more productive, aesthetic and eco-friendly as well as contribute to the socio-economic upliftment of the rural people.

Specific objectives of MLP include sustainable employment to families living below poverty line, enhancing productivity of degraded forests, ensuring income to community, provide basic infrastructural subsistence facilities for sustainable development, promote competence building and skill development among the beneficiaries, ecological amelioration, transfer of technology on cultivation, processing and marketing of medicinal plants and essential grasses.

Thus the programme has dual objectives of rehabilitation of degraded lands and reduction of rural poverty simultaneously.

The Approach

The MLP adopted time bound programme for rehabilitating the degraded forests. For generating income and employment, it promotes scientific harvesting and processing of medicinal and aromatic plants. Initially, MPFD provided support during the early year of implementation whereas in later stages, responsibilities are to be gradually handed over to the beneficiary communities themselves. The role of MPFD will then be of motivator and facilitator.

The Method

Under this scheme, the poor and landless families in the villages are identified through participatory techniques such as wealth ranking and interactions with the local communities. The degraded forest areas are allocated to identified poor and landless families for assisting them to enhance and diversify the livelihood opportunities. Having provided irrigation facilitates to the degraded lands, multi-layer cropping are promoted to ensure optimum utilization of the land. Timber, Bamboo and Aonla plants in upper canopy and herbs, shrubs and tubers of medicinal or aromatic value in under storey are planted in the degraded forests. The MLP has promoted six different combinations of species, which include Aonla and Aromatic grass model, Aonla and Ashwangandha model, Bamboo and Ashwangadha model.

Bamboo and aromatic grass, Teak and *Safed Musli*, and Teak and *Ashwagandha* model. For timber and fruit bearing plants and aromatic grasses, MPFD assists to raise clonal saplings. Central Institute of Medicinal and Aromatic Plants provides seeds of medicinal plants to beneficiary communities as per their needs. The MPFD attempts to ensure that each of the products has well established market linkages either at local markets or through linkages with the industries or marketing agencies. MPFD provides technical skills and knowledge on cultivation practices, cultural operations, harvesting, processing and value addition of the products. The linkages and coordination with the knowledge institutions are established in delivering specific activities or programmes such as monitoring and evaluation by external agencies, namely State Forest Research Institute, Jabalpur, Tropical Forests Research Institute, Jabalpur or any other Non-Governmental Agency, supply of seeds by Central Institute of Medicinal and Aromatic Plants, Lucknow. The major activities under the MPL are as follows:

- Selection of Village and Formation of Village Forest Committees
- Establishment of Committees' Bank Account and Selection of Committees' Coordinator.
- Wealth Ranking
- Selection of Beneficiary Families
- Signing of Memorandum of Understanding (MOU)
- Training President and Coordinator
- Enumeration, Demarcation and Allocation of Forest Land
- Opening of Beneficiary Bank Account Capacity Development of Families on Plantation

- Field Preparation and Establishment of Irrigation Facilities
- Plantation
- Irrigation
- Fertilizer Application
- Weeding and Soil Working
- Insecticide Application
- Protection
- Harvesting

Management Interventions

With a view to analyse management system practised in the field, Table 2.2 summarizes technological interventions carried out by few Forest Divisions in managing MLP. The table shows that MLP has followed the high-technology plantation, by giving due care to management aspects such as selection of species, spacing, application of irrigation, fertilizer, pesticides etc. Though these were applied at regular intervals, variations were observed among different Forest Divisions based on site conditions, year of plantations and requirement of species. In all the studied Forest Divisions, the scheme was started during the 2002-03. The most common species promoted across include Aonla, Eucalyptus and Bamboo followed by medicinal and aromatic plants such as lemon grass, CNS grass etc. The irrigation was done in all the Divisions. Both organic and inorganic fertilizers were applied at early stages of plantation as well as during the soil working. The most commonly used organic fertilizers were farmyard manure, cowdung and compost where as inorganic fertilizer included Urea, DAP, Potash, NPK etc. Soil working was carried out in most of the divisions for providing early support for establishment of seedlings.

Specialities

MLP is innovative and participatory scheme that seems to address all the three key elements of sustainable livelihoods. These key elements are undermining pressure on natural resources, maintaining or enhancing assets and capabilities and recovering or maintaining from the shocks and stresses. The scheme follows result demonstration methods, which is said to be one of the best extension approaches for transforming new technologies. Some of the specialities of the MLP are as follows:

Optimum Utilization of Land: The MLP has promoted multi-layer plantation with timber, Bamboo and Aonla plants in upper canopy and herbs, shrubs and tubers in lower layer.

Table 2.2: Technological Interventions

Parameters	*Forest Divisions*				
	Chindwada			*North Seoni*	*South Betul*
	East	*West*	*South*		
Initiated Year	2002/03	2002/03	2002/03	2002/03	2002/03
Number of Demo plots	4	5	4	3	2
Area (ha)	65	75	40	30	80
Major Species	Aonla, Eucalyptus, Bamboo, Aloe Vera, Kalmegh	Bamboo, Lemon Grass	Aonla,Aloe Vera	Aonla, Lemon Grass	Bamboo
Number of seedlings planted	284600	3282000	100150	6762	24000
Irrigation (Total No)	73	95	71		24
Application of Fertilizer	Yes	Yes	Yes	Yes	Yes
Chemical fertilizer	Urea, DAP, Potash, Themate, Phosphorus	DAP, Urea, Phosphate, NPK	Grow Mix RPI	DAP, Urea, Phosphate, NPK	Urea, NPK
Organic fertilizer		Farm yard manure	–	Farm yard manure	Cowdung
Soil working/treatment (Number)	37	57	42	12	6
Any other inputs	Neem Cake, Indane, Copper oxide, Chloride etc.		Farm yard manure, Non-Lindane	Ti-termite Pesticide, Lindane, Monochrotophas, BSC Powder	
Expense (Rs. lakh)	36.67	48.26	10.76	18.50	5.04

Reinforce Existing Legislations and Policy: This programme has reinforced on the government's commitments towards participatory forestry. It has given due recognition to principle and preamble of forest policy and MP JFM resolution, which intends to empower forest dwelling community both socially and economically.

Addressing the Poverty: The programme identifies poor and landless communities as its target group through wealth ranking exercise. This has created direct access and opportunities to poor and landless communities in diversifying their livelihoods through sustained local employment opportunities and sharing of revenue and benefits from forestry development activities.

Use of Modern Technologies and Inputs: The programme uses better clonal and planting materials for rehabilitating degraded forests. Modern inputs such as chemical fertilizer, pesticides and irrigation are used at regular intervals to obtain better results or performance of forestry crops.

Dissemination of Technologies: The Programme promotes six different combinations of woody and non-woody species types for the economic benefits. It has adopted result demonstration to show impacts of the scheme. Beneficiaries are trained on cultivation, collection, processing, value addition and marketing of medicinal plants and essential grasses. The skills and capacities are enhanced for better management and then the management responsibilities are handed gradually over to the beneficiaries themselves.

Partnership with Local Community/Improvised Community Forestry Concept: The programme is implemented in partnership with the local communities. MPFD provides modern technologies and inputs where as local communities provide labour contribution as well as watch and ward of the forest area. The revenue is shared as per the contribution, i.e. 30 per cent of voluntary labour inputs are provided by farmers and hence they are entitled for 30 per cent share in final harvest. This is the improvised way of community forestry in sharing the cost as well as benefits from management of the forests.

Coordinated and Collaborative Linkages with Stakeholders: The Programme is being implemented in coordination and collaboration with different stakeholders with defined roles and responsibilities. The roles of higher level to lower level management staff along with communities has been clearly spelt out. Apart from this, role of different institutions in managing MLP has been identified with their key support area or their potential stake in managing the MLP.

An Innovative and Participatory Benefit Sharing Mechanism: The MLP has an innovative way of benefit sharing arrangement. Communities are provided with free Nistar and 100 per cent share from minor forest products and intermediate harvest where as 30 per cent share from final harvest of timber and Bamboo. This corresponds to their voluntary labour contribution to protecting the forest.

Employment for Poor: Families living below poverty line whether land holders or landless get employment under the project. Landless families having minimum income and at the bottom of the other families are considered in ascending order based on the results of the wealth-ranking list. Thus the poor and vulnerable groups have direct access to the project benefits specially on local employment.

Immediate Return: Compared to other schemes, MLP has very short gestation period. The local community starts getting benefits from the second year of plantation due to multi-storey management and hence the farmers can enjoy the benefits from the second year of their work. Hence, it is more likely to empower local community as well as on the sustainability of the project due to immediate return.

Economics of MLP

As discussed earlier, MLP has promoted six different models on multiplayer plantation, which primarily include the multi-storey crop management. The plantations are completed in two years and start yielding from thc lower storey crops. After two years, maintenance of the plantations are done through the income accrued from cultivation of medicinal/ aromatic plants and other silvicultural operations.

Aonla (Emblica Officinalis) and Aromatic Grass Model: Aonla starts giving benefits from 5th year whereas the aromatic grasses fetch profits from second year onwards. The net profit in the second year is Rs. 25,000 whereas the profits accruing in third year onwards is Rs. 50,000 ha.

Aonla and Ashwangandha (Withania Somnifera) Model: The benefits from Aonla would be same as above. From third year onwards, committee will get about Rs. 24,000 on a sustainable basis from Ashwagandha.

Bamboo and Ashwangadha Model: There will be no profit from Bamboo up to four years. But Ashwagandha will provide profit of about Rs. 24,000 from the second years onwards.

Bamboo and Aromatic Grass (Lemon Grass) Model: There will be no profit from Bamboo up to 4 years. But lemon grass will fetch approximately Rs. 50,000 in third and subsequent years. This ensures sustainability of employment.

Teak (Tectona grandis) and Safed Musli (Chlorophytum spp.*) Model*: There will be no profit from Teak up to five years but Safed Musli will fetch approximately Rs. 25,000 from third onwards.

Teak and Ashwagandha Model: There will be no profit from Teak up to five years but Ashwangandha will fetch profit of about Rs. 24,000 from third year onwards.

Since most of the models are labour intensive and demand intensive management, these ensure sustainable employment to the local communities. Table 2.3 summarizes costs and returns under different combinations of MLP.

Table 2.3: Cost and Return from Different Models upto 5 Years of Plantation

Sl. No.	*Combination*	*Cost (Rs.)/ha*	*Return (Rs.)/ha*	*Benefit cost ratio*
1.	Aonla and Aromatic grass	116,000	250,000	2.2
2.	Aonla and Ashwangandha	94,000	148,000	1.6
3.	Bamboo and Ashwangadha	104,400	121,200	1.2
4.	Bamboo and aromatic grass	106,600	223,860	2.1
5.	Teak and Safed Musli	150,500	250,000	1.7
6.	Teak and Ashwagandha	104,000	158,000	1.5

The accrued cost and benefits are calculated on the basis of cash flow analysis taking five years duration though some of the upper storey crops would provide benefits up to 20 years only. All the models show benefits exceeding the cost. In short, the programme appears very useful for diversifying the livelihoods of the people due to its short duration as well as high financial gain from the model. It has provided an excellent programme to attack the poverty of poorest of the poor and landless.

Strengths and Weaknesses

The strengths and weaknesses of project have been assessed on the basis of field observations and interactions with the stakeholders. The strengths of the project reflected from its origin itself as it attempts to reduce poverty through sustainable management of forests. Major strengths and weaknesses of the project are highlighted in Table 2.4 As discussed earlier, it would be too early to assess the impacts of the project, since it has not completed even a five years of duration. But it seems that this scheme has been evolving as a promising approach to address the poverty of the rural people. Some JFM committees have already started getting benefits from the schemes, e.g. Kalapani FPC from North Betul has earned Rs. 88,000 from grass seeds and Rs. 12,094 from *Mahua* collection. Similarly, MLP has provided sustainable rural employment to the landless and poor families for their diversifying livelihood opportunities. Apart from this, support to the local infrastructure and community development has further supported to improve livelihoods of the people.

Although programme has contributed significantly to biodiversity conservation, increasing extent of the forest cover and improvement of site quality, its advantages/benefits could have been further increased through proper planning, and timely implementation.

Table 2.4: Strengths and Weakness of MLP

Strengths	*Weakness*
▪ Regular employment to landless and poor family ▪ Participatory, flexible and innovative ▪ Promoted involvement of the villagers in JFM ▪ Diversify the livelihoods ▪ Intensive Management ▪ Enhanced Forestry Productivity ▪ Improved site quality ▪ Improved moisture regime ▪ Improved microclimatic conditions ▪ Increased grass production ▪ Developed community infrastructures ▪ Reduced biotic pressure ▪ Enhance capacity development of the poor people ▪ Socio-economic upliftment of poor and marginalized communities	▪ Poor planning and selection of beneficiaries ▪ Untimely implementation and flow of funds ▪ Inadequate technical guidance and support ▪ Less monitoring ▪ Poor linkages with the market ▪ Not able to cover the area as well as its intended target of production

Some of the weaknesses of the programme, which were observed in the fields, included poor planning and selection of beneficiaries, untimely implementation and flow of funds, inadequate technical guidance and support, and linkages with market. Despite the modern techniques and inputs were provided at regular intervals, monitoring seemed quite weak.

Lessons Learnt

MLP is an innovate scheme for rehabilitating the degraded forests. It has maintained harmony with conservation and socio-economic upliftment of rural people. Shifting from watch and ward approach to multi-story management, this scheme has shown how livelihoods of local communities could be diversified through rehabilitation of the degraded forests. It has recognized poor and landless people as a key partner or stakeholder in forestry development activities since their livelihood revolves around the forests. MLP has succeeded to diversify livelihood opportunities by involving poor and landless in forestry development activities and through developments of the village infrastructures and enhancing opportunities for local employment. MLP has used participatory techniques in reaching the rural poor, which seems quite effective to address the underlying causes of poverty in rural domains. Adopting result demonstration methods for promoting use of high tech plantations to disseminate different agroforestry models seem appropriate. Likewise, giving priority to capacity development rather than to technical packages or incentives is one of the key factors to contribute towards the success of MLP that should be considered in designing similar innovations. There is need to empower poor and landless people both socially and economically through their active participation as has been practised in MLP.

MLP has demonstrated that use of modern techniques such as clonal seed planting, multi-layer management and intensive care and management can be useful in enhancing or maintaining forest conditions, health and vitality. For this, partnership and sharing responsibilities among beneficiaries and other relevant stakeholders are necessary. This would not only help to share the responsibilities but also ensure the sustainability of scheme/programme.

NOTES

1. MoEF, 1988. *National Forest Policy Resolution, 1988*. No. 6-21/89-P.P. Ministry of Environment and Forests, Paryavaran Bhavan, Lodi Road, New Delhi.
2. MPFD, 2003. *JFM Statistics*. JFM Cell, Madhya Pradesh Forest Department, Bhopal.
3. Bhattacharya, A.K. and Basnyat, B., 2003. *Empowerment through Joint Forest Management: A Study from Madhya Pradesh*, Unpublished Study Report, M.P. Forest Department, Bhopal, India.

3

Can Agroforestry Contribute to Livelihoods of People?

The Nepalese Experience

Bijendra Basnyat and *A.K. Bhattacharya*

Introduction

In Nepal, agroforestry, the practice of integrating trees into farm systems to enhance agricultural production, is a traditional practice. However, it is regarded as a means to reduce the dependency of people on natural forests and its resultant socio-economic benefits are often largely ignored. And little efforts are made to assess the contribution of agroforestry towards the livelihood as well as economic returns. The Study is an attempt to fill this gap by assessing the different agroforestry practices and their contribution to the livelihoods of people. With the brief review of the Nepalese agroforestry, the paper describes the existing situation of agroforestry practice in the western Terai of Nepal and assesses the different agroforestry options in relation to their contribution towards the improvements on the livelihoods of people.

Conducted in three western Terai districts of Nepal, namely Nawalparasi, Rupandehi and Kapilbastu, the study has used exploratory research design. The survey was conducted in two stages. At the first stage, it was attempted to find the incidence of agroforestry practices. In the second stage, detailed survey was carried out with farmers practising agroforestry to assess the extent of agroforestry practice and its contributions to the livelihoods of the people. 150 farmers were interviewed and in-depth investigation was carried out to assess the way they had been managing their farms through participatory processes such as focus group discussions, transect walk, seasonality analysis etc.

Agroforestry in Nepalese Context

History of Agroforestry

Agroforestry is not a new practice for farmers. Traditionally, they have been growing trees, fuelwood and fruit trees on the bunds of terrace fields, marginal lands and along the stream banks, and raising animals. They have been integrating trees with their agriculture crops to meet their diverse need of the forest products as well as insurance to their agriculture crops. Of the several types of agroforestry, the most famous agroforestry practice is growing trees on farmland and homestead kitchen gardening.

Different models of agroforestry were developed and disseminated during the mid-70s to reduce the pressure on the natural forest. In 1976, Taungya system was introduced in Bara district of Nepal to tackle the encroachment problems. Later on, this system was adapted by Sagarnath Forestry Development Project (SFDP) to reforest the area. Encouraged with the success of SFDP, other projects too started to introduce agroforestry as one of the major components in their programme. In the mid-1980s, Terai Community Forestry Project introduced the agroforestry with intercropping in afforestation or plantation, which included the free distributions of the seedlings. This speeded up the expansion of agroforestry in Nepal, particularly in the Terai.

Agroforestry in Nepal's Farming System

Forest is the integral part of Nepalese farming system. Farmers depend heavily on forests for supply of fodder, fuelwood and construction materials. The pressure on natural resources or forest resources has increased by many folds due to rapid population growth, and lack of alternate livelihood opportunities over-exploitation of forest for fulfilling daily subsistence needs of forest products such as fuelwood, fodder, animal bedding and small timber. The improper land use system, heavy pressure on forest, improper cultivation practices are the main problems of these areas. Some specific problems, which the farmers are facing these days, are as follows:

- Dependency on natural forests for daily use needs like fuel wood, fodder and timber.
- Declining productivity of the land as well as cultivation of traditional crops.
- Diminishing wood and forest resources, which has led to an acute shortage of fuelwood and other essential wood products.
- Large number of cattle population and poor productivity of the cattle.
- Shortage of fodder sources for livestock.
- Land degradation, both in terms of loss of fertility and aggravated erosion.
- Lack of income and employment opportunities.
- Fragmented land holdings.

Above scenario suggests for the need to promote non-farm agroforestry to sustain rural livelihoods. Similarly, use of agricultural residues for cooking has substituted the demand of fuelwood and animal feed. Agroforestry as a multiple land use option for optimum biomass production per unit land can play an important role in fulfilling the demand of tree products and agricultural crops as well as in soil conservation.

Not only agroforestry supplements the tree products based demand of people but also it is one of the financially viable options of optimum land use. Kanel (1995)[1] has reported that the share of the trees to the total net present value is 51 per cent.

Thus the scope of agroforestry is in increasing trend due to decrease in per capita land holding size and access to forests which are caused as a result of the rapid population growth, increasing poverty and declining of agricultural productivity.

Policies and Legal Environment

This chapter reviews policy, periodic plans and legal framework of the forestry sector in Nepal with reference to private forestry and agroforestry to understand the commitments of Government on the promotion of agroforestry. Nepalese policies and legislations have no special clauses or provisions related to the promotion of agroforestry, and it is included within the private forestry. Private forestry programme refers to trees on private agricultural or marginal land used to supplement animal fodder, fuelwood and other basic resources or simply to provide saleable produce. Since private forest refers to trees in small woodlots or trees integrated with agricultural crops as agroforestry, the policy and legislations related to private forestry can be applicable in context of agroforestry as well.

The Master Plan for the Forestry Sector, 1988[2] aims to meet the people's basic needs for fuelwood, fodder, timber, and other forest products and contribute to food production through encouraging people to establish tree farms on their uncultivated lands and by distributing of free or subsidized seedlings of desirable tree species. Similarly, the revised forest sector policy (2000)[3] suggests to expand agroforestry techniques by inter-cropping of fruit trees with medicinal and aromatic plants as well as by other multiple land use techniques adapted to various farming systems through effective interaction between forestry and farming practices. Likewise, the Agriculture Perspective Plan (APP, 1995-2015)[4] urges the government to encourage farmers to commercialize farming operations on environmentally more robust lands in order to relieve pressure on limited natural resource base for achieving higher economic growth through improved productivity in agriculture. In addition, promotion of intercropping with medicinal herbs and other cash crops for providing economic benefits and generating off-farm employment to the rural poor has also been mentioned in the APP. The Tenth Five-Year Plan (2002-2007)[5] also emphasizes the greater role of the forest resources in reducing poverty through various forest development activities. In order to achieve above objectives, the plan gives a priority to agroforestry programme.

For encouraging farmers to practise agroforestry, the Forest Act (1993)[6] allows farmers to utilize, sell or distribute the forest products by fixing their price according to his/her will when they develop, conserve and manage the woodlots in their farm lands. For this, the District Forest Office is to provide necessary technical assistance.

Understanding Agroforestry Practice

Farmers Practising Agroforestry

For the purpose of the study, 150 households (HHs) from three districts were selected randomly to find incidence of farmer's practising agroforestry. Most of farmers (80%) reported that they are either having trees on their farm land or practising the agroforestry. The study reveals that number of trees per HH is 9.8.

The number of trees per ha is 13.5 (Table 3.1) which is almost similar to Kanel's study which reported 11 trees per ha of cultivated land in the Terai. Of 1179 trees enumerated from the HH survey, 64.6 per cent accounted for the Sisso (*Dalbergia sisso*) followed by Aam (*Mangifera indica*) (18.6%) and other tree species as reported in Table 3.3.

The number of Sisso trees was quite high in the study area because Terai Community Forestry Project had promoted this species for tree planting by providing seedlings free of cost for on-farm planting.

The average number of seedlings planted per HH was around 199 (Evans, 1989)[7].

Table 3.1: Agroforestry in Western Terai of Nepal

S. No.	*Response*	*Attributes*
1.	No. of farmers practising agroforestry	120 (80%)
2.	Total operated land (ha)	86.76
3.	Operated land per HH (ha)	0.72
4.	Total number of trees	1179 (100%)
	• Sisso	762 (64.6%)
	• Mango	219 (18.6%)
	• Others	128 (16.8%)
5.	No of trees per HH	9.8
6.	Number of trees per ha	13.5
7.	Place of plantation	
	• Khet land (irrigated)	59 (49.8 %)
	• Khet land (Unirrigated)	31 (25.8 %)
	• Waste land	27 (22.5 %)
	• Homestead land	24 (20.0 %)
	• Others	8 (6.7 %)

Of the total planted trees, about 50 per cent are planted on the borders of "khet" land (irrigated land). This is consistent with Evan's study, which reported 46 per cent plantation in similar type of land.

Reasons for Initiating Agroforestry

Table 3.2 presents farmers' perception for practising the agroforestry. Seventy two per cent farmers reported that they practised agro-forestry to meet their demand of forestry products. This was followed by the support received from external agencies (50%) which refers to the free distribution of seedlings. This shows that majority of farmers are unaware of the resultant benefits of the agroforestry as the objective of agroforestry is to optimize production and economic returns per unit area, while respecting the principle of sustainable development.

Table 3.2: Reasons for Practising Agroforestry

Sl. No.	*Response*	*Number*	*Per cent*
1.	Meet the demand of forest products	86	71.7
2.	External support/free seedlings	60	50.0
3.	Secure higher income	42	35.0
4.	Multiple use of the land	27	22.5
5.	Increase the productivity of land	17	14.2
6.	Conserve the land and water	11	9.2
	Total	120	100.0

Note: Total does not tally due to multiple responses.

Agroforestry Models

As discussed earlier, agroforestry is a land use system that involves deliberate retention, introduction or mixture of trees or other woody perennials in crops/animal production to benefit from the resultant ecological and economic intersections (Nair, 1984)[8]. Based on the field observations and interactions with the farmers, the patterns of tree planting on farmland or agroforestry include following categories (LFP 2003)[9]

- Trees and agricultural crops;
- Trees and grass;
- Trees and Non-Timber Forest Products (NTFP);
- Trees planted around pond or fish pond;
- Trees on bund;
- Trees and crops in rows (Alley cropping);
- Scattered trees on farm land;
- Row or line plantation; and
- Block plantation.

The above categories can be broadly classified into three groups, trees planted in blocks, inter-cropping/agroforestry and trees on farmland.

Trees planted in blocks: Trees are planted in the block with poor soil conditions or even in patches of the farmland. The species planted were *Dalbergia sisso, Mangifera indica.* Orchards of mango trees in the Terai farmlands are intercropped with agricultural crops. The crops grown under mango orchards are not reported. However, these orchards are also surrounded by one or two rows of *D. sissoo* and *Dendrocalamus strictus*. When the fruit yield declines, trees are felled for timber and fuelwood. The land is then cultivated for agricultural crops presumably for some time, and then reverted back to the orchard.

Intercropping/agroforestry: For intercropping between trees and crops farmers are practising different models as discussed below. The common agroforestry models found in the district are agri-silviculture (crops + tree); silvo-pastoral (trees + pasture); Agri-horticulture (Crops + fruit trees); Agri-silvi-pastoral (crop+fruit trees+ pasture); and homestead agroforestry (multiple combination of various components). The common species planted is *Dalbergia sissoo* and *Mangifera indica. D. sisso* covers about 90 per cent of the total trees on farm (LFP, 2003).

- *Crops in between tree rows*: Tree species (*Dalbergia sisso, Mangifera indica)* are planted in rows and crops are grown in between them. The rows width varies from 10 to 20 metre which also serves as the farmer's field boundary in most of the case. The agriculture crops include paddy, wheat, lentil, pea etc.
- *Paired tree rows*: Trees are planted in the paired rows with the spacing between 1.5m × 1.5 m to 3m × 3m. The width of the space varies from 10 to 20 metres and the empty spaces between the rows are used for cropping. This system enables farmers to use mechanical/bullocks ploughing as well as helps them to grow crops for longer duration. The most common species planted are *Dalbergia sissoo* followed by *Mangifera indica.*

Silvofishery: This is a system where trees are used in conjunction with fish farming. This system is currently gaining momentum in the Terai parts of the country where *Dalbergia sissoo* and other trees are being planted on the risers of the pond along with some banana, pineapple and papaya. In some cases, ducks and pigs are also included.

- *Trees and NTFPs*: Under this system the trees are used in conjunction with the NTFPs

Trees on farm land: The trees are grown without specific spacing and scattered along the farm. The crops are grown in the blank spaces. In this system, tree population is kept low to avoid shading of the crops. The common species planted on farm land are again *Dalbergia sissoo* and *Mangifera indica.* The list of species planted on farm land observed in this survey is given below in Table 3.3.

Table 3.3: Species Reported under Different Agro forestry Systems

Species	*Local*	*Major Use*
Albizia species	Siris	Fodder, fuel
Artocarpus lackoocha	Badahar	Fodder, fruit
Dalbergia sissoo	Sissoo	Timber, fuel, fodder
Eucalypstus camaldulensis	Masala	Timber, fuel, oil
Ficus semicordata	Khanu	Fodder
Leucaena leucocephala	Ipil Ipil	Fuel, fodder
Mangifera indica	Amp	Fuel, timber, fruit
Populus species	Lahare peepal	Timber, fodder, fuel
Sesbania grandiflora	Dhaincha	Fodder, green manure

Management Practices

Agroforestry system involves the close interaction of trees with crops, livestock or both. Hence, tree and agriculture crop management is essential for effective and efficient management of agroforestry. Similarly, different agroforestry models or systems need to be managed differently and management will depend upon the objectives of the agroforestry. The management of agroforestry begins with the selection of species, managing spacing between trees, protection, pruning and thinning etc. However, farmers give priority to crop management as compared to the tree management. This could be mainly because of long gestation period of forest crops, increasingly attack of disease in *Dalbergia sissoo* and less management support or assistance. Table 3.4 reflects that farmers are either unaware or have ignored the tree management aspects of agroforestry though they are giving due considerations in crop management. More than half of the farmers (58.3%) reported that the management of trees mainly included protection and use of forest products. One third (32.5%) of farmers reported that they were carrying out lopping and pruning operations for fuelwood, fodder and animal bedding materials as well as to reduce the shading effects on the crop. Very few farmers have informed to have applied fertilizer, irrigation etc. in the agroforestry. This reflects the lack of technical knowledge on agroforestry management. The application of insecticides and pesticides was high (32.5%) which was due to widespread of disease in *Dalbergia* trees.

Table 3.4: Management Practices

S. No.	Response	Number	Per cent
	A. Crop management		
1.	Selection of crops	102	85.0
2.	Weeding	108	90.0
3.	Application of fertilizers, manures etc	112	93.3
4.	Application of pesticides/insecticides	68	56.7
5.	Intercropping	79	65.8
	B. Tree management		
1.	None/ Left as it is (Use)	70	58.3
2.	Soil working/weeding	18	15.0
3.	Thinning	26	21.7
4.	Pollarding	18	15.0
5.	Pruning and lopping	39	32.5
6.	Application of fertilizers, manures	17	14.2
7.	Application of pesticides/insecticides	39	32.5
8.	Protection from grazing, fire etc.	25	20.8
	Total	120	100.0

Note: Total does not tally due to multiple response.

Return from Agroforestry System

It was extremely difficult to collect financial details of agroforestry as farmers could not provide necessary details. This in turn led to another important discussion, whether farmers recognize or realize financial benefits from the system. They hardly cared about species planted or on the financial returns from the trees. When asked about the return from the agroforestry, most common answers were as follows:

- How can I remember the value of the amount of fuelwood, fruits or fodder that I have collected and what I spent on them?
- What benefits did I get from this farm? It hardly provides food to feed my whole family?
- What economics! I will be happy if I could just return my costs?
- I'm just not able to recollect the exact amounts on yearly basis, but yes, I can certainly tell you what efforts I had to make at the time of planting.

Hence, cash flow analysis showing the flow of cash expenditures and receipts resulting from the introduction and maintenance of an agroforestry system was used to assess the return from the agroforestry system. Table 3.5 compares the return of agroforestry model for one ha of land. This analysis does not include the value of the standing trees because, it only attempts to calculate the annual benefits derived by the farmers under different agroforestry model. The highest ratio of benefits (2.84%) has been found for the silvo-fishery practice because farmers are getting multiple benefits from it such as pigs raising, duck farming, banana cultivation, fishery and along with fodder and fuelwood and that too in a short duration of time, i.e. within a year.

Table 3.5: Cost and Benefits of Different Agroforestry Models

Sl. No.	*Models/practice*	*Cost (Rs.)*	*Return (Rs.)*	*Benefit cost ratio*
1.	Trees with cereal crops	13,689	26,715	1.95
2.	Trees with horticulture crops	14,315	35,512	2.48
3.	Fruit trees with cereal crops	13,456	29,319	2.18
4.	Fruit trees with vegetables	15,619	42,689	2.73
5.	Trees scattered on farm land	11,110	20,989	1.89
6.	Silvo-fishery	19,236	54,653	2.84

Note: Benefits were calculated without valuing the standing tree.

This model can be promoted as one of the most viable options for improving the livelihoods as majority of the farmers have either their own or public ponds in their vicinity. Similarly the combination of fruit trees with vegetable cultivation also yielded high benefits to farmers since they can get regular income. Though most of the farmers have planted trees on their farm land, benefit cost ratio was low compared with the other models. This was mainly because trees were planted for their subsistence use and farmers often ignored the return from woody components. However, the return was high compared to the agriculture crops alone as shown in Table 3.6.

Problems Faced

When respondent farmers were asked about the problems that they had been facing in management of the agroforestry farms, majority of them (72.7%) mentioned lack of technical knowledge as their main problem followed by the management assistance from the concerned agencies (Table 3.7). Though District Forest Office is responsible for monitoring

Table 3.6: Comparative Advantage of Agroforestry Compared with Agriculture Crop

Particulars	*Agriculture*		*Agroforestry*	
	Expense	*Income*	*Expense*	*Income*
Cost of cultivations	12,215		11,726	10,202
Return from crops		21,021		15,555
Management of tree			3,021	
Tree products				4,721
Value of sanding tree				37,450
Total expense (Rs.)	12,215	21,021	14,747	67,929
Return from agroforestry	1.72		4.61	

Note: Value to standing tree is estimated at market price of current year

woody components, agroforestry has never been a priority programme of the DFO. Farmers have complaints for both quality and quantity of the services available through the DFO. The other problems faced by farmers in managing and developing the agroforestry includes dead and dying of *Sissoo* trees, training and extension support, lack of management assistance, marketing problems of the products, knowledge about policies, acts and regulations etc.

Table 3.7: Problems Faced by Agroforestry Farmers

S. No.	*Response*	*Number*	*Per cent*
1.	Lack of technical knowledge	86	71.7
2.	Dead and dying of sissoo trees	71	59.2
3.	Training and extension support	68	56.7
4.	Lack of management assistance	59	49.2
5.	Marketing problems of the products	39	32.5
6.	Dying of trees in plantation	31	25.8
7.	Knowledge about policies, acts and regulations	23	19.2
8.	Taxation on forest products	17	14.2
9.	Lack of irrigation, particularly for vegetable crops	17	14.2
10.	Limited opportunities for forest enterprises	16	13.3
11.	Lack of quality and appropriate species	12	10.0
12.	Legal complications and formalities	9	7.5

Impacts of Agroforestry

The agroforestry is economically and financially viable option of optimum use of land. The comparative analysis of agriculture and agroforestry in one ha of land has been presented in Table 3.6. The table reflects that return of agriculture is 1.72 where as that of agroforestry is 4.61 which is almost double. Hence, agroforestry may not only reduce pressure on natural forests but also supports farmers to raise additional income through the management of same unit of land.

The impact and implication of agroforestry on livelihoods is assessed in terms of production of crops, fruits, self-sufficiency on fodder and fuelwood, reduction of time in collection of forest products, generation of income and other benefits from agroforestry, no of livestock and its productivity. Table 3.8 summarizes impacts of agroforestry. 55.8 per cent farmers felt that crop production had decreased after agroforestry as a result of shading effect or competition for nutrients where as 45.8 farmers considered that increase in fruit production was due to improved soil fertility, use of improved seeds or seedlings along with upgraded knowledge and skills of agroforestry practices. Majority of the farmers (76.5%) credited agroforestry for helping them to achieve self-sufficiency on fuelwood/fodder due to different silviculture operations carried out in the woodlots. According to the farmers (70.8%), agroforestry has significantly contributed to reduction in the time in collection of forest products. Although 53 per cent respondents informed that number of livestock has remained nearly the same, about 61 per cent confirmed increase in productivity of their livestock due to availability of highly nutritious fodder and palatable species at farm itself. Majority of the farmers (74.2%) reported that their income had increased from agroforestry as their income source had been diversified. Farmers have mixed perceptions on impacts of agroforestry on soil conservation and soil fertility as they were not able to measure any tangible benefits from it despite a few reported increase in production of crops as a result of agroforestry.

Lessons Learnt

As stated earlier, Nepal's forestry sector policies and periodic plans have identified agroforestry as one of the best options to enhance livelihood opportunities and reduce the dependency on natural forests. However, adaptation and application of agroforestry has not taken place as expected. The study shows that farmers were not able to perceive actual benefits of agroforestry. These are mainly due to weak research and development in

Table 3.8: Impacts of Agroforestry

Impacts	*Increase*	*Similar*	*Decrease*	*Total (N = 120)*
Crop production	17.5	26.7	55.8	100.0
Fruit production	45.8	38.3	15.8	100.0
Sufficiency in forest products	76.7	20.8	2.5	100.0
Time spent on collection of forest products	5.8	23.3	70.8	100.0
Number of livestock	24.2	53.3	22.5	100.0
Productivity of livestock	60.8	22.5	16.7	100.0
Income	74.2	20.0	5.8	100.0
Soil conditions/fertility	18.3	64.2	17.5	100.0
Soil Erosion	14.2	46.7	39.2	100.0

agroforestry, lack of awareness among farmers, traditional beliefs and poor market linkages and coordination. Except free distribution of seedlings, there seems very little promotional activities on agroforestry techniques. Extension services are practically absent at the moment.

The farmers have planted tree species because they got free seedlings or to meet demand of forest products though it has complementary relationships on natural resource base as well as on productivity of land.

Farmers are not able to harness the benefits of agroforestry due to lack of technical skills and management assistance required as well. Though, District Forest Offices are responsible for monitoring woody components, agroforestry has never been their priority programme. Farmers have complaints with regard to both the quantity and quality of the services provided by them.

The study also reveals the absence of scientific management of agroforestry. The management of agroforestry includes mainly protection and use. The lopping and pruning operations are carried out for fuelwood, fodder and animal bedding materials. Very few farmers apply fertilizer, irrigation and herbicides in the agroforestry. Whenever, they prepare fields for seasonal cultivation, management of woody components gets due care.

Agroforestry is not simply mixing of a few trees and crops. For promoting viable agroforestry system, understanding tree-crop interaction and its relationships is crucial. But this has not received priority from both the farmers and the development practitioners. Scarce resources need to be allocated in the right way.

Agroforestry is still at a very rudimentary stage in the country. An appropriate and properly managed species mix will bring about the sustainable land use system. Lessons learnt from the Terai community forestry have also indicated for appropriate documentation of agroforestry system.

Issues and Actions

The study has identified the following crucial issues which need to be addressed for practising agroforestry as a vehicle to improve the people's livelihoods as well as for the scientific management and expansion of the agroforestry system in the country.

- *Induced* vs. *need based*: In the past, expansion of the agroforestry system did little to understand farmers' needs and priorities. Species were identified and introduced in localities without detailed understanding of local situations. For sustainability of the system and enabling farmers to diversify their income, agroforestry system should cater to farmers' needs and their priorities. As a prelude to promoting agroforestry, it is necessary to investigate opportunities for management of resources, which fit in with the local situations and culture.
- *Promotion* vs. *management assistance*: The purpose of the agroforestry should be to assist farmers to manage their lands to meet management objectives through educational and technical support rather than forcing or encouraging them for agroforestry through incentives or any other benefits such as tax exemption. The resource centres might be established to share information as well as for disseminating different extension materials. Credit and insurance facilitates must be available to farmers and linkages should be established with bank, forest based industries and other money lending institutions.
- *Incentive* vs. *capacity development*: Past efforts concentrated on push strategy of agroforestry and provided incentives such as free seedlings, subsidies etc. and did little to develop human skills and capabilities required for the management. As sought by many farmers, capacity development should prevail over incentive system to push the agroforestry system. Incentive system is not sustainable.
- *Technical package* vs. *technical support*: Considering the different site conditions and needs of the farmers, it will not be

possible to develop a model or provide same technical packages applicable to all the farmers. Hence, educational and technical support must be provided to farmers interested in agroforestry for multiple use and benefits as to their management objectives. Potential service provider should be identified for the development of agroforestry for having comparative advantage and complementary in delivering the services.

- *Adoptive* vs. *adaptive approach*: In the past, agroforestry models were developed and disseminated without detailed understanding of the local situations and conditions. Hence the focus should be on identification and documentation of effective, efficient and potential agroforestry practices which not only fit in with the needs of the farmers but also are compatible with their farming system. Demonstration plots for different management objectives (Farmers' Field School) should be established so that results of different management objectives can be seen and replicated in other areas as well.
- *Land use* vs. *landscape*: The purpose of the agroforestry should be to increase the number of trees in the landscape to meet the demands of the forest products and not changing the land use patterns.
- *Subsistence* vs. *market oriented approach*: Earlier main purpose of agroforestry development was to meet household demand of forest products. As a result, market assistance and market linkage development were amongst the most neglected components. However, management of agroforestry is likely to be unsustainable if due attention on marketing aspects is not given. Thus, marketing assistance is required to provide practical marketing assistance to farmers by providing information related to market trends and conditions, current prices, marketing methods, logging and grading of timber and establishing institutional linkages with the forest based industries.
- *Unilateral* vs. *coordinated efforts*: Agroforestry does not mean cultivation of crops and trees together but it is concerned with interactions between trees and crops. Therefore, neither forestry nor agriculture department can alone contribute to the development or promotion of the agroforestry. Hence close collaboration and coordination among district line agencies such as District Agriculture Development Office, District Forest Office, District Livestock Development Office, District Soil Conservation

Office and District Small Enterprise Development Office is necessary.

NOTES

1. Kanel, K.R., 1995. *Farmer and Tree Linkages in Terai of Nepal*. A Dissertation Presented for the Degree of Doctor of Philosophy. University of Minnesota, USA.
2. HMGN/ADB/FINNIDA, 1988. *Master Plan for the Forestry Sector Nepal*, Main Report, Ministry of Forest and Soil Conservation, Kathmandu, Nepal.
3. HMGN, 2000, *Revised Forestry Sector Policy 2000*, Ministry of Forests and Soil Conservation, Kathmandu, Nepal.
4. APROSC/JMA, 1995, *Nepal Agriculture Perspective Plan*, Agricultural Project Service Centre/John Mellor Associates, Kathmandu, Nepal.
5. HMGN, 2002, *The Tenth Plan* (*2002-2007*), National Planning Commission. His Majesty's Government of Nepal, Kathmandu, Nepal.
6. HMGN, 1995, *Forest Act 1993 and Forest Regulation 1995* (*Official Translation*). Ministry of Forests and Soil Conservation, Kathmandu, Nepal.
7. Evans Ptrick T., 1989, *Farm Forestry Survey 1989*, Terai Community Forestry Development Project. Hetauda, Nepal.
8. Nair, P.K.R., 1993, *An Introduction to Agroforestry*, in cooperation with International Centre for Research in Agriculture. Delhi, India
9. LFP, 2003, *Draft Programmes of Participatory Strategic District Forest Sector Plan*, Livelihoods and Forestry Programme—Terai Component.

4

Technical and Financial Appraisal of Agroforestry System

A Study from Villages Around Bhopal, Madhya Pradesh (India)

A.K. Bhattacharya and *Bijendra Basnyat*

Background

In India, agroforestry has been recognized as an important farming system since time immemorial. In recent years, scope of agroforestry has increased significantly as an ecologically based natural resources management system which diversifies and sustains production for increased social, economic and environmental benefits (Nair, 1984)[1]. Apart from this, decrease in per capita land holding, rapid population growth, increasing trend of poverty, conversion of forestland to agriculture uses has created lot of opportunities to promote agroforestry. Though agroforestry has been recognized as an effective means to improve the livelihoods and energy demands, its contribution has not been documented scientifically. Past studies reflect on inadequate information related to management aspects of different agroforestry models, specially on technical and financial aspects. The situation of Madhya Pradesh is not exceptional to this (Bhattacharya, 2003)[2].

Farmers in MP have been practising many innovative systems to meet their daily demands of forest products as well as increase their productivity of land. However, these systems have not been scientifically documented (Bhattacharya *et al,* 2003)[3]. Very few studies had been carried out to document agroforestry practices being practised in MP, specially the indigenous ones. Technical and financial appraisal is one of the most ignored aspects in these studies. The detailed appraisal of technical and financial aspects of the various agroforestry models is essential to evolve a proper strategy for the promotion of the agroforestry and identifying the capacity building needs of farmers. The paper documents different agroforestry models being practised around the villages of Bhopal and estimate on financial returns of different agroforestry models.

Study Approach and Methodology

Considering the nature of study, the exploratory research design was followed. After reviewing the concepts and past efforts on agroforestry development in India and in Madhya Pradesh in specific, knowledge institutions working for the cause of promotion of forestry or agroforestry in the MP were consulted to know about different agroforestry systems being practised, implemented or promoted by them as well as to understand the problems and constraints of agroforestry development. The review and consultations call for operationalizing the definition of agroforestry since it is a traditional practice. Trees are spontaneously grown and no management interventions is being carried out. Hence, with a view to narrow down the scope of study, as well as for better documentation of agroforestry practice, the study adopted the agroforestry definition as proposed by Nair (1984)[1]

> "*Agroforestry is a land use system that involves deliberate retention, introduction or mixture of trees or other woody perennials in crops/ animal production to benefit from the resultant ecological and economic intersections*". Even if a farmer has deliberately planted or retained a single tree, he/she has been considered as the agroforestry farmer.

The study adopted two approaches in identifying the agroforestry farmers, which include the knowledge institutions consultations and rapid field survey. Knowledge institutions were consulted to identify the pocket areas or location, where agroforestry has been promoted and identify individual farmers or groups who have been practising agroforestry or has been either approached or are getting support. The list obtained from those institutions were verified with the concerned farmers and detailed survey was carried out. Rapid survey was conducted in 15 villages around

Bhopal following the direction of road network. Transect walk was done with Sarpanch and Forest Beat Guard to understand about village farming system and identify the potential farmers for detailed survey. The farmers were then consulted for the detailed discussions with the help of checklists and schedules. After completion of field survey, focus group discussion was carried out to supplement the findings of study as well as to identify potential areas of interventions for managing the agroforestry. This approach not only helps to triangulate the findings of study, but also helps to make farmers aware about the situation of agroforestry development in the village. Dual approach was used in analyzing the data and information collected during the study. This implies use of both quantitative and qualitative methods of analysis. Simple statistical parameters such as frequency and percentages were used, as applicable. The financial analysis was carried out by adopting cash flow analysis, and discounted cost/benefit analysis (Mathur *et al., 2003*)[4].

Results and Discussion

Agroforestry Practice

Trees with annual crops or agri-silviculture are perhaps the most easily recognized and most widely employed agroforestry practice in the studied villages. Agri-silviculture was found to be more popular with the farmers followed by agri-horti-silviculture. It was observed that limited tree species are grown with staple food crops. *Eucalyptus, Phoenix* and *Acacia nilotica* are the most commonly found species in combinations with wheat, soyabean, gram and maize. The major function of the woody component was demarcation of boundaries and commercial wood production. Fences not only provide fencing for livestock, protection and privacy but also serve as windbreaks that produce wood and foliage products. Thorny species such as *Acacia* and *Prosopis* spp. were the most common ones on boundary. Commercial trees like Eucalyptus and Mango were always found in the form of boundary plantations protecting the various horticultural species. The different agroforestry models and tree-crop combinations found in the study areas include following combinations:

Eucalyptus + Wheat + Soyabean

One of the most popular and common tree-crop combinations is

Eucalyptus and wheat. Eucalyptus trees have been found in the form of boundary plantations and are mainly planted for commercial purposes.

Mango + Wheat + Soyabean

Mango, being an evergreen, perennial tree, is preferred significantly. Mango trees are found in either separate patches on croplands or in the form of rows between crop areas. Wheat and soyabean are cultivated in different seasons annually while fruiting in mango trees starts around the fifth year.

Gram + Wheat + Orange + Mango

Horticultural trees too are gaining popularity among farmers as fruiting starts early and returns are good. They are grown generally with wheat and gram.

Wheat + Khejari

Khejari, a small to medium sized thorny tree, is grown with wheat. Farmers are fully convinced that this tree species improves fertility beneath its canopy and has no competition with arable crop for soil moisture and nutrients.

Soyabean + Jamun + Aonla + Neem

Although this combination was not a popular one, it was found to be existing in a small area.

Eucalyptus + Wheat + Teak

This combination was found at some places and was one of the few successful ones.

Other tree species most commonly found and integrated with agriculture crops included *Emblica officinalis (Aonla)*, *Tectoria grandis (Teak)*, *Azadiracta indica (Neem)*, *Madhuca indica (Mahua)*, *Diospyros melanoxylon (Tendu)*, *Tamarindus indica (Imli)*, *Terminalia arjuna (Koha/Arjun)*, *Zizyphus mauritiana (Beri)*, *Pongamia pinnata (Karanj)* and *Acacia nilotica (Babul)* etc.

Technical Appraisal

This section documents technical or management interventions carried out by farmers. The farmers are more concerned and providing high inputs on crop management compared to tree management. On intensive enquiry about these issues the most common answers of the farmers were as follows: "We have planted trees not to supplement the farm income but to fence our boundary. We are less concerned about it because we did not know when we would get returns from those trees." This statement itself reflects on level of management interventions practised by farmers. Trees wherever existing on croplands, planted or natural, are mainly grown to demarcate the croplands and act as live fences. Food crops invariably have priority, followed, if possible, by other seasonable crops that can generate cash easily.

Crop Management

The farmers are cultivating grams, wheat and soyabean along with woody component. They are cultivating these crops on the basis of their traditional knowledge and skills. None of the farmers in the study areas have been cultivating improved varieties of wheat or soyabean. Most of the farmers use bio-fertilizers. Very few farmers reported of applying chemical fertilizers. Application of fertilizers and irrigation is almost negligible for both crops. Weeding is generally done for soyabean. No intensive care is given for either of the crops as these are planted. Once the crop is ready to harvest, the farmers harvest it. The crops are either stored for subsistence use or sold in the local market.

Tree Management

Tree management practices are quite traditional with very little technical inputs. Though, tree management starts from the selection of species to its final harvest, farmers are ignorant of the scientific management practices. When asked about the tree management, *most of the farmers put back the questions to the researchers and asked what exactly it meant. For him the tree management is just the planting and collecting the product out of it.* The study reflects that farmers have less technical knowledge and skills in managing the tree component. Table 4.1 presents the different management interventions carried out by farmers. There exists no difference on management interventions though they have practised different agroforestry models. The differences were observed

Table 4.1: Management Interventions Practised by Farmers

Interventions	*Agro-Silvo-horticulture*	*Agri-silviculture*
Site preparation	Manually and ploughing up to 10 cm suitable for cultivation of crops	Manually and ploughing up to 10 cm suitable for cultivation of crops
Planting materials	Seedlings available at local nurseries	Seedlings available at local nurseries
Manuring	Use of organic fertilizer and chemical fertilizer during plantation only	Use of organic fertilizer and chemical fertilizer during plantation only
Pit size	30-45 cu. cm	30-45 cu. cm
Pit preparation	Mechanical means	Mechanical means
Layout of plots	Row	Row / Block
Soil working	Up to the third year of plantation	Up to the third year of plantation
Irrigation/Watering	Rain fed	Rain fed
Soil and moisture conservation	None	None
Weeding	As required based on weed growth and intensive care is given until establishment of seedlings	As required based on weed growth and intensive care is given until establishment of seedlings
Silviculture operations	None	Thinning and prunning rarely followed
Grazing	Not allowed	Not allowed
Insect pest/ diseases	Use of chemicals as and when required	Use of chemicals as and when required
Fire	Fire control lines	Fire control lines
Harvesting	Annual for crops and selection system for trees	Annual for crops and selection system for trees

on harvesting of products. Farmers get fruits and fodder annually whereas timber are harvested once the rotation age is completed or gets severely attacked by insects.

Selection of Species

Most of the farmers had selected the species on the basis of their traditional knowledge and beliefs. The choices of species were governed by the purpose of planting trees, for example, commercial trees like Eucalyptus and Mango were planted on boundary to protect the various horticultural species whereas thorny species such as *Acacia* and *Prosopis* were planted for fencing. Specific tree characteristics received high priority compared to its multiple purpose utility.

Spacement and Tree Density

This is governed by different sets of factors such as objectives of management, agriculture crops to be grown and vehicular or machinery movement such as tractor, cart to be taken during the land preparation and species itself. Based on the specific requirements, farmers keep different spacement and tree density. The common patterns adopted in the study are as follows:

- *Eucalyptus spp.*: The plantations are done mostly on the boundaries. It is most preferred for growing on farmlands due to its high production of wood and sparse shade. Eucalyptus, usually, occupies 40 per cent of the cropland leaving rest of the area for agricultural crops. The spacing varies from 10 to 20 ft.
- *Mangifera Indica*: Mango trees have always been found in separate patches along with agricultural crops and have not been found to exist in a spatially mixed form. Wherever Mango trees were found, they were in large numbers occupying almost an equal area. Spacing is usually kept as 30 × 30 ft. Mango is mainly cultivated for its fruits and firewood.
- *Prosopis Cineraria*: Khejari, as it is commonly called in the region, is a small to medium sized tree which is grown on the boundaries only as it is considered to be a very effective live fence. Leaves are collected, dried and stored to feed livestock. Trees are not normally thinned, as they are usually not grown for wood purposes.
- *Citrus spp.*: The most popular species for horticultural purposes is citrus species such as Oranges and Lemon. Planting is mainly in the form of intercropping with rows of trees between agricultural crops.
- *Syzygium Cumini*: Jamun, a common name, is an evergreen fast growing tree. The spacing is usually wide, around 30 × 30 ft. Trees are usually not trained and cultivated mainly for fruits.

Preparation of Land

Woody perennials usually demand more thorough land preparation than seasonal crops. Farmyard manure, chemical fertilizers are usefully incorporated in the soil before planting. Planting pits are dug with varying depths and widths. Farmers are not aware of pit size or nutrient requirement of different tree species. They prepare the land on the basis of their local traditional skills and knowledge and most of them have not approached concerned departments or offices for the management of species.

Early Plant Protection

For the early plant protection, most of the farmers carried out timely weeding, application of chemical herbicides and pesticides, and control of the grazing. Some of the farmers have also been conducting soil working depending upon the weed growth. Almost all the farmers have controlled or prohibited grazing to protect the plant. The source of the chemical herbicides and pesticides are the local agrovets present around the vicinity of their villages.

Thinning / Pruning

Thinning and pruning procedures are probably the least understood practices. They did not know what exactly both these terms meant. Transect walk reflected that farmers had been practising thinning and pruning operations in their fields. But for them, removal of the crowded trees or cutting of lower branches is just to increase the sun penetration on the field rather than from tree improvement aspects. Though farmers are to some extent aware of possible tree crop interactions and their effects, they hardly practise tree management. Most of the farmers are only collecting the dead wood or dried branches of the trees. This clearly reflects that farmers have their own traditional knowledge in tree management but do not understand technical terms of forest management.

Pests and Diseases

Though farmers apply regularly insecticides or pesticides as a preventive measure to their agriculture crops, it is hardly applied in case of forestry crops. They are applied only when the tree is severely infected by diseases. Most common methods of the pest and disease control are to just remove the infected trees and use for the fuelwood. Farmers hardly approach any institutions if their tree crops get diseased. Most of the farmers either render support or prefer to go to the local agro-vet centres rather than to different institutions working for the cause of agroforestry promotion. For them visiting these institutions would not be effective from all the aspects such as cost, time and suggestions or inputs provided.

Logging and Harvesting

Logging and harvesting operation is one of the most neglected aspects of tree management. As the purpose of planting of tree was either to fence the agriculture land or to provide shade to horticulture crops, farmers are least concerned for its harvesting. They are not practising any scientific harvesting and logging operations. Trees are only harvested when they are dead or severely infected by diseases. Trees are generally felled when the land is fallowed.

Financial Appraisal

Technical appraisal of the agroforestry systems revealed the absence of any technical or scientific management of agroforestry. Under this situation, it was extremely difficult to collect economic details and estimate the cost and benefits of agroforestry. Hence the study adopted cash flow analysis methods to calculate the financial appraisal of different agroforestry models. Benefit cost ratio (BCR) based on farmers' recall methods was used to calculate on the yield over last five years by different crops and woody component. BCR of different agroforestry models has been summarized in Table 4.2. The BCR is maximum in fruit trees with soyabean followed by fruit tree and gram. The return from agri-silviculture model is low because farmers are not getting benefits from tree economically though it might have contributed ecologically by preventing erosion or increasing the productivity of soil.

Issues and Problems

A farmer who plants trees or other woody perennial plants instead of seasonal crops has a different set of commitments. Trees, however, are not a panacea for all ills. To achieve the full potential, a farmer must be able to carry out all those operations that control how an agroforestry system functions in a timely and skilful way. But it was found that farmers were not really aware of any new technological or management systems. This can be one of the reasons why agroforestry is still struggling. Some of the specific issues or problems which agroforestry should address are as follows:

Table 4.2: BCR of Different Agroforestry Models after Five Years of Plantation

(*Rupees per acre*)

S.No.	Model	Cost	Return	Benefit	BCR
1.	Orange and Soyabean	9540	26000	16460	2.7
2.	Eucalyptus and Soyabean	8640	24000	15360	2.7
3.	Lemon and Soyabean	9540	24000	14460	2.5
4.	Orange and Gram	11165	26000	14835	2.3
5.	Eucalyptus and Gram	10265	24000	13735	2.3
6.	Orange and Wheat	10765	24000	13235	2.2
7.	Eucalyptus and Wheat	9865	22000	12135	2.2
8.	Guava and Soyabean	8940	20000	11060	2.2
9.	Lemon and Gram	11165	24000	12835	2.1
10.	Mango and Soyabean	12210	26000	13790	2.1
11.	Guava and Gram	10565	20000	9435	1.9
12.	Mango and Gram	13835	26000	12165	1.8
13.	Guava and Wheat	10165	18000	7835	1.7
14.	Mango and Wheat	13435	24000	10565	0.5
15.	Lemon and Wheat	10765	22000	11235	0.4

Technical Knowledge and Skills

Tree management practices are quite traditional with very minimal technical inputs. Farmers have given high priority in crop management compared to tree management. Most of the farmers implement different management inventions till trees are established. The farmers are familiar with a limited number of trees, which can be incorporated in the agricultural fields. Hence, present study reflects that farmers have less technical knowledge and skills in managing the tree component.

Technical Support

When asked about the key institutions providing the support to farmer or in the village for the promotion of agroforestry, most of the farmers were not able to even list the name of those knowledge institutions consulted earlier. For them, the private nursery or seed suppliers from where they either purchased seed or approached them for any technical interventions is the key institution for the promotion of agroforestry.

They hardly recall even the name of Forest Department. Thus, there is a complete absence of extension activities for agroforestry.

Marketing System

Of the agriculture and forestry products, agriculture crops are sold in village markets or at Bhopal. Very few farmers have sold forest products specially the timber to the local people. Though there was highest return from the combination of soyabean with orange, farmers are not able to get the direct benefits from the marketing of the products. There exists no special marketing channel for the promotion of the marketing aspects of agroforestry.

Operated Land Size

Agroforestry, a means of minimizing risk, is actually associated with a lot of risks. Farmers associate risks with agroforestry and try to avoid the consequences of making a wrong decision. Farm size is an important issue because farmers with access to small areas of land want to protect their families' subsistence needs and plant food and other preferred crops. Thus agroforestry options were found to be more readily accepted as the farm size increased. Thus the aim of the agroforestry would be to increase more trees on landscape rather than changing the farming system.

Risk Diversification

Agroforestry is not covered by the extension services, and is never considered as a potential intervention. Moreover farmers are more interested in the tangible economic benefits from agroforestry and care little for the long term ecological benefits derived from incorporating trees on farmlands. Younger farmers are more inclined to plant crops, while tree planting is favoured by the older generation.

Research and Development

Very little has been done in terms of agroforestry research because of various complexities like methodological, ecological and socio-economic aspects associated with agroforestry research. Of these, tree crop interaction and its possible effects are one of the most neglected one. Agroforestry is not simply mixing a few trees and crops together and hoping for the best. Different tree-crops and their interaction with crops have to be understood, and this hasn't received any research inputs. There is a critical need to allocate scarce resources in the right way.

Potential Areas for Intervention

The review of above situation clearly reflects the need to promote agroforestry to maintain and enhance per unit productivity of land by multiple-use management and effective utilization. Thus the agroforestry support programme should be designed to assist farmers to manage their land to meet their management objectives, such as timber production, fuelwood and forage production, recreation, livestock grazing, soil and water conservation, fruit and small wood etc. through educational and technical support while providing environmental, economic and social benefits. The aim of the agroforestry should be to incorporate more trees into the farmland rather than changing the farming system. Some of the potential areas, where the agroforestry programme should focus on, are as follows:

Extension and Publicity

Extension and publicity programme is essential to increase awareness about agroforestry. The development and implementation of participatory approaches towards public awareness and sensitization is an important activity for the promotion of agroforestry. Extension is carried to increase awareness about planting trees in existing agriculture system as well as on community land. Some of the possible ways are as under:

- Develop and disseminate information related to agroforestry in local languages through use of different extension media.
- Organize exposure visits and study tours to understand agroforestry and tree crop interaction and its management carried out by peer groups.
- Provide assistance (material and financial) for establishing information resource centre at each Panchayat for dissemination of different extension materials.

Training and Capacity Development

Agroforestry farmers have limited skill and technical knowledge in managing their forests or for initiating multiple use forestry activities. This requires strengthening of their skills in managing farms for sustainable utilization rather than protection. Thus capacity building should be taken up as a continuous process through identifying the priority areas that are likely to have the greatest overall impact and as identified or needed by the farmers themselves. Some of the possible ways are as follows:

- Enhance agroforestry farmers' capacities through effective information network, exposure visits, study tours, trainings, workshops at village and district levels to provide them a platform for exchange of ideas, experience and knowledge.
- Provide skill-based trainings on silvicultural operations, multi-storey management, plantation techniques, disease and pest control, NTFP's management, tree crop management, criteria for selection of species and crops, diagnosis and design of agroforestry projects etc.
- Provide follow up and post training support for effective implementation of learned skills in the field as well as to disseminate information and technical advice.

On-Farm Demonstration Plots for Different Agroforestry Models

Agroforestry is still at a very rudimentary stage. An appropriate and properly managed species mix will result into sustainable land use system that can produce as well as conserve. Hence existing agroforestry practice needs to be documented and among these best and efficient ones need to be identified and replicated in other similar domains. Some of the possible methods to achieve above are explained below:

- Establish different *on-farm agroforestry demonstration plots* of different agroforestry systems to show the tree crop interactions and its resultant benefits.
- Encourage and help farmers to establish on-farm agroforestry demonstration plots and plantations.
- Disseminate and share information/experiences of different agroforestry systems to concerned stakeholders.
- Document different management systems followed in each demonstration plot and calculate its financial rate of return to assess the financial feasibility of the activity.

- Develop detailed guidelines for practising different agroforestry systems based on results of different demonstration plots and farmers' field school practice.

Management Assistance

The management of agroforestry includes mainly protection and use. Very few farmers have reported to apply fertilizer, irrigation and herbicides. There is lack of technical knowledge for management of agroforestry. Hence, management assistance is necessary to meet their personal management objectives while providing environmental, economic and social benefits. The main thrust of the programme should be to provide technical assistance and support farmers interested in agroforestry for multiple resource benefits rather than providing technical packages.

- Identify potential service provider for agroforestry and agree on roles and responsibilities based on comparative advantage and complementary in delivering the services.
- Provide information on organizational support for tree planting, credit facilities available from different banking institutions, and potential service provider for management of agroforestry.
- Support farmers to meet their management objectives by suggesting methods on proper management of species and appropriate selection of species based on local conditions and situation.
- Provide technical advice and on-farm visits to assist farmers in management of land according to their management objectives.
- Explore the local market places, demands of forest products and market price for different products and provide market information.
- Promote and strengthen the *Lok Vaniki (Private Forestry) concept* in agroforestry management as well.

NOTES

1. Nair, P.K.R., 1984, *Soil Productivity Aspects of Agroforestry*, ICRAF, Nairobi.
2. Bhattacharya, A.K., 2003, *Financial and Economic Analysis of the*

Agroforestry Models and the practices and the technological Interventions in the Agroforestry models. Project Report, DPIP, MP. Unpublished.

3. Bhattacharya, A.K., Basnyat, B., Chaudhary, B. and Pandit, M., 2003, *Agroforestry benefits are not fully realized. Appropriate Technology*, 30 (1), 70-71.
4. Mathur, P.K., Kumar, H., and Lehmkuhl, J.F., 2003. *Terrai Grasslands-Diversity, Management and Conservation Perspective*, ENVIS bulletin on *Grassland Ecosystems and Agroforestry:* 1(1) 1-28.23.

5

Bamboo in Agroforestry

A Study from Madhya Pradesh, India

P.B. Gangopadhyay and *A.K. Bhattacharya*

Introduction

Bamboo is a critical element of the economy. Bamboo and its related industries already provide income, food and housing to over 2.2 billion people worldwide. There is a 3-5 year return on investment for a new Bamboo plantation *versus* 8-10 years for Rattan. Governments such as India, China and Burma with 19,800,000 ha of Bamboo reserves collectively, have begun to focus attention on the economic factors of Bamboo production.

Bamboo is a renewable resource for agroforestry products. Bamboo is a high-yield renewable natural resource; ply Bamboo is now being used for wall paneling, floor tiles; Bamboo pulp, for paper making, briquettes for fuel, raw material for housing construction, and rebar for reinforced concrete beams.

Bamboo Resource in Madhya Pradesh

On Forest Land

Madhya Pradesh is the largest State of India having 14 per cent of the total geographical area of the country. Out of 44.3 million ha of total geographical area, the State has 15.45 million ha area under forests. Thus about 35 per cent of the geographical area of the State is under forests.

According to the available estimates, the area under Bamboo forests is about 1.5 million ha. Normally, Bamboo grows either as understorey or with other compatible species in the forests. Area designated as Bamboo forests usually has about 200 clumps (clump is group of Bamboos) per hectare. The first and second year Bamboo culms (culm is an individual Bamboo) are not harvested. Culms, which are three years old and more are silviculturally available for harvest.

On an average, a natural sound Bamboo clump is capable of yielding about six Bamboos, each of about 5 metres in length on a four-year cycle, i.e. once in four years. Thus if one hectare of Bamboo forests has 200 clumps, 1.5 million ha Bamboo forests of the State have a potential of yielding 4500 lakh Bamboos every year.

However, in the recent past, due to tremendous biotic pressure, the Bamboo forests are getting degraded at a very rapid pace. The actual production of Bamboo in the State is, therefore, hardly 900 lakhs, i.e. about one-fifth of the potential. These forests need an urgent attention; otherwise the Bamboo that is regarded as poor man's timber will soon go beyond the reach of a common man. Because of the variety of uses of Bamboo, it has aptly been described as poor man's timber. The Bamboo is not only a source for manufacturing paper and pulp but it is a friend of villagers, tribals and forest dwellers due to its multiple uses and fast rate of growth. In villages it is used largely in construction of houses. It plays a significant role in the upliftment of rural economy, especially for *Basod*s, who are totally dependent on Bamboo for their livelihood.

Bamboo Area and Degraded Bamboo Forests: Table 5.1 gives an overview of the Bamboo area and the extent of degraded Bamboo forests in the forest divisions of the erstwhile MP (MP + Chhattishgarh).

On Non-Forest Land

Since there is huge demand of Bamboo, and supply from government forests is dismally low, a pressure has been generated for development of Bamboo resource outside forests also. Villagers have raised Bamboos on the bunds of their agricultural land and also in their homestead. Demand for Bamboo is partially met with from this domestic and agroforestry source. Looking to the vast expanse of the State, it was not feasible to take up detailed study regarding assessment of non-forest Bamboo resource. However, during the study an attempt has been made to assess the quantum of Bamboo contributed by the non-forest source.

Table 5.1: Division-wise Bamboo Area

Forest Division having bamboo area	*Bamboo Area (in ha)*	*Degraded B. Area (in ha)*
Balaghat North	10837.413	5419
Balaghat South	142310.633	71155
Bastar Central	63576.226	31788
Bastar South	49412.000	24706
Bastar West	1047779.00	52389
Betul North	44111.600	2206
Betul South	9216.00	4608
Betul West	Data not available	
Bhanupratappur	84168.00	42084
Bhopal	No. Bamboo	(482)
Bilaspur	24355.542	12178
Bilaspur North	21614.109	10807
Burhanpur	33083.500	16542
Chhatarpur	8240.780	4120
Chhindwara East+West+South	37951.200	18976
Dewas	9624.930	4812
Durg	22494.350	11247
Guna	26902.220	13451
Harda	59402.331	29701
Hoshangabad	9991.90	4996
Jabalpur	964.874	483
Kanker	19392.00	9696
Kawardha	41687.910	20844
Khandwa	48696.216	24348
Khargone	32055.720	16028
Korba	9733.176	4867
Korea	64178.564	32090
Mandla East	Data not available	
Mandla West	13446.500	6723
Narayanpur	47822.71	23911
Narsinghpur	7474.978	3738
Panna North	18708.340	9354
Panna South	7289.010	3645
Raipur East	6536.500	3268
Raipur North	46024.648	23012
Raisen + Obaidullaganj	37244.300	18622
Rajanadgaon	32025.020	16012
Sagar North + South	68813.616	34407
Sarguja North	48507.215	24254

(Contd.)

(Contd.)

Forest Division having bamboo area	Bamboo Area (in ha)	Degraded B. Area (in ha)
Satna	44540.600	22270
Sehore	41945.752	20973
Seoni North	12427.500	6214
Seoni South	20838.896	10419
Shahdol North	87465.070	43733
Shahdol South	4118.84	2059
Sheopur	5882.900	2941
Sidhi East	34997.720	17499
Umaria	67519.110	33760
Vidisha	Data not available	(870)

Primary Survey

Primary survey was carried out in the five selected districts, namely Betul, Bilaspur, Mandla, Panna and Sehore, one from each stratum. The details of the selected villages are given in Table 5.2.

Table 5.2: Study Area

S.No.	Division	Range	Village
1.	Mandla	Ghori	Guwara
		Nanipur	Dhanipur
2.	Betul	Multai	Sonara
		Multai	Khedamla
3.	Panna	Kalda	Mahuadol
		Piparia	Saleha
4.	Bilaspur	Korba	Patrapali, Thakruketa, Cheetapali
		Balco	Bela, Dondrao
5.	Sehore	Sehore	Sonkacchh
		Ashta	Bhimkhedi

Direct survey was carried out in the selected villages by contacting the villagers and collecting information from them in a redesigned questionnaire. The head of the family, number of family members, number of clumps on their private land, number of culms in each of these clumps, total number of culms, and average number of culms per clump have been recorded. The demographic data of the village, total families living in the village, number of families having Bamboo on their private land, total number of culms on their land have been computed. The district-wise and village-wise total number of families, number of families having Bamboos on their private lands, number of culms, total number of culms per family have been computed and shown in Table 5.3.

Results

In 13 sample villages of the five districts, 2340 families were surveyed. Out of these, 354 families (15.13%) are having Bamboo on their fields. The total number of culms on their field is 26,041 and the average number of culms per family is 73.56.

Production of Bamboo from Private Land

The State has been divided into five strata. For each stratum number of rural families is shown in Table 5.3. The percentage of rural families having Bamboos on their private lands has been projected from the sample data. On the basis of the number of culms per family, computation has been done to find out the total number of Bamboo culms present in each stratum. Taking average utilizable length of Bamboos as five meters, the total growing stock has been calculated by dividing the product of number of culms and five metres by 2400 (1 NT = 2400 running metre). On a four year felling cycle, 25 per cent of the growing stock will be available as annual yield. Hence the annual available production of Bamboo from private holdings has been calculated for each stratum and shown in Table 5.3.

Table 5.3: Projection of Bamboo Resources on Private Lands

Stratum	*Sample district*	*Total no. of rural families in the stratum*	*% of families having Bamboos on their lands*	*Rural families having Bamboos on their lands*	*Per family no. of culms*	*Weight of growing stock (in NT)*	*Annual yield @ 25% of growing stock (in NT)*
First	Betul	14,10,151	8.46	119,299	61.12	15,191	3,798
Second	Bilaspur	19,00,506	16.14	306,742	66.30	42,369	10,592
Third	Mandla	19,78,991	3.21	63,526	218.00	28,851	7,213
Fourth	Panna	11,76,399	46.73	549,731	81.88	93,775	23,444
Fifth	Sehore	51,46,414	10.60	545,520	96.63	109,820	27,455
Total		116,12,461	15.13	1,756,965	73.56	2,69,255	67,314

Note: According to this projection, 67,314 NT of annual production of Bamboos is available from non-forest source, which is 30 per cent of the production from forest source.

Production Potential of Bamboo

From the analysis of the data it is clear that there is a yawning gap between the requirement of Bamboos and the available supply. It is also clear from the statistics of Bamboo supply and family income of *Basods*, Nistaris and Pan Barejas that if various levels of satisfaction could be achieved than the corresponding improvement in the economic level of villagers will also be higher.

On the basis of the area estimation, Bamboo plantations in the State cover an area of 1,08,306 ha considering the plantations, under Five Year Plans by the various Departments, Forest Development Corporation, and Tree Cooperative Scheme. Besides, there are other areas under private plantations and those raised by various other departments which could not be accounted for in the present study. However, annual available yield of Bamboos from the plantation area of 1,08,306 ha will be 54,153 tonnes, on a four year felling cycle at the rate of 2 tonnes per ha as estimated in the study.

Estimation from non-forest Bamboo resource was also carried out and according to this, Bamboo resource from the farmers' field has been estimated as 67,314 tonnes per year, which is about 30 per cent of the annual yield from Forest Department coupes. Thus the total available Bamboo yield on an annual basis can be summarized as under:

From government forests	2,25,384 NT
From plantations areas	54,153 NT
From non-forest land	67,314 NT
Total	3,46,851 NT

In financial terms the value of the annual yield is Rs 235.85 crore at the rate of Rs. 6800 per NT as computed in the subsequent para. The study also indicates that Bamboo resource available on non-forest lands is also a significant part of the total production system. This necessitates a more focused attention to regenerate and manage scientifically the Bamboo plantations in homestead and agroforestry sector by adequate technical support of the Forest Department.

State Level Projection of Bamboo Demand Supply for Basods

On the basis of sample district data projection has been done for the *Basod* families of the same stratum to which the sample district belongs. At the rate of sample supply, sample demand and 1500 Bamboos per family, the quantity of Bamboo required for the entire *Basod* families of the State have been calculated. The corresponding figures are 95.56 lac, 268.92 lac, and 461.25 lac Bamboos respectively. Against this, the actual average supply of last five years is 73.82 lac Bamboos. The demand supply gap in the three scenarios has been calculated for each stratum. This shows a shortfall of 21.74 lac Bamboos, at the rate of sample supply, i.e. 248 Bamboos per family, 195.10 lac Bamboos, at the rate of sample demand, i.e. 908 Bamboos per family and 387.43 lac Bamboos, at the rate of 1500 Bamboos per family. The shortfall is more acute in third stratum where *Basod* families are more and in the 5th stratum where Bamboo availability is less. The demand supply gap is shown in Fig. 5.1.

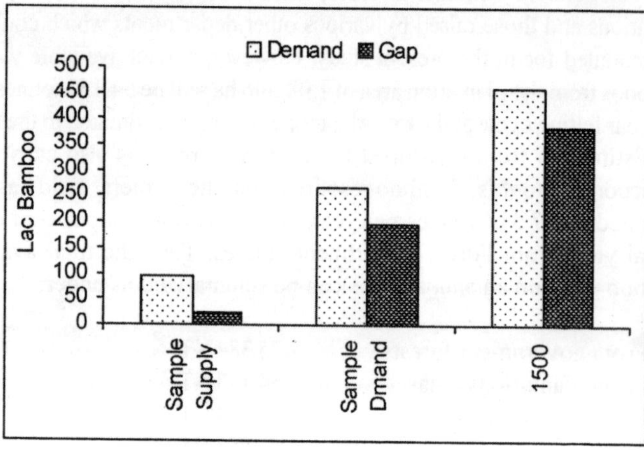

Fig. 5.1: Based Bamboo Demand Supply Gap

On the basis of poverty improvement scenario for the three types of supplies, the contribution of Bamboo to the income of *Basod* families, for the whole State will be as follows:

Rs. 5.57 crore @ sample supply i.e. 248 Bamboos.
Rs. 40.21 crore @ sample demand i.e. 908 Bamboos.
Rs. 68.22 crore @ 1500 Bamboos per family.

In the corresponding event, out of 30,750 *Basod* families 27,288, 15,832 and 7,062 families only will remain below poverty line respectively for three types of supplies. This quantifies the present and potential contribution of Bamboo for increasing economic status of *Basod* families.

Nistar Demand and Income Projection

The demand of Bamboo for Nistar use has been projected for the whole State on the basis of the results obtained in sample districts. The number of rural families around forests, entitled for Nistar, i.e. within 5 km of the forests, is 48.77 lac. Stratum-wise calculation for Bamboo requirement will be as follows :

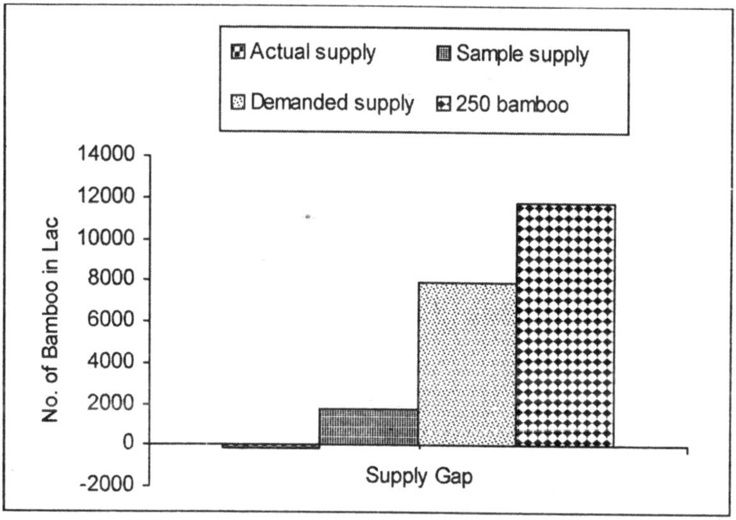

Fig 5.2: Nistar Bamboo Demand Supply Gap

Sample supply	-	42 Bamboos
Sample demand	-	167 Bamboos
1500	-	250 Bamboos.

In the above scenario the requirement of Bamboo will be 21.74 crore, 83.44 crore and 121.93 crore respectively. As against this, the average production of last 5 years is only 4.66 crore. After meeting the Basod demand, only 3.92 crore Bamboos remains available for Nistar. This stock when put against the huge demand, the gap is 17.81 crore, 79.51 crore and 118.01 crore Bamboos respectively. As against the projection of sample data, the actual annual supply of Bamboo in the last five years has been 2.31 crore. Thus only 11 per cent of Nistar demand, as calculated from sample data could be met with the actual supply of Bamboo. The demand supply gap is shown in Fig. 5.2.

Probable income supplement, if Bamboo could be provided as per three scenarios mentioned above, works out as Rs.192.84 crore, Rs. 271.34 crore, and Rs. 662.04 crore for supply of Bamboo at the rate of sample supply, actual demand and 250 Bamboos per family respectively. This explains the wide gap in demand supply of Nistar Bamboo and also highlights the quantified financial benefit that could reach villagers by concessional supply of Bamboo.

Bamboo Demand Projection for Pan Barejas

There are 5435 Pan Bareja families in 16 districts of the State. At present they are getting a supply of 2,22,759 Bamboos. On the basis of sample district data at the rate of 100 Bamboos per family, the total requirement is 5,43,500 Bamboos. On the basis of actual demand at the rate of 190 Bamboos noted in the sample district, the total requirement is 10,32,650 Bamboos. According to the *Nistar* policy provision of 1000 Bamboos, total requirement works out to 54,35,000 Bamboos. The total demand projection is shown in Fig. 5.3.

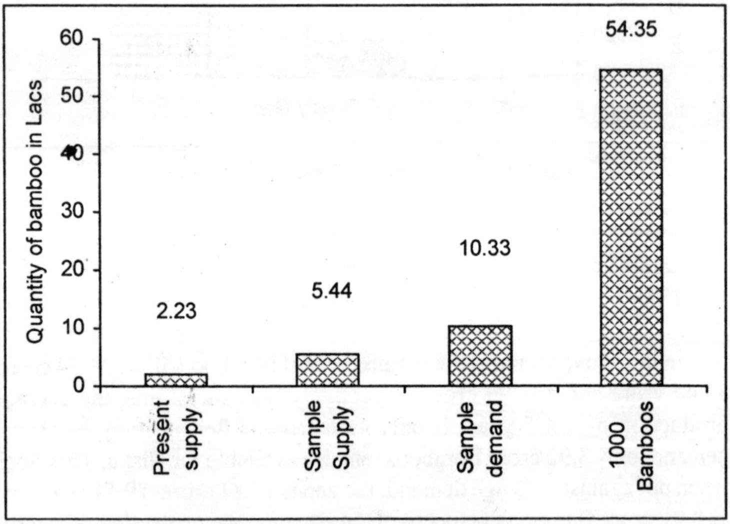

Fig. 5.3: Demand Projection of Bamboo for Pan Barejas

The corresponding income supplement at the rate of Rs. 6.58 per Bamboo will be Rs. 35.76 lac, 67.97 lac and 357.62 lac respectively. Rs. 2.74

lac is however the actual income supplements in the present supply scenario. The income supplement in four scenarios above is shown in Fig. 5.4.

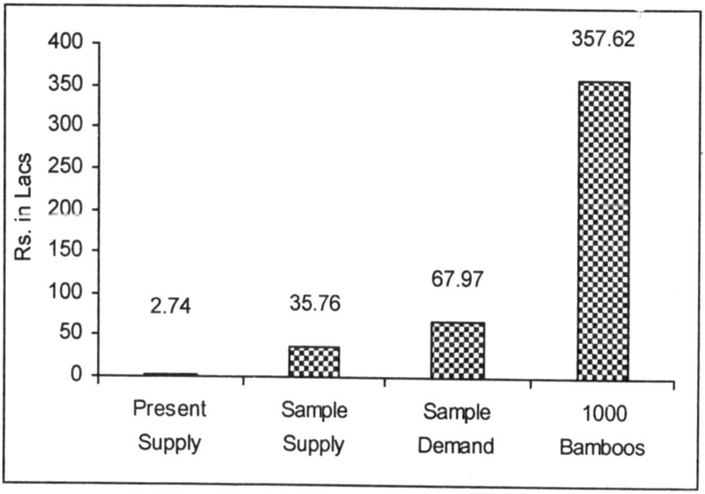

Fig. 5.4: Income Supplement to Pan Barejas

Combined Demand of Commercial Bamboo

The requirement of Bamboo for the three important categories of consumers in the State, namely *Basod*, Nistari and Panbareja have been quantified in the study under four different scenario: (1) at the rate of actual supply, (2) at the rate of supply as per sample data, (3) at the rate of Bamboo demanded by the consumer, (4) at the rate of Bamboo promised by Nistar policy of the government subject to availability (Table 5.4).

Table 5.4: Commercial Bamboo Combined Demand (*No. in Lacs*)

Consumer	*Actual supply*	*Supply as per sample data*	*Supply as per consumer demand*	*As per nistar policy*
Basod	73.82	95.56	268.92	461.25
Nistar	231.00	2174.00	8344.00	12193.00
Panbareja	2.23	5.43	10.33	54.35
Total	307.05	2274.99	8623.25	12708.60

This shows that at present 307 lac Bamboos is being supplied to the consumers of three different categories while the actual departmental production is 466.24 lac Bamboos. Thus almost 70 per cent of commercial Bamboo production is supplied to the three categories of consumers under Nistar supply, thereby leaving only 30 per cent for other consumers and open market disposal. If the data of actual supply to the sample families are projected for the total rural families around forests who are eligible for Nistar supply, the total requirement works out to be 2275 lac Bamboos which is seven times higher than the actual supply. Thus for every person benefited from Nistar supply, there are six others who are deprived. The deficiency is mainly in Nistar supply. The requirement of Bamboo to meet the actual demand of the *Basod* and also the target of Nistar policy is astronomically higher, as is clear from the demand projection in the last two scenarios. The graphical representation of the combined demand scenario is in Fig. 5.5.

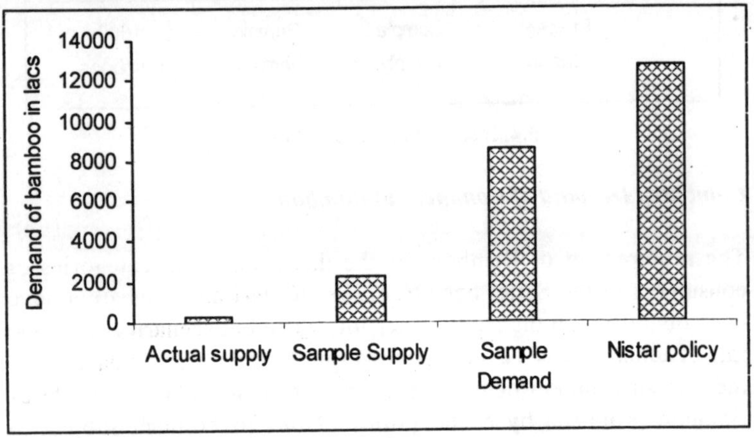

Fig. 5.5: Combined Demand of Bamboo in Three Different Scenarios

Industrial Bamboo Demand

The State presently is producing industrial Bamboo to the extent 1,31,970 NT per year. The demand of paper mills in M.P. for utilizing the current installed capacity of 2,63,550 tons is 6.85 lac tons of raw material. Even if the present production is committed entirely for the Paper Mills within the State still a gap of 5.53 lac tons of industrial Bamboo demand will remain unsatisfied. Production of Bamboo from non-forest areas needs to be activated to meet this gap in raw material supply.

Combined Income Supplement

The study has quantified the contribution of Bamboo in the economy of *Basod, Nistari* and *Panbarejas*. The addition to their family income under four different scenarios can be summarized as in Table 5.5. This is diagrammatically explained in Fig. 5.6.

Table 5.5: Combined Income Supplement

(*In Lac Rs.*)

Consumer	Actual supply	Supply as per sample data	Supply as per consumer demand	As per nistar policy
Basod	73.82	95.56	268.92	461.25
Nistar	231.00	2174.00	8344.00	12193.00
Panbareja	2.23	5.43	10.33	54.35
Total	307.05	2274.99	8623.25	12708.60

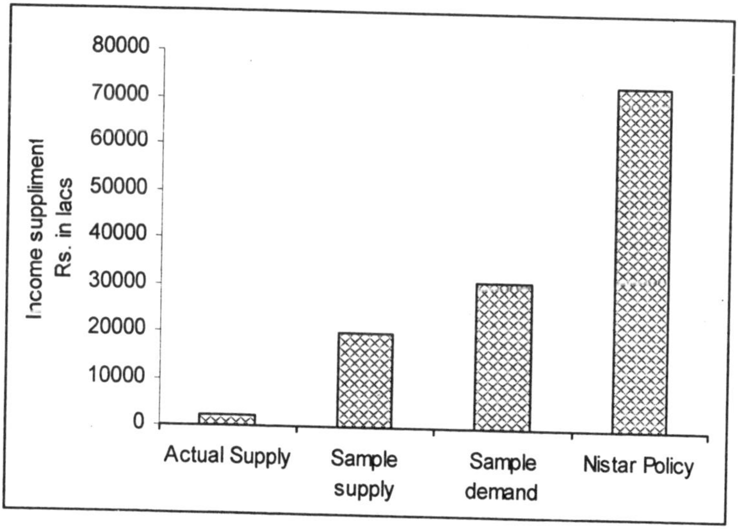

Fig 5.6: Combined Income Supplement in Three Different Scenarios

Thus in MP even at the present rate of Bamboo supply which is a small fraction of their requirement a huge amount of Rs. 21.24 crore is flowing to the beneficiaries as income supplement. If Bamboo could be provided to the consumers of three categories mentioned above at the same rate as have been provided to the consumers in the sample districts the income supplement could have gone up to Rs. 198.77 crore. This could have increased to a very high figure of Rs. 312.25 crore and Rs. 733.83 crore if it would have been possible to supply Bamboo to the consumers as per their actual demand and as per Nistar policy commitment respectively.

The combined demand supply (DIS) and income supplement has been summarised in the following diagram:

		Actual Supply	Sample Supply	Demand Supply	Supply as per policy
Basods	Requirement in lac Bamboo	73.82	95.56	268.92	461.25
	D/S gap in lac Bamboo		21.74	195.10	387.43
	Income Supplement (Rs. In Lac)	108.00	557.00	4021.00	6822.00
+					
Nistar	Requirement in lac Bamboo	231.47	2174	8344	12193
	D/S-gap in lac Bamboo	–161	1781	7951	11801
	Income Supplement in Crore Rs.	2014.00	19284.0	27137.0	66204.0
+					
Panbareja	Requirement in lac Bamboo	2.23	5.43	10.33	54.35
	D/S-gap in lac Bamboo		3.20	8.10	52.12
	Income Supplement in Crore. Rs.	2.74	35.76	67.79	357.62
↓					
Combined	Requirement in lac Bamboo	307.05	2274.99	8623.35	12708.6
	D/S-gap in lac Bamboo	–161	1805.94	8154.20	12240.5
	Income Supplement in Crore Rs.	2124.74	19876.7	31224.9	73383.8

This amply proves the tremendous contribution of Bamboo for poverty alleviation of rural forest dependent families and its potential to achieve a much higher limit in case the Bamboo resource augmentation becomes possible.

Plantation Economics

One of the most important ways to increase Bamboo resource is to take up large-scale plantations. Bamboo with diverse utility in the field of domestic and industrial consumption gives better return than most favoured species, e.g. Teak, Eucalyptus, etc. Many Forest Development Corporations and Forest Departments selected Teak for raising commercial plantations. It is realized that Bamboo has not got its due place in the country's afforestation and reforestation programmes. Even in social forestry plantation programme, Bamboo is not given the place, which it deserves. Bamboo is more economical than Teak, which is considered by and large the most valuable commercial forest species. Comparative economic parameters of Teak and Bamboo plantations are shown in Table 5.6.

Table 5.6: Results of Economic Analysis

Model	*IRR %*	*NPV at 12% in Rs./ha*	*B C RATIO at 12 %*
Teak			
1. Model I (II site quality)	16.5	+2100	1.733
2. Model II (III site quality)	11.5	-23	0.980
3. Model III (IV site quality)	8.5	-1257	0.560
Bamboo			
Model I (good site quality)	20.7	+3294	2.621
Model II (Medium site quality)	16.7	+1456	1.717
Model III (Poor site quality)	11.0	-193	0.905

Source: Anon, 1981[1]

Above analysis amply proves that Bamboo (*Dendrocalamus strictus*) plantation is more profitable and economically more viable than Teak plantations. Economic returns from Teak plantations are greatly influenced by the value of thinned material and specially from earlier thinning. The marketability of thinning material from first and second thinning are generally low in remote areas and areas which are agriculturally and industrially less developed. Bamboo on the other hand is always in demand in areas, which are less developed and remote. The other species of Bamboo viz., *Bambusa bambos, B. vulgaris, B. nutans, B. tulda*, etc. which are frequently grown by villagers, are believed to give more yield as compared to *Dendrocalamus strictus*. Therefore, if these species are planted in suitable localities, yield will be much higher than estimated for *Dendrocalamus strictus*.

State Government has already initiated a massive tree plantation programme under various schemes. It is necessary that Bamboo should be planted in admixture with other species. The following factors would fully justify its large share:

- Bamboo is easy to grow and manage. Villagers are well versed with plantation and management techniques.
- Bamboo starts yielding from 8 to 10 years of age. Repeated yield is obtained on short cycles of 3 to 4 years up to a period of 30 to 40 years, till it flowers.
- It has diverse utility both in domestic consumption and industrial use. Bamboo yields basic raw material for a number of cottage industries. It is principal raw material for paper and pulp manufacture in India. With the increasing demand for paper, it is imperative that Bamboo plantation is taken up on a large scale.
- Bamboo species are not very exacting in nature and can be grown in a variety of agro-climatic conditions.
- Bamboo plantations are economically more paying than most commercial tree species.
- Bamboo easily grows as under crop with many tree species. It can be grown with Teak and other tree species on the same site and thus augment the resource generating potential of the forests.

Suggestions

To mitigate the acute shortage of Bamboo resource and also to improve the socio-economic condition of the rural people particularly those dependent on Bamboo the following measures are suggested:

- Looking to the economics of raising Bamboo plantation, those high market price of Bamboo, the decreasing annual production of Bamboo from areas within forests, agroforestry system and community plantations can fetch encouraging price. The local people within the villages can utilize the commercial Bamboo at remunerative price, while the industrial Bamboo can be supplied to the paper mills who are looking for such supply sources these days, after expiry of the industrial lease agreement.

- It is also suggested to popularize *Bambusa vulgaris* and *Bambusa bambos* among farmers, as Bamboosa is known to establish fast, recruitment rate of new culm is relatively more and is easy to work. Although the species is common in eastern MP, it can be grown throughout the State. Productivity of these two Bamboos on field bunds and in village wastelands is believed to be more than *Dendrocalamus strictus*.
- Genetic improvement of existing Bamboo resources is an important step. Bamboos are highly cross pollinated and selection of superior seedling after gregarious flowering as a result of crossing followed by genetic recombination, should be easier. Early vigour and growth habit may be criteria for seedling selection. Superior clones based on phenotypic characters like number of culms/clump, height, thickness, girth of culms, length of internodes, fibre length, resistance to disease and pests can be selected from the adult wild populations and plantations (Balasubramanyam *et al.*, 1992)[2].
- It is necessary to encourage tree planting on non-forest land and attract private investment for wasteland afforestation. The increase in Bamboo production from farmers' field can in a very big way augment the supply situation. Removal of the hurdles from private plantation will go a long way in improving the supply situation along with socio economic development of rural people.
- Suitable tax exemption on income from Bamboo plantation should be considered to encourage private entrepreneurs for growing Bamboo.
- Bamboo based Cottage industry should be encouraged so that the advantage of value addition goes to the local craftsmen. This will improve their economic condition. Training in Bamboo handicraft should be imparted to rural artisan in general and *Basod* community in particular.
- Marketing support to derive the best benefit from Bamboo articles should be organized, may be in the cooperative sector, in the pattern of Kerala Bamboo Corporation Ltd. The corporation takes care of the marketing of Bamboo and cane articles, manufactured by their members. Sericulture Department requires large quantity of Bamboo for manufacturing "chandrika" and "chhabra" which are used for silk worm rearing. Supply of these commodities could be tagged with *Basod* societies (which could be formed) for their benefit.

With these efforts, Bamboo is likely to play a more important role in the socio-economic development of rural population in general and Bamboo dependent communities in particular.

NOTES

1. Anon, 1981, *Prosperity through Bamboo planting*, Forest Survey Organisation, MP. Bhopal. April, 1981.
2. Balasubrahmanyam, V.R. and Kumar, N. 1992, *Problems and Prospects of Bamboo Cultivation in India*, International Symposium of Industrial Use of Bamboo.

6

Farm Forestry and Agroforestry Initiatives of ITC, Bhadrachalam
A Case Study

S.N. Rao and *A.K. Bhattacharya*

Background

It is ITC's endeavour to continuously explore opportunities for growth by synergising and blending its multiple core competencies to create new opportunities of growth. The employees of ITC are inspired by the vision of growing ITC into one of India's premier Institutions and are willing to go the extra mile to generate value for the economy, in the process creating growing value for the shareholders.

Paperboards and Speciality Papers Division a core sector environment and ecology conscious enterprise, one of the important divisions of ITC having a turnover of around Rs.600 crores becoming instrumental in the development of a backward scheduled tribal area around Bhadrachalam (temple town), district Khammam in the state of Andhra Pradesh, India.

To achieve self-sufficiency in cellulosic raw materials (present need 4.00 lac tonnes per annum and it is likely to grow to 6.00 lac tpa and then 8.00 lac tpa) launched plantation programme (Farm Forestry). In the early stages encouraged 6185 farmers to cover 7441 hectares with Eucalyptus seed route plantations from 1138 villages in the districts of Khammam, West Godavari, Krishna, Guntur, Prakasam, Nalgonda and Warangal of Andhra Pradesh.

To make this programme more attractive by increasing the levels of productivity from 6-10 CuM (MAI) to 20-58 CuM (MAI) and to show to the conventional agri farmers a best alternative land use option, initiated Research and Development for tree improvement and developed 'Bhadrachalam' clones.

Farm Forestry: For commercial and sustainable tree growing, developed a package of scientific silvicultural practices and transferring to the farmers (details given in Annexure-I) to raise and maintain highly scientific and successful plantations with a buy back arrangement at prevailing market price. So far encouraged 6372 farmers participation in plantation programme and covered an area of around 10,000 hectares.

In addition to achieve self-sufficiency to improve the livelihoods of people below poverty line ITC had envisaged social forestry programme to alleviate poverty in support of National Poverty Alleviation Programme during 2001-02. This model simultaneously tackles problems of endemic poverty of tribal and meets part of the needs of woods for our business. Multi species plantations along with inter crops are promoted on private wastelands by providing long term loans to resourceful households.

Outcome

Achievements in all three initiatives have been very encouraging.

(a) *Research and Development*: 86 fast growing and high yielding, disease resistant clones are being produced on a commercial scale. The productivity of 'Bhadrachalam' clones ranges between 20-58 m^3/hectare/year, which is 3 to 9 times more productive than normal seedlings. 23 site-specific clones (includes 86) adapted to problematic alkaline and saline soils have also been developed so far.

(b) *Farm Forestry*: The commercial viability of these clones and subabul selected varieties is evident from the fact that since 1992, 6,372 farmers have become our partners and planted over 10,000 hectares.

(c) *Social Forestry*: Since its inception in 2001-02 covered 2,500 poor tribal farmers, converted 1730 hectares of private wastelands into productive farmlands and planted 4.5 million saplings of different species.

(d) *Environment*: Apart from other well-known consequences of such a large-scale greening effort, these plantations have the potential to sequester 0.5 million tonnes of carbon, thus mitigating GHGs.

(e) *Farm Incomes*: Average net gains to farmers is about Rs. 25,000/ hectare/year (US$ 520) under rainfed condition and Rs.40,000/ hectare/year (US$ 833) with irrigation on a 4-year rotation cycle which is significantly higher than most other cash crops in our operational area, and at much lower risks.

Plantation for Pollution Prevention

By promoting the planting of nearly 30.86 million plants, the project has increased the green cover in the country over 10,000 hectares. Apart from the innumerable benefits of such large-scale afforestation, it directly contributes to *in situ* moisture conservation, groundwater recharge and significant reduction in topsoil losses due to wind and water erosion.

This plant population has the potential to sequester 0.5 million tonnes of carbon thus helping in the reduction of Green House Gas (GHG) and conservation of natural forest resources. This 0.5 million tonnes of carbon is approximately worth US$ 1.5 million at the rate of US$ 3 per metric tonne. Year-wise the volume of carbon sequestration is as follows:

Year	*No. of farmers*	*No. of clonal saplings (lakhs)*	*Area (ha.)*	*Carbon Sequestration (MT)*
1992	5	9613	4	200
1993	6	25857	15	750
1994	47	92817	39	1950
1995	187	299919	183	9150
1996	127	428075	220	11000
1997	181	948426	489	24450
1998	286	1643934	838	41900
1999	376	1825111	980	49000
2000	423	2211354	1178	58900
2001	647	2458681	1110	55500
ITC SF Project	967	917396	560	28000
2002	800	2936770	1322	66100
ITC SF	1543	1657000	1170	58500
Project Subabul	777	15406904	1914	95700
Total	6372	30861857	10022	501100

Over 6,000 rural households having access to their own woody biomass, they can meet most of their fuel wood requirements in-house through lopping, cuttings and pruning. To this extent, therefore, existing public forests are protected from wanton destruction by rural households in search of fuelwood.

As a result of the leaf-litter from multi-species plantations and the promotion of leguminous inter-crops between rows, depleted soils are constantly getting enriched, making these farmlands more productive. In the near future, the increase in soil fertility will lead to a decline in fertilizer and pesticide consumption, thus reducing the pollution of groundwater sources through leaching of such chemicals.

The details like year-wise spread of plantations, clones suitability to soils, clones performance on different soils, yields and standard performance levels of 'Bhadrachalam' clones are as follows:

Participant Farmers and Their Levels

Year of Plantation	*Size of Plantation*					*Total No of Farmers*
	< 1 ha	1-5 Hareli	5-10 ha	10-50 ha	> 50 ha	
1992	2	3				5
1993	3	2	1			6
1994	30	16				46
1995	149	37	1			187
1996	61	54	10	1		126
1997	42	73	11	5		131
1998	110	120	15	10	1	256
1999	148	237	22	16		423
2000	156	214	35	15	2	422
2001	224	383	23	10		640

Clones Suitability to Soils

S.No.	*Type of Soil*	*Adaptable Clones*
1.	Alluvial	1, 3, 6, 7, 10, 27, 105, 115, 122, 130, 266, 274, 284, 286, 288, 292, 316
2.	Black Alkaline	1, 10, 27, 71, 83, 99, 105, 130, 271, 285, 316, 405, 411, 412, 413, 470, 526
3.	Black	1, 3, 7, 10, 27, 71, 72, 83, 99, 105, 122, 266, 271, 272, 285, 290, 316, 405, 411, 412, 413, 417, 439, 526
4.	Red	1, 3, 6, 7, 10, 27, 71, 72, 99, 105, 122, 130, 158, 159, 223, 266, 272, 274, 284, 285, 286, 288, 290, 292, 315, 316, 319, 405, 411, 412, 413, 417, 469, 470, 499, 526
5.	Red Sandy	3, 6, 7, 71, 105, 122, 266, 288, 316
6.	Sandy	3, 6, 7, 10, 288

Clonal Performance on Different Soils

S. No.	*Type of Soil*	*MAI (Cmt)*	
		Maximum	Minimum
1.	Alluvial	35.41	12.08
2.	Black Cotton	41.93	15.61
3.	Black	47.81	15.04
4.	Red	57.37	16.70
5.	Red Sandy	40.54	10.23
6.	Sandy	50.72	18.64

Performance of Clones as Per the Soil and Management Practices Followed

(*In per cent*)

S. No.	*Type of Soil*	*No. of Plantations*		
		Poor	*Normal*	*Excellent*
1.	Alluvial	11	22	67
2.	Black Cotton	11	18	71
3.	Black	9	18	73
4.	Red	10	15	75
5.	Red Sandy	11	23	67
6.	Sandy	9	23	69

Standard Performance Levels

Age of the Plantation (Years)	*MAI (Cmt/ha/yr)*		
	Poor	*Normal*	*Excellent*
1.	< 8	8 to 10	> 10
2.	< 10	10 to 12	> 12
3.	< 12	12 to 15	> 15
4.	< 15	15 to 18	> 18
5.	< 18	18 to 20	> 20
6.	< 20	20 to 22	> 22
7.	< 22	22 to 24	> 24

Yields (Mai) as Per Holding Levels of Participant Farmers

S. NO.	*Size of Plantation*	*MAI (Cmt)*	
		Minimum	Maximum
1.	Less than 1 Hareli.	16.70	57.37
2.	1 ha. to 5 ha.	18.64	49.38
3.	5 ha. to 10 ha.	10.23	39.65
4.	10 ha. to 50 Hareli.	12.08	33.71
5.	Above 50 ha.	–	–

Village Case Study

Name of the village	: Kavuluru
Mandal	: G. Konduru
District	: Krishna
State	: Andhra Pradesh

Village Kavuluru is located 28 kms away from Vijayawada city, 10 kms from Ibrahimpatnam. (3 kms interior on Vijayawada, Tiruvuru State Highway).

Geographical Location

Latitude	: 17° 15′ North
Longitude	: 80° 40′ East

Total size of the village is around 2000 hectares of cultivable land having 4000 voters (1100 families) and catagorized as follows:

Big farmers (above 8 hectares)	: 50
Medium farmers (between 4-8 hectares)	: 200
Small farmers (between 2-4 hectares)	: 150
Marginal farmers (land holding below 2 hectares)	: 200
Agriculture labourers (land holding below 2 hectares)	: 100
Landless farmers	: 100
Landless agriculture labourers	: 200

Major economic activity of the villagers is agriculture, dairying and getting casual employment through surrounding industries. Land use pattern is dominated by cotton, chilli and pulses (black gram and green gram) followed by paddy, sugarcane and maize. Around 100 members are working in banks, schools, industries and village administration as permanent employees. Average annual income among big farmers is Rs.80,000 per family and Rs.40,000 for small farmers. Women folk from landless families and agri labourer categories are employed in farm activities. The prevailing wage rates for male and female members are Rs.70 and Rs.30 respectively. The variation mainly is due to the industries existing around.

One of the typical farmers Shri G. Venkata Rao is of the opinion that ITC had taken an excellent initiative for developing high yielding clones for the sake of farmers to increase the productivity per unit area. Many farmers of the village are eagerly waiting to witness the higher yields and actual returns from clonal plantations.

Eleven farmers of the village have taken-up Eucalyptus 'Bhadrachalam' clonal plantations and the total area is 7.8 ha. These farmers have planted Eucalyptus during the years 2000 and 2001. The farmer-wise details are as follows:

(a) Name of the Farmer : Shri G.Venkata Rao, S/o Chowdary
Father's Name : Chowdary
Age : 35 years
Qualification : Intermediate (10+2)
Family : Joint Family
Total family members : 7 (wife, son, parents, brother and his wife)
Total Land Holding : 7.2 hectares
Status of Land : 1.2 hectares—ancestral
6.0 hectares—purchased
Area of Plantation : 1.2 hectares
Planting Date : 16.8.2000
Species Planted : Eucalyptus 'Bhadrachalam' Clones
Irrigation Facility : No assured irrigation, but back waters of
Thermal power station is available.
Phases of Plantation : Planted in one phase
Other Crops under cultivation : Chilly—1.6 ha.
Cotton—0.8 ha.
Black gram & Green gram—2.8 ha.
Paddy—0.8 ha.

Labour Employed	: Family members—35 man days Hired laborers—60 man days (male) 80 person days (female)
Financial Resource	: Bank loan
Intercrops: (1st year only)	: Black gram
Yield obtained	: 20 bags/1.2 ha.
Income	: Gross—Rs.48,000 (US$ 1000) Net—Rs.23,000/ (US$ 479)
Motivation	: Exposure visit to other farmer plantation.
% of income thru tree farming	: Crop is growing. It may reach 50%.

(b) Name of the Farmer	: Shri D.Chandrasekhara Rao
Father's Name	: Venkata Subbaiah
Age	: 48 years
Qualification	: 10th Class
Family	: Single Family
Total family members	: 4 (wife and two children)
Total Land Holding	: 3.6 hectares
Status of Land	: 2.5 hectares—ancestral 1.1 hectares—purchased
Area of Plantation	: 1 hectares
Planting Date	: 20.9.2000
Species Planted	: Eucalyptus 'Bhadrachalam' Clones
Irrigation Facility	: No assured irrigation, but back waters of Thermal power station is available.
Phases of Plantation	: Planted in one phase.
Other Crops under cultivation	: Chilly—1 ha. Cotton—0.8 ha. Paddy—0.8 ha.
Labour Employed	: Family members—35 man days Hired labourers—45 man days (male) 70 persondays (female)
Financial Resource	: Bank loan
Intercrops: (1st year only)	: Black gram
Yield obtained	: 15 bags/1 ha.
Income	: Gross—Rs.35,000 (US$ 729) Net—Rs.20,000 (US$ 416)

	Motivation	: Faced labour problem with other crops and fluctuations in market price of agricultural commodities. Also exposure visit to other farmer plantation.
	% of income thru tree farming	: Crop is growing. It may reach 25%.
(c)	Name of the Farmer	: Shri M. Venkateswara Rao
	Father's Name	: Subbaiah
	Age	: 55 years
	Qualification	: Illiterate
	Family	: Joint Family
	Total family members	: 6 (wife, son, daughter-in-law and two grandchildren)
	Total Land Holding	: 4.2 hectares
	Status of Land	: 0.8 ha.—ancestral 2.8 ha.—purchased 0.6 ha.—leased
	Area of Plantation	: 0.8 hectares
	Planting Date	: 20.9.2000
	Species Planted	: Eucalyptus 'Bhadrachalam' Clones
	Irrigation Facility	: No assured irrigation, but back waters of Thermal power station is available.
	Phases of Plantation	: Planted in one phase.
	Other Crops under cultivation	: Sugarcane—2.4 ha. Paddy—1 ha.
	Labour Employed	: Family members—25 man days Hired labourers—50 man days (male) 65 person days (female)
	Financial Resource	: Bank loan
	Intercrops (1st year only)	: Black gram
	Yield obtained	: 14 bags/0.8 ha.
	Income	: Gross—Rs.28,000 (US$ 583) Net—Rs.18,000 (US$ 375)
	Motivation	: Exposure visit to other farmer plantation.
	% of income thru tree farming	: Crop is growing. It may reach 20%.

(d) Name of the Farmer : Shri Ch Venkata Rattaiah
Father's Name : Venkaiah
Age : 49 years
Qualification : 8th Class
Family : Single Family
Total family members : 4 (wife and two daughters)
Total Land Holding : 7.2 hectares
Status of Land : 0.8 ha.—Ancestral
2.4 ha.—Purchased
4.0 ha.—leased
Area of Plantation : 0.8 hectares
Planting Date : 26.10.2000
Species Planted : Eucalyptus 'Bhadrachalam' Clones
Irrigation Facility : No assured irrigation, but back waters of Thermal power station is available.
Phases of Plantation : Planted in one phase.
Other Crops under cultivation : Chilli—3.6 ha.
Sugar cane—0.8 ha.
Black gram—0.8 ha.
Maize—1.2 ha.
Labour Employed : Hired labourers-55 man days (male)
85 person days (female)
Financial Resource : Bank loan
Intercrops (1st year only) : Cotton
Yield obtained : 20 quintals/0.8 ha.
Income : Gross—Rs.36,000 (US$ 750)
Net—Rs.8,000 (US$ 166)
Motivation : By seeing adjacent plantations
% of income thru tree farming : Crop is growing. It may reach 20%.

(e) Name of the Farmer : Shri D. Nageswara Rao
Father's Name : Venkateswara Rao
Age : 47 years
Qualification : Intermediate (10 + 2)
Family : Single Family
Total family members : 4 (wife and two children)
Total Land Holding : 1.6 hectares
Status of Land : 1.6 ha.—ancestral
Area of Plantation : 0.4 hectares

Planting Date	:	20.9.2000
Species Planted	:	Eucalyptus 'Bhadrachalam' Clones
Irrigation Facility	:	No assured irrigation, but back waters of Thermal power station is available.
Phases of Plantation	:	Planted in one phase.
Other Crops under cultivation	:	Gave on Lease—1.2 ha.
Labour Employed	:	Hired labourers—31 man days (male) 42 person days (female)
Financial Resource	:	Savings
Intercrops (1st year only)	:	Black gram
Yield obtained	:	5 bags/0.4 ha.
Income	:	Gross—Rs.10,000 (US$ 208) Net—Rs.5,000 (US$ 104)
Motivation	:	By seeing adjacent plantations
% of income thru tree farming	:	Crop is growing. It may reach 15 per cent.

(f)

Name of the Farmer	:	Smt. D. Suseela
Husband's Name	:	Ranga Rao
Age	:	35 years
Qualification	:	5th Class
Family	:	Single Family
Total family members	:	4 (wife and two children)
Total Land Holding	:	3.2 hectares
Status of Land	:	1.6 ha.—ancestral 1.6 ha.—leased
Area of Plantation	:	0.4 hectares
Planting Date	:	20.9.2000
Species Planted	:	Eucalyptus 'Bhadrachalam' Clones
Irrigation Facility	:	No assured irrigation, but back waters of Thermal power station is available.
Phases of Plantation	:	Planted in one phase.
Other Crops under cultivation	:	Chillis—0.8 ha. Black gram—0.8 ha. Green gram—0.8 ha. Jowar—0.4 ha.
Labour Employed	:	Family Members—10 man days Hired labourers—21 man days (male) 42 person days (female)

	Financial Resource	: Bank Loan
	Intercrops (1st year only)	: Black gram
	Yield obtained	: 5 bags/0.4 ha.
	Income	: Gross—Rs.10,000 (US$ 208) Net—Rs.5,000 (US$ 104)
	Motivation	: By adjacent farmers
	% of income thru tree farming	: Crop is growing. It may reach 20%.
(g)	Name of the Farmer	: Shri J. Venkateswara Rao (Scheduled Caste farmer)
	Father's Name	: Panakalu
	Age	: 63 years
	Qualification	: Illiterate
	Family	: Joint Family
	Total family members	: 13 (wife, three sons, three daughters-in-law and grandchildren)
	Total Land Holding	: 4.4 hectares
	Status of Land	: 1.2 ha.—purchased 3.2 ha.—leased
	Area of Plantation	: 0.4 hectares
	Planting Date	: 20.9.2000
	Species Planted	: Eucalyptus 'Bhadrachalam' Clones
	Irrigation Facility	: No assured irrigation, but back waters of Thermal power station is available.
	Phases of Plantation	: Planted in one phase.
	Other crops under cultivation	: Paddy—0.8 ha. Cotton—1.2 ha. Black gram—0.4 ha. Chilli—1.6 ha.
	Labour Employed	: Family Members—25 man days Hired labourers—20 man days (male) 20 person days (female)
	Financial Resource	: Bank Loan
	Intercrops (1st year only)	: Black gram
	Yield obtained	: 6 bags/0.4 ha.
	Income	: Gross—Rs.11,000 (US$ 229) Net—Rs.4,000 (US$ 83)
	Motivation	: By adjacent farmers
	% of income thru tree farming	: Crop is growing. It may reach 15%.

(h) Name of the Farmer : Shri G. Pichaiah
Father's Name : Venkaiah
Age : 55 years
Qualification : Illiterate
Family : Joint Family
Total family members : 5 (wife, son, daughter-in-law and one granddaughter)
Total Land Holding : 4 hectares
Status of Land : 0.8 ha.—ancestral
1.6 ha.—purchased
1.6 ha.—leased
Area of Plantation : 0.4 hectares
Planting Date : 20.9.2000
Species Planted : Eucalyptus 'Bhadrachalam' Clones
Irrigation Facility : No assured irrigation, but back waters of Thermal power station is available.
Phases of Plantation : Planted in one phase.
Other Crops under cultivation : Chilly—1.2 ha.
Paddy—0.8 ha.
Cotton—1.6 ha.
Labour Employed : Hired labourers—25 man days (male) 45 person days (female)
Financial Resource : Bank Loan
Intercrops (1st year only) : Black gram
Yield obtained : 6 bags/0.4 ha.
Income : Gross—Rs.12,000 (US$ 250)
Net—Rs.6,000 (US$ 125)
Motivation : Exposure visit to other farmer plantation
% of income thru tree farming : Crop is growing. It may reach 15%.

(i) Name of the Farmer : Shri P. Satyanarayana
Father's Name : Sivudu
Age : 45 years
Qualification : I.T.I.
Family : Joint Family
Total family members : 4 (wife and parents)
Total Land Holding : 9.2 hectares
Status of Land : 9.2 ha.—ancestral

Area of Plantation	:	1.6 hectares
Planting Date	:	19.7.2001
Species Planted	:	Eucalyptus 'Bhadrachalam' Clones
Irrigation Facility	:	No assured irrigation, but back waters of Thermal power station is available.
Phases of Plantation	:	Planted in one phase.
Other Crops under cultivation	:	Cotton—3.2 ha. Black gram—1.2 ha. Paddy—0.8 ha. Chilly—2.4 ha.
Labour Employed	:	Hired labourers—45 man days (male) 90 person days (female)
Financial Resource	:	Savings
Intercrops (1st year only)	:	Black gram
Yield obtained	:	15 bags/1.6 ha.
Income	:	Gross—Rs.37,500 (US$ 781) Net—Rs.22,500 (US$ 469)
Motivation	:	Exposure visit to other farmer plantation
% of income thru tree farming	:	Crop is growing. It may reach 15%.

(j)	Name of the Farmer	:	Shri A. Venkateswara Rao
	Father's Name	:	Tirupataiah
	Age	:	45 years
	Qualification	:	7th Class
	Family	:	Single Family
	Total family members	:	4 (wife and two children)
	Total Land Holding	:	4 hectares
	Status of Land	:	1.6 ha.—ancestral 2.4 ha.—leased
	Area of Plantation	:	0.4 hectares
	Planting Date	:	19.7.2001
	Species Planted	:	Eucalyptus 'Bhadrachalam' Clones
	Irrigation Facility	:	No assured irrigation, but back waters of Thermal power station is available.
	Phases of Plantation	:	Planted in one phase.
	Other Crops under cultivation	:	Cotton—1.2 ha. Chilli—1.2 ha. Paddy—1.2 ha.

Labour Employed	:	Family members—7 mandays Hired labourers—13 mandays (male) 25 persondays (female)
Financial Resource	:	Bank Loan
Intercrops (1st year only)	:	Black gram
Yield obtained	:	3.5 bags/0.4 ha.
Income	:	Gross—Rs.6,300 (US$ 131) Net—Rs.2,500 (US$ 52)
Motivation	:	Exposure visit to other farmer plantation
% of income thru tree farming	:	Crop is growing. It may reach 20%.

(k)	Name of the Farmer	:	Shri D. Subba Rao
	Father's Name	:	Chandraiah
	Age	:	50 years
	Qualification	:	Illiterate
	Family	:	Single Family
	Total family members	:	3 (wife and daughter)
	Total Land Holding	:	4 hectares
	Status of Land	:	2.0 ha.—ancestral 2.0 ha.—purchased
	Area of Plantation	:	0.4 hectares
	Planting Date	:	19.7.2001
	Species Planted	:	Eucalyptus 'Bhadrachalam' Clones
	Irrigation Facility	:	No assured irrigation, but back waters of Thermal power station is available.
	Phases of Plantation	:	Planted in one phase.
	Other Crops under cultivation	:	Blackgram—0.4 ha. Gave on lease—3.2 ha.
	Labour Employed	:	Hired labourers—20 mandays (male) 35 persondays (female)
	Financial Resource	:	Savings
	Intercrops (1st year only)	:	Black gram
	Yield obtained	:	3.5 bags/0.4 ha.
	Income	:	Gross—Rs.6,300 (US$ 131) Net—Rs.2,500 (US$ 52)
	Motivation	:	Exposure visit to other farmer plantation
	% of income thru tree farming	:	Crop is growing. It may reach 20%.

Case Studies on Extent of Plantation Area

Less than 1 ha

(1) Name of the farmer : Shri M.N.V. Prasad
Father's Name : Chalapathi
Age : 46 years
Village : Nimmalagudem
Mandal : Buttaigudem
District : West Godavari

With agriculture family background, Mr. M.N.V. Prasad has a small family and has three daughters. First daughter, B.Com, second daughter B.Sc. and third daughter post-graduate in Chemistry. He had planted Bhadrachalam clonal saplings in the folloiwng manner:

1995	2 acres
1996	2 acres
1997	2 acres
1998	3 acres
2001	7 acres.

Seen Cousin's Plantation and Motivated by ITC Managers, for the Following Reasons:

Less labour intensive, no pest problems, doesn't need intensive management and regular irrigation, less problems from biotic interference and natural calamities like floods and drought. Above all, there is an assured market from ITC Limited Paperboards and Specialty Papers Division.

Economics

Clones Planted : 3 & 7
Month of Planting : August, 1996
Area : 0.85 ha.
No.of saplings planted : 2000
No.of saplings survived : 1870

	Nature of Activity	Amount Rs.
1st Year	Ploughing	800.00

	Planting stock @ Rs.8.50 including transportation	15895.00
	Pitting and Planting	1800.00
	Neem Powder	200.00
	BHC-50 kgs	900.00
	FYM 8 tractor loads	2000.00
	Weed Management (application of Glycil 4 ltrs)	1600.00
	Irrigation twice	800.00
	Total	**23995.00**
2nd Year	Ploughing (4 times)	1600.00
	Fertilizer	2120.00
	Irrigation	400.00
	Total	**4120.00**
3rd Year	Ploughing (4 times)	1600.00
	Fertilizer	2120.00
	Total	**3720.00**
4th Year	Ploughing (4 times)	1600.00
	Fertilizer	2120.00
	Total	**3720.00**
	Cutting and Transportation expenses:	
	Cutting, debarking, loading and transportation @ Rs.380/-per MT	49400.00
	Expenses—Grand Total	**84955.00**
	Expenses per hectare	99947.00
Income	Yield—130 MT @ Rs.1450/-per MT	188500.00
	Net Returns	103545.00
	Net Returns per hectare	121817.65

(2) Name of the farmer : Shri K. Janakiramaiah
Age : 42 years
Qualification : Intermediate (10+2)
Village : Dippakayalapadu
Mandal : Koyyalagudem
District : West Godavari

With agriculture background, Mr. K. Janakiramaiah has a small family with wife Smt. Achyutavalli, one son Yasasvi studying 9th class and one daughter Sudhishna studying 7th class, four members in total. Planted 'Bhadrachalam' clonal saplings in the following manner:

1994	5 acres
1995	5 acres
1996	4 acres
1999	5 acres
2000	36 acres
2001	3 acres.

Motivated by ITC Managers for the following Reasons:

- Less labour intensive.
- Minor fluctuations in market.
- Can cut and sell the plantation produce on need basis.
- Number of irrigations are less.
- Doesn't need intensive management unlike tobacco and sugarcane.

Economics

Clones Planted	: 3 & 7
Month of Planting	: January, 1995
Area	: 0.64 ha.
No.of saplings planted	: 1348
No.of saplings survived	: 1266

Nature of Activity		Amount Rs.
1st Year	Ploughing	700.00
	Planting stock @ Rs.6.50 including transportation	8230.00
	Pitting and Planting	1500.00
	Ploughing	1500.00
	Irrigation	2500.00
	Total	**14430.00**
2nd Year	Ploughing (5 times)	1500.00
	Fertilizer	900.00
	Irrigation	2500.00
	Total	**4900.00**

3rd Year	Inter-crop—Sunnhemp	200.00
	Ploughing (5 times)	1500.00
	Irrigation	2500.00
	Total	**4200.00**
4th Year	Fertilizer	700.00
	Ploughing	1500.00
	Irrigation	2500.00
	Total	**4700.00**

Cutting and Transportation Expenses

Cutting, loading @ Rs.110/- per MT	13090.00
Transportation charges	23800.00
Miscellaneous expenses	5950.00
Total	42840.00
Expenditure—Grand Total	71070.00
Expenditure per hectare	111046.00

Income

Yield—119 MT @ Rs.1450/-per MT	172550.00
Net Income	101480.00
Net Income per hectare	158562.00

1 to 5 hectares

Name of the farmer	: Shri Guglothu Eerya Naik
Father's Name	: Parthya
Age	: 35 years
Village	: Tekulabanzar
Mandal	: Chandrugonda
District	: Khammam

With agriculture background, Mr. Guglothu Eerya Naik has a joint family with wife Lakshmi, one daughter studying in 9th class, two sons studying in 7th class and 5th class, brohter, brother's wife and brother's children. He was motivated by ITC managers to take-up Bhadrachalam Eucalyptus clonal plantation because of the following reasons:

- Less expenditure compared to commercial crops

- Less involvement, enabling him to concentrate on other activities.
- Less problems from biotic interference and natural calamities like floods and drought.
- Assured market.

He had planted 'Bhadrachalam' clonal saplings in the following manner:

Clones Planted	: 3, 7 & 27
Month of Planting	: August, 2001
Area	: 2.40 ha.
No.of saplings planted	: 6000
No.of saplings survived	: 5400

Economics

	Nature of Activity	Amount Rs.
1st Year	Ploughing	1800.00
	Planting stock @ Rs.4/-	24000.00
	Saplings transportation to field	1500.00
	Pitting and Planting	6000.00
	Anti-termite (Chloropyriphos)	1800.00
	Fertilizer	1260.00
	Ploughing (2 times)	3000.00
	Total	**39360.00**
2nd Year	Ploughing (3 times)	4800.00
	Fertilizer	2200.00
	Total	**7000.00**
3rd Year	Ploughing (3 times)	4800.00
	Fertilizer	2200.00
	Total	**7000.00**
4th Year	Ploughing (3 times)	5000.00
	Fertilizer	2500.00
	Total	**7500.00**
5th Year	Ploughing (3 times)	5000.00
	Fertilizer	3000.00
	Total	**8000.00**
	Cutting and Transportation Expenses	
	Cutting, loading and transportation @ Rs.370/- per MT	111000.00

Expenditure—Grand Total	**179860.00**
Expenditure per hectare	74942.00

Income

Yield—300 MT @ Rs.1400 per MT (@ 125 tonnes per hectare)	420000.00
Net Income	240140.00
Net Income per hectare	100058.00

5 to 10 hectares

Name of the farmer	: Shri G. Veerabhadra Rao
Father's Name	: Satyanandam
Age	: 33 years
Village	: Kapavaram
District	: Khammam

With agriculture background, Mr. G. Veerabhadra Rao has a small family with wife Smt.Vijayabharathi and son Abhiram. He had planted Bhadrachalam clonal saplings in the following manner:

2001	5 acres
2002	10 acres

Mr.Veerabhadra Rao is an employee of ITC Limited, Paperboards and Speciality Papers Division and was motivated by plantation managers to take-up clonal plantation after showing him the farmers' plantations in the nearby village.

Month of Planting	: December 2001 and July 2002
Area	: 5.80 ha.
No.of saplings planted	: 13050
Espacement	: 3 × 1.5 M.

Economics

	Nature of Activity	Amount Rs.
1st Year	Mechanical Uprootal (proclainer) including loading into lorry and transportation charges	48000.00

	Site preparation i.e. balance uprootal, filling of gaps etc.	22000.00
	Fencing charges (with iron barbed wire)	38000.00
	Ploughing	13000.00
	Alignment charges	500.00
	Pitting and Planting	20000.00
	Anti-termite treatment	3200.00
	Fertilizer	5820.00
	Planting Stock cost including transport	58725.00
	Irrigation	3500.00
	Water pipes, electricity charges	10000.00
	Ploughing (3 times)	21000.00
	Soil Working	2000.00
	Watch and Ward Rs.1200/-p.m.	14400.00
	Total	**260145.00**
2nd Year	Ploughing (3 times)	21000.00
	Fertilizer	27300.00
	Irrigation	6000.00
	Watch and Ward	14400.00
	Total	**68700.00**
3rd Year	Ploughing (3 times)	21000.00
	Fertilizer	27300.00
	Irrigation	6000.00
	Watch and Ward	14400.00
	Total	**68700.00**
4th Year	Ploughing (3 times)	21000.00
	Fertilizer	27300.00
	Irrigation	6000.00
	Watch and Ward	14400.00
	Total	**68700.00**
5nd Year	Ploughing (3 times)	21000.00
	Fertilizer	27300.00
	Irrigation	6000.00
	Watch and Ward	14400.00
	Total	**68700.00**
6th Year	Cutting and Transportation expenses	
	Cutting, loading and transportation @ Rs.250/-per MT	362500.00
	Expenses—**Grand Total**	**897445.00**
	Expenses per hectare	154732.00

Income

Yield—1450 MT @ Rs.1400/- per MT	2030000.00
Net Income	1132555.00
Net Income per hectare	195268.00

10 to 50 hectares

Name of the farmer	: Dr. P.V. Ramana
Father's Name	: P. Sundara Rao
Age	: 45 years
Qualification	: M.B.B.S.
Village	: Sundernagar
District	: Khammam

Dr. P.V. Ramana, MBBS from a well educated family, worked as Medical Officer in M/s. Singareni Collieries. Since 1983, he is occupied with development of his 40 hectares of land. He has taken-up the clonal plantation in his land in the following manner:

1999	28 hectares
2000	12 hectares

Motivated to take-up the clonal plantations because of the following reasons:

- Convinced with the productivity of clonal plantations.
- Fluctuations in market price of cashew, at the same time assured market with buyback guarantee for clonal plantation produce.
- Pest menace in cashew brought down yields.

Economics

Month of Planting	: July 1999 and July 2000
Area	: 40 ha.
No.of saplings planted	: 64000

Expenditure

	Nature of Activity	Amount Rs.
1st Year	**Mechanical**	
	Mechanical uprootal of cashew stumps (100/ha.) including loading into lorry and transport	120000.00
	Site preparation i.e. uprooting balance root System, picking of of bark etc.,	

	stacking and burning	28000.00
	Filling of excavated soil	16000.00
	Total	**164000.00**
	Manual	
	Manual uprootal of cashew stumps (100/ha)	
	Including filling of excavated soil	140000.00
	Loading and transport	28000.00
	Total	**168000.00**
	Ploughing	40000.00
	Staking—2.5 × 2.5 M.	6400.00
	Cost of stakes	16000.00
	Mechanical pitting	102400.00
	Planting	22400.00
	Internal transport and basal dressing	18000.00
	Cost of anti-termite treatment (aldrin and thimmet)	20000.00
	Fertilizer	16000.00
	Planting stock @ Rs.4.50 including transport	288000.00
	Anti-termite treatment 6 times (with application)	80000.00
	17:17:17 NPK fertilizer @ 50 gm with appln.	28800.00
	Harrowing in first year—3 times	80000.00
	Weeding—twice	32000.00
	Soil working	32000.00
	Shoot cutting—3 times	6000.00
	Watch and Ward	18000.00
	Total	**970000.00**
2nd Year	Harrowing criss cross—3 times	80000.00
	Fertilizer	66000.00
	Shoot cutting	6000.00
	Watch and Ward	18000.00
	Total	**170000.00**
3rd Year	Harrowing criss cross—3 times	80000.00
	Fertilizer	100000.00
	Shoot cutting	6000.00
	Watch and Ward	18000.00
	Total	**204000.00**
	Yield for 40 ha. @ 75 MT/ha. = 3000 MT.	

Cutting and Transportation expenses

Cutting, loading, transportation and misc. expenses @ Rs.400/-per MT	1200000.00
Expenditure—**Grand Total**	**2544000.00**
Expenditure per hectare	63600.00

Income

3000 MT @ Rs.1400/- per MT	4200000.00
Net Income	1656000.00
Net Income per hectare	41400.00

Annexure-1

Silvicultural Packages for Better Management of Plantations

A number of criteria need to be made exigent for better management practice, which are mentioned below:

Selection of Entrepreneur: Farmers, absentee landlords, businessmen. The participant should look at diversification to achieve sustainability through agriculture activity. He should be resourceful, economically sound, to cope up with the high expenses and long gestation of the crop. Estimated cost benefits are mentioned in Table 6.1.

Site: It is mandatory to study Soil Profile. *p*H should be less than 8.5, and Electrical conductivity should be less than 2 milli mhos/cm. Water logging and highly eroded sites are to be avoided. 6 ft. deep neutral soils are preferred.

Site preparation: Very good site preparation by mechanical means is required to facilitate good aeration in the soil, which allows maximum percolation of rain-water.

Planting stock: Genetically superior, fairly disease resistant and well hardened quality clonal planting stock ensures optimum survival and growth rate which results in better productivity. Minimum age of the planting stock should be 6 months from the date of setting.

Planting: Spacing should be 3 × 2 metres and pit size—30 × 30 × 30 cm. The appropriate planting time is the beginning of the monsoon season. Pit should be filled with soil, leaving a space of 7.5 cm in red soils and 5 cm in black soil. Pot watering should be continued for 7 to 10 days. Phorate @ 2 gm per pit should be applied at the time of planting to protect the sapling from root grub. 15 days after planting the saplings should be treated with anti-termite chemical (Chloropyriphos) @ 2 to 3 ml per litre water.

Operations

Weed Management:Timely and proper weed management avoids competition from weeds and allow the saplings to pickup fast growth. Timely weeding operations would improve soil aeration.

Table 6.1: Cost Benefits Per Hectare of Bhadrachalam Clonal Plantation

Spacing: 3 × 2 M

S.No.	Operation	Unit	Rate	Qty	Years 1	2	3	4	5	6	7	Total Rs.
1.	Ploughing	HA	2400	1	2400	2520	2646	2778	2917	3063	3216	19541
2.	Alignment/Staking	LS	150		150							150
3.	Digging of Pits & Planting	PLANT	1.5	1666	2499							2499
4.	Weeding/Cleaning/Soil Working	HA	833	2	1666	1749						3415
5.	Cost of Fertilizers/Green Manure	HA	2250	1	2250	2363	2481	2605	2735	2872	3015	18320
6.	Cost of Anti-termite Treatment	HA	800	2	1600							1600
7.	Provision for Fencing/Maintenance	LS	2000		2000	200	200	200	200	200	200	3200
	Sub-Total				12565	6832	5327	5583	5852	6135	6431	48725
8.	Contingencies			5%	628	342	266	279	293	307	322	2436
	Sub-Total				13193	7173	5593	5862	6145	6441	6753	51161
9.	Cost of Plants		8	1750	14000							14000
	Sub-Total				27193	7173	5593	5862	6145	6441	6753	65161
10.	Insurance premium			1.25%	340	430	499	573	650	730	815	4036
	Total Cost (per Ha)				27533	7603	6092	6435	6794	7172	7568	69197
11.	Gross Returns	Tons	1400	175	(average yield)							
	Net Returns											175803

Note: Land lease amount, interest on investment not taken into consideration.

Water Conservation measures enhance the growth almost two times. Such measures, taken up in APFDC plantations boosted up the yields to almost double. It is an actual example happening around Hyderabad.

Fertilizer Application: Timely application of fertilizer has been helping the plants to put on more growth. Application of fertilizer during first year for the Eucalyptus 'Bhadrachalam' clonal plantation is as follows:

Well Decomposed FYM (Farm Yard Manure): should be applied as basal dose. FYM should be spread properly throughout the field and incorporated thoroughly in the soil well before planting. One month after planting NPK in the form of complex @ 20-25 gm per plant should be applied by pocketing method (minimum 6″ away from the plant and at 4″ depth). Proper moisture should be ensured whenever fertilizer is applied. Subsequently, every year NPK should be applied in two split doses during the rainy season with a gap of 1 to 1-1/2 months.

Irrigation: Timely and on need basis provision of irrigation to the growing plants will enhance the growth by two times to normal growth.

Ploughing: Soil working by means of ploughing in between the rows of technology based plantations is very essential which facilitates proper aeration and conservation of moisture. These conditions are essential for the luxuriant growth of the saplings. Deep ploughing is to be done in either direction followed by harrowing. Ploughing twice a year is beneficial.

Protection of Plantations: Protection is the foremost thing. Protection from mechanical damages—wrong ploughing practices, biotic interference—cattle, white ants, fire and human beings and chemicals—improper application of phyto-toxic chemicals.

Monitoring of Plantations: Samples of 25 trees are measured in the standing population to assess the survival and growth rates. This practice ensures proper accountability of the participating individual or agency.

7

Agroforestry Model Designing for Specific Areas

With Special Reference to Mountainous Region of Nepal

Bijendra Basnyat and *A.K. Bhattacharya*

Background

The mountain region comprises land between 4,877 and 8848 m above the sea level in the Himalayan range, between 1500 and 3000 m above sea level in the Mahabharat Mountain, of the total land area, 147,481 sq. km, 35 per cent are in the mountains. Approximately 7.8 per cent of the population lives in this area; Areas with less steep slopes and narrow valleys are used for cultivation in the mountains. The upper limits of cultivation are 4200 m elevations. These high regions can support only the crops like buckwheat, barley or potatoes a year or once every two years. High Himalayan areas are mostly rocky with snowfields and glaciers. Agricultures activities are limited to a minimal tilling of land. Raising of sheep, goats and yaks is common. Meadowlands in the area are used for grazing livestock. Rotational grazing is characteristics of the high altitude areas.

The per capita land of people in mountain area is also decreasing day by day. People are cultivating more land, and converting forest in agriculture land to ensure their food security. Due to the sloppy terrain, and improper cultivation practices the quality and productivity of land are also decreasing very rapidly. Forest areas are also under pressure due to the heavy dependency of people on forest for fodder, fuelwood, and timber. Number of cattle holding per household is also high and people use to leave their cattle in forest for grazing. The main livelihood source in high altitude is livestock rather than cultivation of crop. Decreasing availability and gradual deterioration of quality of fodder are the major problems associated with these people. Soil erosion is the another problem in these areas. To over come with these types of problems Agroforestry is the best option.

Agroforestry as a dynamic, ecologically based, natural resources management system that, through the integration of trees on farms and in the agricultural landscape, diversifies and sustains production for increased social, economic and environmental benefits for land users at all levels. Agroforestry is a land use system that involves deliberate retention, introduction or mixture of trees or other woody perennials in crops/animal production to benefit from the resultant ecological and economic intersections (Nair, 1984). Actually Agroforestry is new name for a set of old practices. The major objective of agroforestry is to optimise production and economic returns per unit area, while respecting the principle of sustainable development. The main components of agroforestry systems are trees and shrubs, pasture and livestock, together with the environmental factors of climate, soils and landforms. There are a lot of models available, designed and developed by various institutions for the high altitudes regions based on land condition, climatic condition, elevation, and soil type etc. Agroforestry models are designed on the problem based like fodder, fuel wood, timber based or grasses based, and erosion control based, rehabilitation and reclamation models. Taking the above factors into accounts the agroforestry model was developed for the farmer for his socio-economic upliftment and livelihood enhancement of the farmer of the mountain region.

Designing Agroforestry Model

Characteristics of Farmer

Location

This model has been designed for 10 ha of land in the mountain region of Nepal. This model was developed for Mr. Ram Prasad, who is one of the progressive farmer in the village but due to various problems, he was not able to secure the food security of the family due to following regions. Box 1 below summarizes the overview of the model.

Box 1		
S. No.	*Name of Farmer*	*Ram Prasad*
1.	Location	Lalu VDC, Kalikot District, Far Western Development region, Nepal
2.	Area	10 ha
3.	Physiographic region	Mountain
4.	Farming system	Subsistence
5.	Model Designed for	Mr. Ram Prasad and family
6.	Designed by	Regional Agriculture Research Centre

Land Distribution of the Farmer

Table 7.1: Land Distribution

Land types	*Area (ha)*	*%*
Agriculture fields (plain)	3	30
Upland hills (Bari)	4	40
Forest land	3	20
Total	10	100

The different types of land holding of farmer is presented in Table 7.1. The household and home garden area constitutes for the 1 ha. The agriculture fields is about 2 hours walk and covers the area of 3 ha from his home in the foothills of the valley. The upland area 4 ha and his private forest (2 ha) is about one hrs walk from his house in the uphill side. Total land area for the farmers is 10 ha of land.

Farming System

Livestock based farming dominates in the farming system. The basic unit of farming is the individual farm, the land he cultivates, the livestock it holds and other available area that may include privately owned forests, pastures and other lands. The main cropping patterns are:

Maize–Potato–Wheat
Crop–fallow–barley
Potato—Fallow

He is practicing the maize-potato-wheat in the low land areas and crop fallow barley and potato-barley in the high land areas. The forest is mainly used for the fodder and fuelwood purposes. The use of farmyard manure is quite common though chemical fertilizer is available at village itself. The cattle are mostly unproductive and are kept for the religo-cultural value. The farming system is oriented towards the subsistence level and productivity is declining day-by-day.

Existing Agroforestry Practices

The existing agroforestry practices the farmer is following are:

- Meeting the demand of the basic food requirement from the homestead farms.
- Few trees are planted in the bund and border of the agriculture field. The species mostly includes are *Alnus nepalensis, Juglance regia, Prunus`persica, Morus serrata* etc. These species are planted for the fodder and fuelwood purposes.

The analysis of the existing agroforestry systems reveals that:

- There is no agroforestry system practiced over the entire villages. People are following traditional agroforestry system to bet their livelihoods.
- The species planted are of less economic importance as well as low yielding capacity.
- The soil erosion is one of the major problems in the highland areas. People are not practicing any system and methods for the control of it.
- The forest is kept for the fuelwood and fodder purposes. No other activities are carried out for the improvement of the forest as well as generating income from it.

Need for Agroforestry Intervention

The major problems in the higher altitude have been generated by the high populations, which have exceeded the carrying capacity of the land in most parts of the region. The improper land use system, heavy pressure on forest, improper cultivation practices is the main problems of these areas. Some specific problems which the farmers are facing is:

- Dependency on national forest area for daily use needs like fuel wood, fodder and timber.
- Poor productivity of the land as well as use of traditional crops in the regions.
- Diminishing wood and forest resources, which has led to an acute shortage of fuel wood and other essential wood products. (80% of fuel used for cooking is normally from wood).
- Large number of cattle population and poor productivity of the cattle.
- Shortage of fodder sources for livestock limits livestock integration.
- Land degradation, both in terms of loss of fertility and aggravated erosion.
- Lack of income and employment opportunities .
- Fragmented land holdings.

Thus, there is the lot of scope for the implementation of agroforestry in the farmer's field.

Objectives of the Model

The main objective of the agroforestry model is to optimize the production and economic returns per unit area while respecting the principle of sustainable development. Specifically this model intends to—

- To meet the demand of the fodder, fuelwood and timber from their own land through the practice of agroforestry models hence reduce the pressure in the forests.
- Increase the productivity of the land as well as cattle for the socioeconomic upliftment.
- Ensure the food security of the family.
- To reduce the environmental hazards as erosions, landslides etc. by practicing the different agroforestry models.

Approaches in Designing Model

In order to attain the above objectives, certain agroforestry models have been evolved and standardized combining optimal land use system with the 'agriculture–livestock production system' to give maximum economic returns, simultaneously or sequentially, however, in a manner so as to make them technically feasible, ecologically sustainable, economically viable and socially acceptable. Conforming to the cultural ethos of the people existing agroforestry models must be given due consideration also.

Technical Considerations for Designing of Agroforestry Systems

- Use of nitrogen-fixing trees or shrubs.
- Use of fast growing and deep rooted trees or shrubs for nutrient uptake (near the pump).
- Plant trees or shrubs along the contours.
- Hedge-row intercropping and synchronize the timing of tree pruning/lopping.
- Practice crop rotation for the intercrops.
- Plant cover crop or green manure crops in fallow areas.

Approach

The 3-D approach was followed for designing the agroforestry model. The approach includes:

- *Diagnosis*: Diagnosis of the existing agroforestry practices, problems assessment, scope for future interventions.
- *Design*: Design of the agroforestry models and its expected return.
- *Delivery*: Delivery mechanism of the suggested model and impact assessment.

The agroforestry model is designed in such a way that each model is:

- Technically feasible
- Ecologically sustainable
- Economically viable
- Socially acceptable

The models was designed by giving due consideration to the following factors:

- *Culture Specificity*: The value of the agroforestry model perceived by villagers depends upon his cultural beliefs, values, faiths into delivery system and their past experiences with the existing practices, certain values associated are functional value, futuristic value, social value and status value.
- *Need Specificity*: The needs of various groups of people differ according to their demographic and socio-economic conditions. The needs of the landowners, cattle owners and daily wage earners have different emphasis.
- *Site Specificity*: Different type of the site had different requirements of the crops and models. Thus, it must be proposed according with the site specification and site conditions giving due consideration of the ecological and Site condition.
- *Technology Specificity*: Make the agroforestry activities, an economically viable and environmentally sound proposition, increasing cropping intensity both in space and time dimensions and adoption of multiple cropping, are being resorted to. The tree crop interactions, design innovation and site capacity, phenological characteristics and management interventions are given due consideration while designing model.

Suggested Agroforestry Model

The agroforestry model suggested is as follows. This model has been suggested on the basis of landholding of the farmers and suitability with the local conditions giving due considerations of the need and requirements of the farmers. The specific designs for each type of models and details plans and cost layout are summarized in later sections. For the control of erosions hazards and landslides, no new models has been suggested as it is assumed that practicing these models will lead to the conservation of soil and moisture as well as erosion control.

Area	*Model*	*Species*
Forest	Silvi-horticulture	*Alnus nepalnensis* + apple/citrus/pineapple Quercus + Pineapple
	Agri-silviculture	*Celtis australis/Grevia optiva, Schima wallichii* /Anlus+ Ginger/Termeric/cardamom *Michelia champaca*/Quercus + Ginger/ Termeric
	Pure silviculture	*Schima wallichii*+ *Alnus nepalensis* *Schima* + *Acasia mangium* / A. meansii

(Contd.)

(*Contd.*)

Area	*Model*	*Species*
Low land	Agri-silviculture	*Celtis australis/Grevia optiva, Schima wallichii* /Anlus+ Maize, Millet, Wheat *Michelia champaca*/Quercus + Millet/maize/cabbage
Upland	Agri-silviculture	*Celtis australis/Grevia optiva, Schima wallichii* /Anlus+ Maize, Millet, Wheat *Michelia champaca*/Quercus + Millet/ maize/cabbage
	Agri-Silvi-pasture	*Schima wallichii* + *Setaria aphacelata* (Nandi grass) Alnus/*Michelia champaca* + Setaria/guinea/ *Thysanolaena agrostis* (Broom grass)+ Maize, Millet, Wheat *Schima wallichii* + Setaria/guinea /*Thysanolaena agrostis* (Broom grass) + Millet/maize/cabbage

Suitable Species for Agroforestry

Anthocephalus chinensis	*Poplar*	*Erythriana poeppigiana*
Duabanga grandiflora	Anlus nepalsis	Mimosa scabrella
Aquilaria agallocha	Gmelina arborea	Sesbania sesban
Spondias axillaris	Spondias axillaris	Trema orientalis
Cedrela serrata	Melia composita	Gleditsia triacanthus
Terminalia myriocarpa	Quercus serrata	Albibizia lebbeck

Grass Species

Paspalam notatum	Lespedeza striata	Cynodon dectylon
Axonopus affinis	Trifolium repens	Paspalam dilatatum

Species for Agroforestry in High Altitude with Specific Uses

Species	*Uses*
Alnus nepalnensis	Timber, Fuelwood, Soil conservation
Juglans regia	Timber, Carving, fruit
Morus serrata	Fodder, Sports article, Furniture, Toys
Populus nigra	Timber, Match wood, Pulp, Fuel wood, Ornamental
Prunus persica	Fruits, Timber, Fuel
Quercus incana	Agriculture implements, Fuel, Fodder, Tussar silk
Robinia psendoacasia	Fuel wood, Fodder, Soil conservation
Salix alba	Bats, Match, Tools, Fuel wood, Fodder

The model was designed on the basis of the parameters discussed above.

Forest Area

The models for the forest area have been suggested on the basis of following assumptions:

- Farmers are willing to manage the forest under the agroforestry system.
- There are no marketing problems of products, the farmer produces.
- All the materials and technology are available in the village itself.

Based on the following assumptions, three models have been suggested in for the management of the forest land.

Model Overview

The model for plantation is summarized in Table 7.1 below. The total forest area is divided into three plots where three different models are followed for the better yield as well as for enhancing the productivity of the crops.

Table 7.1: Overview of the Model

	Silvi-horticulture	*Agri-silviculture*	*Pure silviculture*
Location	Forest	Forest	Forest
Year of Plantation	1995	1995	1995
Area	0.75 Hareli	0.75 ha	1.5 ha
Combination	*Alnus nepalnensis* + apple/citrus/pineapple Quercus + Pineapple	*Celtis australis*/Grevia optiva, Schima wallichii/ Anlus+ Ginger/ Termeric/cardamom *Michelia champaca*/Quercus + Ginger/ Termeric	*Schima wallichii*+ *Alnus nepalensis* *Schima* + *Acasia mangium* / *A. meansii*
Purpose	Income and soil conservation	Income and soil fertility maintenance	Income, erosion control, soil conservation
Rotation	5 years of tress and no rotation for agriculture crops	5 years of tress and no rotation for agriculture crops	5 years
Felling system	Selection felling for trees	Selection felling for trees	Selection felling for trees

Management Techniques

The plantation techniques for each type of the models are summarized in Table 7.2. The trees already exist in the forests so no methods for planting trees are suggested here. The planting techniques for the crops are dealt in this section.

Table 7.2: Management techniques

Name of Plantation	*Silvi-horticulture*	*Agri-silviculture*	*Pure silviculture*
Site preparation	Manually and plaguing upto 10 cm suitable for cultivation of crops	Manually and plaguing upto 10 cm suitable for cultivation of crops	None
Planting materials	Seedlings		
No of seedlings/stumps	500		
Spacement	2m × 2m		
Manuring	No use of organic fertilizer Use of chemical fertilizer during plantation only	No use of organic fertilizer Use of chemical fertilizer during plantation only	None
Pit size	45 cu. cm	45 cu. cm	
Pit preparation	Mechanical means	Mechanical means	
Layout of plots	Row	Row	
Soil working	Upto the third year of plantation, deep mechanical ploughing	Upto the third year of plantation, deep mechanical ploughing	
Watering	Rainfed	Rainfed	Rainfed
Soil and moisture conservation	None	None	None
Weeding	As required based on weed growth	As required based on weed growth	None
Silviculture operations	None	None	None
Harvesting	Annual for crops and selection system for trees	Annual for crops and selection system for trees	Selection system for trees

(Contd.)

(Contd.)

Name of Plantation	*Silvi-horticulture*	*Agri-silviculture*	*Pure silviculture*
Protection			
Grazing	Not allowed	Not allowed	Not allowed
Insect pest/ diseases	Use of chemicals as and when required	Use of chemicals as and when required	Use of chemicals as and when required
Fire	Fire control lines	Fire control lines	Fire control lines

Cost of Production

The cost of production and estimated return from each type of model is summarized in Table 7.3 below. Site preparation includes initial preparation of the crops. The figures below are estimated figures calculated on the basis of production potentiality of site and estimated return. Input cost includes the price of seeds, chemical fertilizers, pesticides and insecticides cost. Labour cost is taken as the wage labour required for the different operations and tending activities. Net return from this model will be Rs. 11,400 per annum where the domestic consumption is not taken into accounts while calculating revenue part.

Table 7.3: Estimated Cost and Return in Rupees

	Silvi-horticulture	*Agri-silviculture*	*Pure silviculture*	*Total cost*
Costs				
Site preparation cost	1500	1000		3000
Inputs cost	500	700	200	1400
Labour cost	4500	4000	1500	11000
Marketing and transportation cost	500	800	2000	3300
Other costs	400	500	800	1700
Total cost	7400	7000	4500	20400
Crops	4800	4500	0	9300
Trees	5000	6000	10000	19000
Total	9800	10500	10000	28300
Return	2400	3500	5500	11400

Note: Estimated Figure.

Upland and Low Land Area

Different models have been suggested for the upland and low land area. The model and species compositon will remain same for all the areas except the plantation techniques for the tree. In the low land area, the trees are planted in blocks whereas in upland area, the trees are planted in rows or in scattered way. Few fruit trees species as apple, citrus will be planted also.

Following models are suggested for the agriculture fields.

Low land	Agri-silviculture	*Celtis australis / Grevia optiva Schima wallichii* / Anlus + Maize, Millet, Wheat Michelia champaca / Quercus + Millet / maize / cabbage
Upland	Agri-silviculture	*Celtis australis / Grevia optiva Schima wallichii* / Anlus + Maize, Millet, Wheat *Michelia champaca* / Quercus + Millet / maize / cabbage
	Agri-Silvi-pasture	*Schima wallichii* + *Setaria aphacelata* (Nandi grass) Alnus / *Michelia champaca* + *Setaria guinea* /*Thysanolaena agrostis* (Broom grass)+ Maize, Millet, Wheat *Schima wallichii* + *Setaria guinea*/ *Thysanolaena agrostis* (Broom grass) + Millet/ maize/cabbage

The plantation techniques, management interventions and expected return are summarized in following sections. This models has been suggested on the basis of following assumptions:

- Farmers is willing to practice this models and is economically feasible.
- The use of the tree species will help to reduce the soil erosions as well as grass species are used as soil binding as well as fodder for animal.
- There are no marketing problems of products, the farmer produces.
- All the materials and technology are available in the village itself.

Table 7.4: Overview of the Model

	Agri-silviculture	*Agri-silviculture*	*Agri-Silvi-pasture*
Location	Low land	Up land	Up land
Year of Plantation	1995	1995	1995
Area	3 ha (0.5 ha tree and 2.5 crop land)	3 ha (0.5 ha tree and 2.5 crop land)	
Combination	*Celtis australis/Grevia optiva* *Schima wallichii* /Anlus+ Maize, Millet, Wheat *Michelia champaca*/Quercus + Millet/ maize/cabbage	*Celtis australis/Grevia optiva* *Schima wallichii* /Anlus+ Maize, Millet, Wheat *Michelia champaca*/Quercus + Millet/maize/cabbage	*Schima wallichii* + *Setaria aphacelata* (Nandi grass) Alnus / *Michelia champaca* + Setaria/ guinea /*Thysanolaena agrostis* (Broom grass)+ Maize, Millet, Wheat *Schima wallichii* + Setaria/ guinea /*Thysanolaena agrostis* (Broom grass) + Millet/ maize/cabbage
Purpose	Income and wind break, shelter belt.	Income and soil erosion control	Income, erosion control, soil conservation
Rotation	10 years of tress and no rotation for agriculture crops	10 years of tress and no rotation for agriculture crops	10 years of tress and no rotation for agriculture crops
Felling system	Selection felling for trees	Selection felling for trees	Selection felling for trees
Mode of regeneration	Coppicing and plantation	Coppicing and plantation	Coppicing and plantation

Overview of Model

The model for plantation is summarized in Table 7.4. The total forest area is divided into three plots where three different models are followed for the better yield as well as for enhancing the productivity of the crops.

Management Techniques

Following management interventions (Table 7.5) are suggested in the agriculture filed. In both the upland and low land area, there was no irrigation facility hence the irrigation is not mentioned here.

Table 7.5: Management Techniques

Name of Plantation	*Agri-silviculture*	*Agri-silviculture*	*Agri-Silvi-pasture*
Site preparation	Manually and plaguing upto 10 cm suitable for cultivation of crops	Manually and plaguing upto 10 cm suitable for cultivation of crops None	
Planting materials	Seedlings/stumps	Seedlings/stumps	
No of seedlings/ stumps	4000	2000	
Espacement	3m × 3m	1.5m × 1.5m	
Plantation model	Block	Row	
Manuring	Use of organic fertilizer during planting in pits Use of chemical fertilizer during plantation only		
Pit size	45 cu. cm		
Pit preparation	Mechanical means		
Layout of plots	Row	Row and scattered	Row and scattered
Soil working	Upto the third year of plantation, deep mechanical ploughing		
Watering	Rainfed		
Soil and moisture conservation	None		
Weeding	As required based on weed growth		
Silviculture operations	Thinning and pruning in 2nd and 5th year	None	None
Harvesting	Annual for crops and clear felling system for trees	Annual for crops and selection system for trees	

(*Contd.*)

(Contd.)

Name of Plantation	*Agri-silviculture*	*Agri-silviculture*	*Agri-Silvi-pasture*
Protection			
Insect pest/ diseases	Use of chemicals as and when required	Use of chemicals as and when required	Use of chemicals as and when required
Fire	Fire control lines	Fire control lines	Fire control lines

Cost of Production

The cost of production for the different models is summarized in table below. The return form the trees includes the fuelwood and fodder species obtained from the singling and thinning operations. The cost did not take accounts of the indirect benefits obtained from the practicing of models, e.g. soil and moisture conservations. Table 7.6 presents the cost and return from the models. The total return from the agroforestry will be Rs. 10,600 per year.

Table 7.6: Estimated Cost and Return from Practicing Agroforestry in Agriculture Fields

	Agri-silviculture	*Agri-silviculture*	*Agri-Silvi-pasture*	*Total cost*
Costs				
Site preparation cost	2500	2500	3000	8000
Inputs cost	1500	2500	2500	6500
Labour cost	4000	5000	5000	14000
Marketing and transportation cost	1000	1500	1500	4000
Other costs	400	400	600	1400
	9400	11900	12600	33900
Return				
Crops (including fruits)	8000	12000	9000	29000
Trees	4000	3000	4500	11500
Animals			4000	
Total	12000	15000	17500	44500
Return	2600	3100	4900	10600

Homestead

No models have been suggested for the homestead farming. The farmer is practicing the multi-storey crop composition and it is sufficient to meet his daily food requirements as well as household requirement of fuel-wood, fodder along with the soil and water conservation. Thus, no model is suggested for the home stead farming.

Support Mechanism

The farmer will be provided with following support for the initiation of the agroforestry in the village as the model farmers.

- Market linkages will be established to sell the products more easily and at competitive prices. This will help to enhance the income of the farmer.
- Coordination with different organizations working for agriculture, forestry, livestock, soil conservation etc. will be made such that farmer can get the required services at earliest as possible and at his door steps.
- Materials inputs will be provided at subsidized prices at the door steps such that it create the incentives for other farmers to start agroforestry practices.
- Free technical inputs and extension facilities will be provided to the farmer by the research organizations such that farmer can always go for the better technology an cost effectiveness.
- Credit arrangements will be made if farmer desire for it at very subsidized rate from the agricultural and rural development banks.

PART II

ECOTOURISM

8

Ecotourism as a Tool for the Sustainable Nature Tourism Management and Community Development

A Holistic Strategic Approach

A.K. Bhattacharya

Backdrop

The term ecotourism was coined by Hector Ceballos Lascurain in 1983. The term was used to describe the nature-based travel with emphasis on education, management and development of sustainable tourism product and activity.

World Tourism Organization has defined ecotourism as "*tourism that involves travelling to relatively undisturbed natural areas with specified object of studying, admiring and enjoying the scenery and its wild plants and animals, as well as any existing cultural aspects (both of past and present) found in these areas.*"

The Ecotourism Society (TES) has defined Ecotourism as "responsible travel to natural areas that conserves the environment and sustains the well-being of the local people". Its main aims are ecological and socio-cultural integrity, responsibility and sustainability (Cater, 1994)[1]. Ecotourism has developed from the traditional nature tourism and sustainable development concepts.

Over the last one decade ecotourism developed as both an industry and a form of sustainable development that conserves natural areas and sustains the well being of local people. Ecotourism, as a concept, aspires in all cases to achieve sustainable development objectives. But ecotourism products have become numerous enough to be considered as a market segment, which has both sustainable and unsustainable characteristics. However, ecotourism has been widely used and often misused as a market label in a variety of circumstances, nevertheless, a holistic approach and the strong orientation of the ecotourism towards the evolution of principles, guidelines, and certification based on sustainability standards gives it a market niche that is strongly associated with sustainability.

With a holistic approach ecotourism can emerge as an effective tool for the three major challenges, viz. biodiversity conservation, promotion of community development and management of sustainable nature tourism.

Principles and Characteristics of Ecotourism

The analysis of all the definitions and the concepts of ecotourism lead to the identification of following integral components of ecotourism:

- ecological sustainable management (long term conservation of resources)
- conservation education and interpretation of natural environment
- local community participation and benefits

Ecotourism should be:

- small scaled, locally controlled, land sensitive and ecologically oriented.
- based on environmental ethics with a concept of small is beautiful, doing more with less and treated with fostering resource integrity.
- sustainable environmentally, socially, culturally and economically.
- capable to enrich experience and should be educational.
- able to promote the participation of the local communities.
- be bio-centric rather than homocentric in philosophy. In fact ecotourism should accept nature largely on its terms and not trying to significantly transform nature for personal convenience.

Community Based Ecotourism

Humankind is facing an unparalleled challenge when it comes to reserving the planet's biodiversity. Solutions that are most likely to be successful are those that will work at the community level in a decentralized manner. So for the successful efforts in ecotourism promotion, the community or people's participation is vital. Community based ecotourism implies that a community is taking care of their natural resources in order to gain income through operating a tourism enterprise and using that income to better their lives. It is also a better option for the people from the developing countries as an alternate means of livelihood, as it can both protect the resource base if properly managed and can act as an economic source.

Why Ecotourism

- Contributes to conservation of biodiversity.
- Sustains the well being of local people.
- Includes an interpretation/learning experience.
- Involves responsible action on the part of tourists and tourism operators.
- Is delivered to small groups by small-scale businesses.
- Requires lowest possible consumption of non-renewable resources.
- Stresses local ownership and business opportunities, particularly for rural people.
- Avoids negative impacts that can damage or destroy the integrity or character of the natural or cultural environments being visited.
- Educates the traveller on the importance of conservation.
- Directs revenues to the conservation of natural areas and the management of protected areas.
- Brings economic benefits to local communities and directs revenues to local people living adjacent to protected areas.
- Emphasizes the need for planning and sustainable growth of the tourism industry and seeks to ensure that tourism development does not exceed the social and environmental "carrying capacity."
- Retains a high percentage of revenues in the host country by stressing the use of locally-owned facilities and services.
- Increasingly relies on infrastructure that has been developed sensitively in harmony with the environment—minimizing use of fossil fuels, conserving local plant and wildlife and blending with natural environment.

Why Community Based Ecotourism

- Communities are an integral part of ecotourism ecosystem. They can help sustain ecosystem.
- Communities are to be affected first by any influences of ecotourism destination areas.
- For sustainable ecotourism involvement of communities in planning and local management is essential.
- Since ecotourism brings many changes at the community level therefore community should have greater voice in development and conservation of their natural resources.

Ecotourism Planning Principles for Natural Areas (WTO 1998)[2]

- Apply strict conservation measures to the natural areas to protect the flora, fauna and ecosystems and any existing archaeological and historic sites.
- Establish carrying capacity standards so that there is not over-development of the tourist facilities or over use of the environment by the visitors.
- Develop small scale tourist facilities in environmentally suitable locations, with locally based design, use of local building materials, energy saving devices and proper disposal of waste materials. A visitor centre with exhibits about the site and local conservation techniques should be developed.
- Prepare and distribute ecotourism codes of conduct for the tourists and tour operators, and monitor applications of these codes.
- Provide well-trained tour guides who will give accurate information to tourists, educate tourists about biological diversity, conservation techniques and observe good conservation measures during tours.
- Integrate local communities into tourism development by providing them jobs and income from tourism, arrange village tours where appropriate and educate tourists about the local cultures including their economic activities and how to show respect for their cultural traditions.

Guidelines for Sustainable Nature Tourism (WTO)

Nature tourism is of growing importance to countries and regions interested in sustainable tourism. It is one segment of the industry, which is difficult to define because it covers a wide range of activities. Nature tourists can be people casually walking through an undisturbed forest, or scuba divers admiring coral formations, or bird watchers adding birds to their lists. But, it is a segment of the market that will respond to environmental issues. The following guidelines can be used by the local planners to encourage community, environmental, and tourism constituencies to work together towards a common goal.

- The success of nature tourism depends on the conservation of nature. Many parks are threatened, and it is critical for everyone involved with nature tourism to realize that intact natural resources are the foundation.
- Nature tourism sites need revenue for protection and maintenance, much of which can be generated directly from entry fees and sale of products. Many protected areas charge nominal or no entrance fees and provide few if any auxiliary services. Nature tourists also desire gift shops, food services and lodging facilities and expect to pay for them.
- Tourists are a valuable audience for environmental education. In many Parks, opportunities are missed to provide environmental education. Whether "hard-core" nature tourists or "new" visitors with little background in natural history, all tourists can enhance their appreciation of the area through information brochures, exhibits and guides.
- Nature tourism will contribute to rural development when local residents are brought into the planning process. For nature tourism to be a tool for conservation and rural development, a concerted effort must be made to incorporate local populations into development of the tourism industry. In some cases, tourism to protected areas is not benefiting the surrounding population because they are not involved.
- Opportunities are emerging for new relationships between conservationists and tour operators. Traditionally, these groups have not worked together; often they have been in direct opposition. However, as more tourists come to Parks and Reserves, tour operators have the opportunity to become more actively involved with the conservation of these areas through education for their clientele and donations to Park management.

Scaling up Community Benefits through Ecotourism

Community-based ecotourism (CBE) is a growing phenomenon throughout the world. The CBE concept implies that the community has substantial control and involvement in the ecotourism project, and that the majority of benefits should remain in the community.

However, all efforts should be made to maximize benefits to local communities, but it is not realistic to expect that all communities will be able to own and manage their own businesses. Researchers have found that problems result from the communal enterprise approach such as the slowness of democratic decision making and inconsistent quality of services.

The community enterprise model is slowly being adapted by the communities themselves to give more responsibilities to specialized and trained members of the community. Case studies indicate that joint ventures between a community or family and an outside business partner are frequently preferable because of the need for the community to connect its small-scale venture to the outside world (Epler Wood 1998[3]; Stronza 2000[4]). Making such ventures equitable to all parties, and creating an environment of trust between partners is a challenge, but contracts between parties has been shown as the best way to proceed. Outside advice for communities, to ensure that their terms are equitable, is also highly recommended.

Understanding the decision-making process of a community is highly important. Many local communities do not have a top-down decision-making structure but may instead decide by consensus. This can be very slow and painstaking. It must also be understood that if the community is not involved in the initial decisions for the project, such as the needs and priorities of the project, the scope and the objectives, they will likely have less of a feeling of a stake in its success. (Adopted from Fergus Tyler Maclaren 2002)[5]

NGO Ecotourism Interface

The NGO sector has a role to play in assisting local communities with viable sustainable development projects. The problem is that inadequate expertise and understanding of the ecotourism product has led to inappropriate assistance patterns.

Some of the primary problems are as follows:

- Lack of cooperation with the private sector that is already offering ecotourism in the destination country.

- Poor communication between NGOs and the private sector.
- NGOs using their tax-free status to offer products that are in direct competition with the private sector.
- Business and tourism planning coming after communities are targeted by NGOs for ecotourism development.
- Top-down, bureaucratic management of projects where the intent is to foster community management.

The most important new trends in development include better pre-assessment of community needs and community readiness, a full analysis of land tenure and other legal issues relating to long-term access to the political process, training that is more targeted at community needs, research on the market for community ecotourism paired with efforts to develop appropriate linkages to the global marketplace, and developing improved systems to protect communities from competing land-uses.

One of the key issues in ecotourism today, is how much this portion of the global marketplace can actually represent a different kind of responsible commerce. A growing number of articles assume that ecotourism will use the same model as much of the rest of the global marketplace, and that it will, by default, disempower local people for the purposes of selling off their land and culture to the highest bidder.

Ecotourism can be a form of enterprise in which local people actively participate in the formulation of social "contracts" that go well beyond standard business formulae. Given that the large majority of ecotourism businesses are locally owned (Sanders and Halpenny 2001)[6], and socially conscious owners are frequently involved (Christ 1998[7], Stronza 2000[4]), communities can achieve more than the *status quo*. The involvement of the NGOs and development assistance community also does frequently help to reverse top-down business patterns. Participatory decision making processes can make a difference, and increasingly these processes are being used, even in regions where communities have been sidelined in the past (Sproule 1998[8], Drumm 1998[9]).

No new economic development system can completely transform the social and political context of people or region. Finding genuinely equitable formulae for responsible commerce between cultures, and between those in the North and the South, the East and the West can without question be challenging. But honest efforts to find increasingly equitable approaches to the development of ecotourism should continue.

Future Strategy

In order to make ecotourism a more effective tool for sustainable community development, the following issues should be addressed:

1. More market research is necessary to understand and evaluate the international market for community ecotourism (Epler Wood, 1998)[3].
2. Community ecotourism enterprises need separate designation within tourism laws and regulations allowing them to legally manage tourism without the same regulations and tax structures that apply to standard tourism businesses (Drumm, 1998)[9].
3. Partnerships between communities, governments, NGOs and the private tourism sector have to be improved. The strengths of each partner must be used in a more effective way and greater coordination between the partners is needed. The private sector should link the projects to the market and explain the customer expectations. NGOs should provide training and act as intermediary for communities in business negotiations. The government should provide appropriate regulation and guidelines.
4. Partners should also become more responsible in relation to assistance in emergency situations, like natural disasters. Preferably these responsibilities even have to be documented.
5. Individual governments should establish national ecotourism strategies. Specific ecotourism development zones should be created. No unsustainable activities like overexploitation of natural resources should be allowed in these zones. Further, a national register of all community tourism projects should be developed to improve awareness and promotion.
6. The issues of participation and distribution of benefits must be evaluated before starting a project and should be reviewed every year. All agreements between the different parties should be approved by the elected community authority and written down in a simple, yet legally binding, document. Participation from the community should be maximized to guarantee the highest benefits. It should include elders, lower social classes and women.
7. Local guides should be acknowledged and recognized for their special knowledge of local ecosystems and cultures. A special

category needs to be created for native guides when licensing tour guides at the national level (Epler Wood, 1998)[3].
8. More community associations should be established to deal with ecotourism projects from the larger regional perspective.
9. Indigenous/local communities should be made aware of the uniqueness of their traditional culture, for example by teaching local culture and history already at the local school level.
10. International policies on guidelines for community ecotourism should be written in cooperation with indigenous/local communities so that it is possible to implement thc policies at a local level. Governments should be held accountable for the already established international agreements relating to ecotourism and indigenous cultures.
11. Development organizations should provide soft loans, assistance with fund raising and other forms of long-term credit needed by communities to help them establish their own tourism programmes. (Adopted from Fergus Tyler Maclaren 2002)[5]

Operational Framework for Ecotourism

As ecotourism is a community development tool, development must be sensitive to the requirement of many stakeholder groups, including tourism provider, public provider and residents. In order to implement ecotourism a partnership can be developed between various groups. The potential partners include the following:

(i) Organizations within the established tourism industry, particularly tour operators;
(ii) The government tourism bureau and natural resource agencies;
(iii) Non-governmental organizations, especially those involved with environmental issues, small business management and traditional community development;
(iv) Universities and other research organizations; and
(v) International organizations, public and private funding institutions, national cultural committees.

In order to achieve maximum benefits from the co-operative efforts of the stakeholders group, following efforts are desirable:

1. Build on the foundation of the local people.
2. Give responsibility to local people.

3. Consider returning ownership of at least some protected areas to indigenous people.
4. Hire local people.
5. Link government development programmes with protected areas.
6. Give priority to small scale local development.
7. Involve local people in the preparation of management plans.
8. Have the courage to enforce restrictions.
9. Support diversity as a value.
10. Build conservation into the evolving new national cultures.

The profile of service providers is illustrated best through the Fig. 8.1.

Figure 8.1 depicts the profile of the government sector and the private sector and the degree of involvement of the local communities with these activities.

Public-Private Partnership

Government can ask the tourist industry to assume a larger share of the burden of financing destination promotion. However, asking the industry to make greater voluntary contribution will not succeed without some form of government intervention. Because, destination promotion is a public good, individual beneficiaries have the incentives not to contribute to free ride on the contribution of others.

Government should play a leadership role in providing the necessary financing, management skills, and knowledge so that the private sector can operate smoothly and efficiently. The government can do this through following actions:

(i) Facilitating efficient private sector activity by minimizing market interference and relying on competition as a means of control.
(ii) Ensure a sound macro-economic environment.
(iii) Guarantee law and order, and the just settlement of disputes.
(iv) Ensure the appropriate provision of infrastructure.
(v) Ensure the development of human resources.
(vi) Protect the public interest without obstructing private sector activity with too many regulations.
(vii) Promote private sector activity by not competing in the business arena with private enterprise.
(viii) Acknowledge the role of small business entrepreneurs and facilitate their activities.

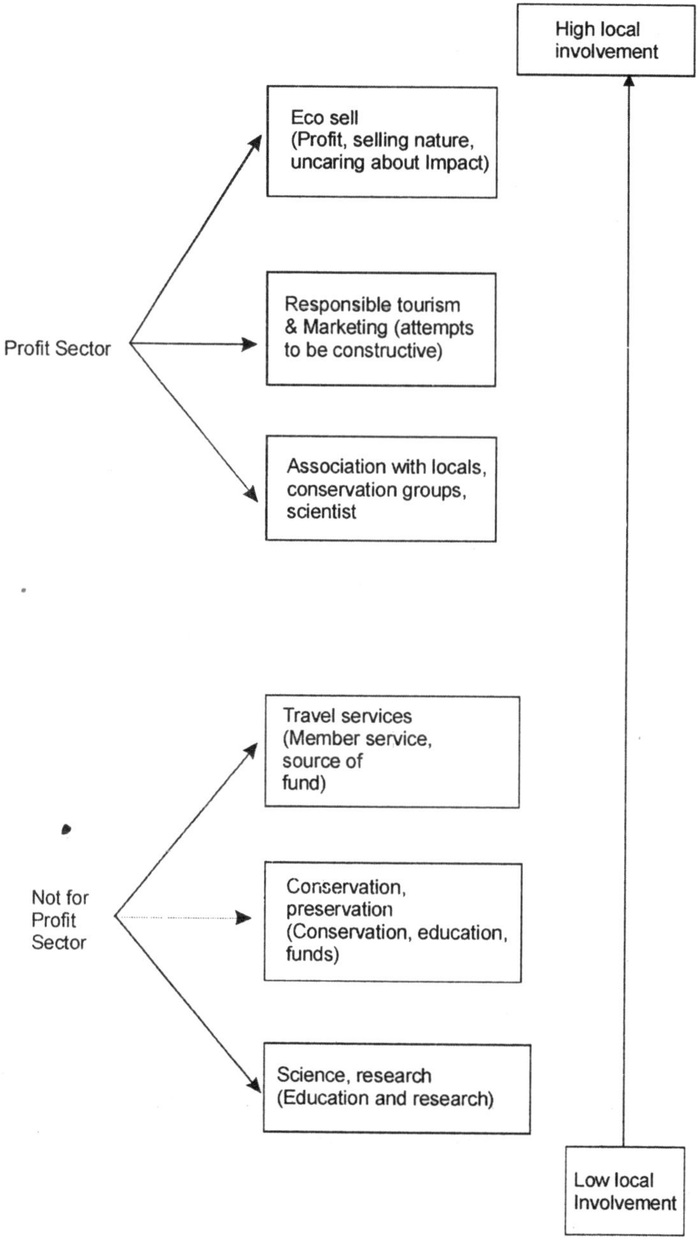

Fig. 8.1: Service provides through Government Sector and Private Sector

Role of Non-Government Organizations

NGOs can play an active role in the development of ecotourism. NGOs can sponsor ecotourism trips for a number of reasons like:

- Providing member service
- Donor rips
- Source of funding
- Education and research

NGOs can indulge in capacity building approach to ensure that ecotourism benefits communities and merges with traditional practices and conservation through the training of local people by organizing workshops.

NGOs can also develop a centre to provide information on ecotourism destinations, tour operators and lodgings, and relevant publications and information.

Role of Tour Operators/Travel Agents

Travel agents and tour operators are key stakeholders in the relationship that exists between the destinations and the tourists. Both have tremendous influence on tourist relative to the choices made and type of experience gained from travel.

The role of tour operators includes following functions:

(i) Preparing the client for trips
(ii) Conflict resolution
(iii) Camps and camp cleanliness
(iv) Legal consideration
(v) Environmental education
(vi) Tour planning

Apart from these, a tour operator performs several other functions. One of the things the tour operator can do is to promote various forms of tourism, like adventure tourism in that area by providing equipments and other specialized services for the public.

Rolc of Community

The involvement of community in ecotourism is must. The communities can appoint their representatives, which will act as a management committee similar to tourism management board. The community can be involved in the following ways—

1. Provisions to allow the community for exclusive use of biophysical resources of the protected area for the subsistence purpose.
2. Provisions for technical and professional training opportunities relating to positions in tourism and in conservation agencies.
3. Priority status in the hiring programmes undertaken by tourism interests and conservation agencies.
4. Priority status in the licensing of business to be operated in the park or protected area.
5. Compilation of traditional knowledge and heritage values of the aboriginal societies, for use both by the communities themselves in strengthening their societal traditions, and by the conservation agency in managing the protected area and in giving to its visitors a heightened appreciation of the traditional societies.
6. Using the services of the local communities in the management of ecolodges and resorts.

These are some of the functions, which can be performed by the various stakeholders and can be used as a guide to operational framework.

Conclusions

Integrated tourism, to be successful, must promote sustainable development by establishing a durable productive base that allows local inhabitants and service providers to enjoy rising standards of living. Ecotourism can be used as an effective tool, which ensures ecological, environmental, economical and cultural friendly tourism. Inappropriate tourism development can result in local people losing access to water, land and community areas, and to the creation of tourist enclaves and to social pollution.

The positive contribution of tourism is significant, but there are a number of challenges to be met if the potential for sustainable local development and poverty reduction, through the localization of benefits, is to be realised. These challenges include issues of ownership, economic leakage (from the local economy and through imports), local employment, benefit distribution, social and environmental impacts and dependency.

These problems can only be effectively addressed at the destination level with the active participation of the local communities, tour operators and government agencies.

Ecotourism businesses remain small, local and fairly well rooted in the context of the places being visited. Ecotourism is the product of a variety of small businesses, NGOs, and communities most of which are making little profit, and come from highly diverse backgrounds that have no relationship at all to big business.

The growth rate of the ecotourism market niche has been inappropriately touted. Despite myths to the contrary, ecotourism remains a small industry operating in remote areas. Some of the worst examples of false labelling of ecotourism seem to result from efforts to exploit its perceived market allure.

Despite these difficult problems, there is reason to believe that ecotourism can continue to improve sustainability standards for enterprise in areas that traditionally have had few development options, that it can help governments diversify their tourism economies, and it can assist communities by finding new formulae for gaining social contracts that respect their culture, intellectual property, and legal rights.

With a holistic strategic approach ecotourism should continue to contribute to make the overall tourism industry more sustainable, by increasing economic and social benefits for host communities, actively contributing to the conservation of natural resources and the cultural integrity of host communities, and by increasing awareness of all travellers towards the conservation of natural and cultural heritage.

NOTES

1. Cater, E., 1994. In: *Ecotourism: A Sustainable Option*? Wiley & Sons, Brisbane. pp. 3-17.
2. WTO, 1998, *Guide for local authorities on developing sustainable tourism*, Published by World Tourism Organization.
3. Epler, Wood M., 1998, *Making the Global Challenge of Community Participation in Ecotourism: Case Studies and Lessons from Ecuador*, The Nature Conservancy, Washington D.C., USA.
4. Stronza, A., 2000. Because it is ours: Community-based ecotourism in the Peruvian Amazon, a dissertation presented to University of Florida, Gainesville, FL, USA.
5. Fergus, Tyler Maclaren, 2002. A Strategic Approach for Community-Based Ecotourism Development.www.recoftc.org/03region/materials conference reports/CBT international.html.

6. Sanders, E. and Halpenny, E., 2001. *The Business of Ecolodges, A Survey of Ecolodge Economics and Finance*, The International Ecotourism Society, Burlington, VT, USA.
7. Christ, C., 1998. *Taking Ecotourism to the Next Level, A Look at Private Sector Involvement with Local Communities, Ecotourism*, A Guide for Planners & Manager, Volume 2, The International Ecotourism Society, Burlington, VT, USA.
8. Sproule, K., 1998. *Guidelines for Community-based Ecotourism Programmes, Lessons from Indonesia*, Ecotourism, A Guide for Planners & Managers, Volume 2, The International Ecotourism Society, Burlington, VT, USA.
9. Drumm, A., 1998. *New Approaches to Community-based Ecotourism Management, Learning from Ecuador,* Ecotourism, A Guide for Planners and Managers, Volume 2, The International Ecotourism Society, Burlington, VT, USA.

9

Ecotourism—Livelihood Security—Biodiversity Conservation Integration

A Holistic Approach

A.K. Bhattacharya and *Yogesh Dubey*

Background

India is among one of the twelve mega bio-diversity countries with vast resources in the form of natural landscapes and bio-diversity areas and heritage sites. These areas provide tremendous potential in terms of promoting Ecotourism. In most of the areas we are faced with at least two major issues in making Ecotourism a practicable tool for sustainable development. One issue is how to develop and manage the 'know-how' for Ecotourism without posing threat to the bio-diversity. The other issue is how to integrate Ecotourism with the biological and the sociological environments. To develop Ecotourism as a tool for such areas it is important that it be developed and managed in a manner whereby all threats to global bio-diversity are minimised and options for sustainable livelihoods are being created. With human pressures all around, bio-diversity conservation is also a question of providing sustainable livelihoods where Ecotourism is a good opportunity for community involvement for outsourcing their livelihoods and a way for bio-diversity conservation. The present understanding amongst the forest policy makers, managers and tour operators about the complexities of Ecotourism is not exhaustive and hence it calls for better understanding and efforts to be made in this direction.

The three basic challenges generally faced in this area relate to bio-diversity conservation, promotion of community development and management of Ecotourism as a viable alternative. Given the status of threatened ecosystems in India, Ecotourism needs to be managed in such a way that the integral values and functions of bio-diversity conservation areas are maintained and enriched and also such activities yield reasonable pay offs at the local level.

Tourism to Ecotourism

Ecotourism is a component of the field of sustainable tourism. Ecotourism aspires in all cases to achieve sustainable development results. However, it is important to stress that all tourism activities, be they geared to holidays, business, conferences, congresses or fairs, health, adventure or Ecotourism, should aim to be sustainable. This means that the planning and development of tourism infrastructure, its subsequent operation, and its marketing should focus on environmental, social, cultural and economic sustainability criteria.

The following basic components are common and integral in all the available definitions of Ecotourism:

- Nature Based Tourism
- Ecological Sustainable Management (Long-Term Conservation of Resources)
- Conservation Education and Interpretation of Natural Environment
- Local Community Participation and Benefits
- Tourists' Satisfaction

The strong orientation of the Ecotourism field toward principles, guidelines and certification based on sustainability standards gives it an unusual position in the tourism field. In the years since the concept was first defined, a general consensus has formed on the basic elements of Ecotourism.

Ecotourism

- Contributes to conservation of bio-diversity.
- Sustains the well being of local people.
- Includes an interpretation/learning experience.

- Involves responsible action on the part of tourists and the tourism industry.
- Is delivered primarily to small groups by small-scale businesses.
- Requires the lowest possible consumption of non-renewable resources.
- Stresses local participation, ownership and business opportunities, particularly for rural people.

Ecotourism is of special interest for its relationship with conservation, sustainability, and biological diversity. As a development tool, Ecotourism can advance the three basic goals of the Convention on Biological Diversity:

- Conserve biological (and cultural) diversity, by strengthening protected area management systems (public or private) and increasing the value of sound ecosystems;
- Promote the sustainable use of bio-diversity, by generating income, jobs and business opportunities in Ecotourism and related business networks; and
- Share the benefits of Ecotourism developments equitably with local communities and indigenous people, by obtaining their informed consent and full participation in planning and management of Ecotourism businesses.

In the field, well-planned and managed Ecotourism has proven to be one of the most effective tools for long-term conservation of bio-diversity when the right circumstances (such as market feasibility, management capacity at local level, and clear and monitored links between Ecotourism development and conservation) are present.

Ecotourism Development to Development Oriented Tourism

"Integrated tourism, to be successful, must promote sustainable development by establishing a durable productive base that allows local inhabitants and service providers to enjoy rising standards of living". This was expressed by Mr. David Barkin in one of his papers in 1996[1] when the concept of Ecotourism was shaping up and word used for "sustainable tourism development" was "integrated tourism". There is no doubt that people consider "Ecotourism" as "Sustainable Tourism". Ecotourism can be used as an effective tool, which ensures ecological, environmental, economical and cultural friendly tourism.

There is no doubt that Tourism has become an important sector for developing countries seeking to maximize foreign exchange earnings, increase employment and secure financial resources to conserve natural and cultural heritage. There is another very important factor in tourism industry, that is its marketing mechanism. Decisions made by tourists and the industry in the originating countries can assist or harm local communities, who are not being involved when tour is being planned, marketed, sold and bought.

Inappropriate tourism development can result in local people losing access to water, land and communal areas, and to the creation of tourist enclaves and to social pollution.

The positive contribution of tourism is significant, but there are a number of challenges to be met if the potential for sustainable local development and poverty elimination, through the localization of benefits, is to be realized. These challenges include issues of ownership, economic leakage (from the local economy and through imports), local employment, benefit distribution, social and environmental impacts and dependency. These problems can only be effectively addressed at the destination level with the active participation of the local communities, tour operators and government agencies.

Rapid and sustained tourism growth and the search for new destinations mean that more and more communities will be affected by the tourism industry. This provides opportunities for economic development, but there are also costs to be minimized. The demand side drives the industry; however, the sustainability of the sector at the destination is dependent upon some public control over the effects of the industry on the environment and socio-cultural structure of the area. It is the natural and cultural heritage of the area and the living culture of the local people that attract tourists. The negative impacts of tourism on the environment and local communities need to be managed and the adverse impacts mitigated in order to maintain the asset.

A tourism monoculture adversely affects the inherent quality of the destination and over-dependence on tourism increases the economic vulnerability of the area to decisions made elsewhere by consumers and investors. Tourism development frequently brings with it demands for goods and services which are not produced in the local economy. These goods and services are then showered outside the local area, often internationally, and only a small proportion of the expenditure remains in the local economy. This is a particular problem in mountain areas. These leakages reduce the development impact of tourism, whereas the development of linkages results in the creation of more jobs and opportunities to locals.

Community Based Ecotourism

In its basic concept, community-based Ecotourism refers to Ecotourism enterprises that are owned and managed by the community. Furthermore, community-based Ecotourism implies that a community is taking care of their natural resources in order to gain income through operating a tourism enterprise and using that income to better their lives. It involves conservation, business enterprise and community development. In any community-based Ecotourism enterprise there will be direct and indirect participants and direct and indirect beneficiaries. It is important that the entire community has some level of involvement and some level of benefit. Direct participants in a community Ecotourism enterprise would be the managing committee and the actual workers involved with producing products or services for sale. In some instances those who are the primary users of a resource might be involved as participants in a project as well. Indirect participants would be the broader community who selected the management committee of a project and those who do not directly use the natural resources involved in an enterprise. Direct beneficiaries would be employees, craft producers, guides, committee members, etc., while indirect beneficiaries would be the wider community as recipients of community development projects funded by tourism revenues would.

Issues to be Aware of When Developing Community-Based Ecotourism

The "community-base" for community enterprises is rarely, if ever, all encompassing. Those community members with some initial disadvantages, such as poor housing, insufficient land or income, tend to be among those excluded from participation with Ecotourism enterprise development. And, depending on how the Ecotourism enterprise is designed, they may be excluded from the benefits of Ecotourism development as well as the issue of participation. While there is an increasing recognition of the need to involve communities in general, there is much less agreement about exactly who should participate and to what extent. One common definition states that participation is "giving people more opportunities to participate effectively in development activities...empowering people to mobilize their own capacities, be social actors rather than passive subjects, manage the resources, make decisions and control the activities that affect their lives." How a community chooses to define participation will be important for determining what level of participation will satisfy the Ecotourism project goals. It's a very tricky subject. For even where attempts at all inclusive community involvement are well thought out, participation has sometimes been decided on the basis of political affiliation, landownership, kinship or gender.

Linking Communities, Conservation and Development

"It has been argued, in a review (WWF 1991)[2] of integrated conservation and development projects (ICDP) in Latin America, Africa and Asia, that for an ICDP to achieve its bio-diversity conservation goals, it is not enough for the development component to foster improved local living standards—a difficult enough task."

The development process must not only be economically and biologically sustainable, but must also conserve the ecosystem of the protected area. To satisfy this exacting requirement, explicit linkages between project development components and conservation objectives are needed. "Ecotourism" is one concept that has been heralded as a means for establishing such linkages. Unfortunately, while there has been a great deal of discussion about the contributions of Ecotourism to local community well being, very little is visible on the ground. There have been numerous efforts to create "guidelines" for Ecotourism development, and the bulk of these assign local participation a privileged position. However, these guidelines have not focused on community-based enterprises nor addressed what, if any, mechanisms exist for nurturing community interest in establishing such enterprises. Rather, they tend to view local participation as something incorporated into the design and implementation of Ecotourism enterprises, which are launched from "outside" the community. Additionally, they tend to focus on the quantitative dimensions of participation (e.g. number of people involved, revenue generated) rather than qualitative aspects of local participation (positions in the Ecotourism enterprise). Whether a community-based tourism enterprise encourages community conservation of natural resources depends on at least four factors:

1. The scale of benefits received by local residents (and whether they outweigh the short term-costs of foregoing resource use or changing resource management);
2. The extent to which the benefits are clearly perceived as dependent on the resource base, and therefore, on sustainable management;
3. Whether benefits reach all resource users; and

4. Whether local institutions are strengthened, so as to increase their capacity for collective resource management.

If the above conditions are not met, massive financial earnings for a few people will not necessarily change communities' approach to resource utilisation. Changes in resource use will also depend on whether communities gain rights/ownership and control over resources and hence a sense of responsibility for their management. Developing mechanisms for local residents to benefit directly from the establishment of Ecotourism enterprises in and around their home areas can help offset any loss of revenue from traditional extractive activities, which may be curtailed, in some instances, by the establishment of the new enterprise. It can also motivate community participation in conservation activities if adequate rewards can be consciously realized. Developing National Community Based Ecotourism Strategies can help focus efforts on achieving such mechanisms. At the least, it's a very good way of beginning the process of forging the partnerships deemed so valuable to the process.

Issues and Options

Ecotourism *per se* in India is identified with only Protected Areas. The very concept of Ecotourism should not only be applied to National Parks and Sanctuaries but should also be extended to areas falling outside the Protected Area network. This would include forested areas outside the PA Network. The need of the day to explore the possibilities of Ecotourism potential in other areas which contain natural character. By doing this we would also be able to overcome the problems related to carrying capacity by evenly or rather distributing the tourism pressure effectively to other areas. The potential to develop the forest fringe areas into recreational sites should also be considered. These areas could be developed for special category of tourists whose prime interest is recreation.

Policy Issues

The national policy on tourism stipulates that tourism should become a unifying force nationally and internationally fostering better understanding through travel. The national policy on Ecotourism also emphasises the need of involvement of local communities for overall development of the area. It also aims at identifying the conflicts between resource use for tourism and the livelihood of local people and attempts to minimize them. The significance of tourism development and its compatibility with environment and socio-cultural characteristics has also been emphasised. Keeping the present framework of policy directives on Ecotourism achieving the goals does not look like a Herculean task. Under the present scenario it is important to do a complete need analysis as to what is required in which all areas considered to have potential of Ecotourism. How and what kind of mechanisms could be used for involvement of community? Ecotourism in the forest areas is generally being regulated by Forest departments of the respective States but it is still not a recognized Programme of the Forest Department. Forest Departments of respective States should take up the Ecotourism as a major Programme and develop policy guidelines in concurrence with the existing national and State level policies. There is also a need to bring in little more element of liberalisation in terms of making policy little flexible to attract more investors in this area.

Strategies for Sustainable Ecotourism

Tourism is one of the most important industries of the world and contributes substantially to the economies of all the countries, with an annual turnover worth US$ 350 billion and employs 11 per cent of the total workforce of the world (PATA) (Pabla 1999)[3]. Even in India, which gets a very small proportion of world's foreign tourist, tourism is the third largest foreign exchange earning activity. However, the share of nature based tourism is very small in the overall earning whereas many of the east African economies are thriving on the nature based tourism industry.

There are many reasons why National Parks and landscapes in India are not the destinations of tourist. One of the main reasons is that most of our nature conservation areas do not have Ecotourism development as one of their prime objectives of management as it is considered to interfere with the conservation objectives of these areas.

Therefore, it is extremely important to evolve a strategy by which ecotourism could be established as an industry on sustained basis. The strategies could be broadly classified into two categories, namely short-term strategies and long-term strategies.

Short-Term Strategy

The first and the foremost to begin with would be to identify potential areas and prepare a detailed inventory of such areas in terms of their tourism potential. The local communities residing in such areas should be taken into confidence right from the beginning and their involvement should be clearly defined. The local communities should be amply motivated to as to sustain their interests in a particular area where Ecotourism is to be planned. Ecotourism planning should be directly linked with the livelihood issue and options that would be generated in any area for the local people. Organizing workshops for local communities to explain them the rationale of Ecotourism and how it is going to better their lives could be part of immediate strategy. These workshops could also be coupled with launch of awareness Programme in such areas. Village level committees could be formed if the motivation levels are found to be satisfactorily high.

Long-Term Strategy

Reexamination of the existing regulations and development of new policy should be done to make Ecotourism more business friendly without affecting the conservation of forest and wildlife. Research and monitoring should be promoted in such areas to find newer possibilities of enhancing and developing tools for effective Ecotourism. Research should also be promoted in these areas to develop best practice guides for regions, which have similar kind of, forest areas and biological diversity. Strategies should be developed so as to ensure the original ecological character of the area on long term basis. Tourism packages should be developed so that they are economically viable and are attractive enough to draw tourists to such areas. A lot needs to be done in terms of infrastructure development. Good interpretation centres should be developed with good audio-visual aids so as to give a synoptic understanding of the area where a tourist is interested to visit. It would also be extremely important to build linkages between various agencies related to the tourism and centralized information centres being created from where information could be easily accessed. A more robust and strong marketing strategy needs to be developed so as to make all the potential areas of tourism visible to the national and international community.

NOTES

1. Barkin David, 1996. www.mtnforum.org / resources / library / harao 01a.htm.
2. WWF Technical Paper, 1991. Community Participation in ICDPs.
3. Pabla, H.S., 1999. Forest Department, Government of Madhya Pradesh, Draft Policy, "A framework for the development of the recreation forests in Madhya Pradesh".

10

Community Based Ecotourism Identification of Issues and Options through Stakeholders Analysis

A.K. Bhattacharya, Suchitra Banerjee and *Vandana Saxena*

Background

Community-based ecotourism refers to Ecotourism enterprises that are owned and managed by the community. Furthermore, community-based ecotourism implies that a community is taking care of their natural resources in order to gain income through operating a tourism enterprise and using that income to better their lives. It involves conservation, business enterprise and community development. In any community-based Ecotourism enterprise there will be direct and indirect participants and direct and indirect beneficiaries. It is important that the entire community has some level of involvement and some level of benefit. Direct participants in a community ecotourism enterprise would be the managing committee and the actual workers involved with producing products or services for sale. In some instances those who are the primary users of a resource might be involved as participants in a project as well. Indirect participants would be the broader community who selected the management committee of a project and those who do not directly use the natural resources involved in an enterprise. Direct beneficiaries would be employees, craft producers, guides, committee members, etc., while indirect beneficiaries would be the wider community as recipients of community development projects funded by tourism revenues would.

Hypothesis

The analysis is based on the following hypotheses:

- Ecotourism is a rapidly growing industry—market forces will play an important role in the growth of ecotourism.
- Ecotourism is an effective tool for sustainable wildlife management.
- Community based ecotourism is the only long-term solution for ecotourism.
- There is tremendous potential of ecotourism at national, state and local levels.
- Ecotourism can play a significant role in the livelihood security of the local communities.
- Sensitive issues like carrying capacity and policy need immediate attention.

Objectives

The specific objectives of the analysis were set to

— identify key emerging issues in community based ecotourism, specially the community benefits.

— generate options to address the key issues.

Approach

— The whole analysis was accomplished through participatory approach at all levels involving all the stakeholders ensuring active participation of the stakeholders.

— The stakeholders included officials of different levels from Forest, Tourism, Energy and Rural Developments, Tourist Operators, NGOs, Professionals, Scientists and Academicians, Community members, Village level institutions including forest committees and *gram sabha* members

— The methodology included intensive and extensive discussions, focus group discussions, village and house survey through schedules, field visits by combined team of major stakeholders followed by brainstorming sessions and presentations of results of the analysis and open house discussions in the workshop.

Ecotourism Defined

About thirty available definitions were reviewed and examined. For the purpose of the analysis following few definitions were considered:

- Responsible travel to natural areas that conserve the environment and sustains the well-being of local people.
- Ecotourism is a significant travel to natural areas to understand the cultural and natural history of the environment, taking care not to alter the integrity of the ecosystem.
- Ecotourism generates economic opportunities that make the conservation of natural resources financially beneficial to local communities.

Basic Components

The analysis of the definitions and the concepts of ecotourism led to the identification of following integral components of ecotourism—

— nature-based tourism.
— ecological sustainable management (long-term conservation of resources).
— conservation education and interpretation of natural environment
— local community participation and benefits.
— tourists' satisfaction.

Why Ecotourism

The review of available literature followed by analysis and discussion identified the following key reasons as to why the ecotourism is significant and be adopted as sustainable tourism:

- Avoids negative impacts that can damage or destroy the integrity or character of the natural or cultural environments being visited.

- Educates the travellers on the importance of conservation directs revenues to the conservation of natural areas and the management of protected areas.
- Brings economic benefits to local communities and directs revenues to local people living adjacent to protected areas.
- Emphasizes the need for planning and sustainable growth of the tourism industry and seeks to ensure that tourism development does not exceed the social and environmental "carrying capacity".
- Retains a high percentage of revenues in the host country by stressing the use of locally owned facilities and services.
- Increasingly relies on infrastructure that has been developed sensitively in harmony with the environment-minimizing use of fossil fuels, conserving local plant and wildlife and blending with natural environment.

Why Community-Based Ecotourism

The key factors to consider 'community-based ecotourism' as essential for sustainable tourism were identified as follows:

- Communities are an integral part of ecotourism ecosystem. They can help sustain ecosystem.
- Communities are to be affected first by any influences of ecotourism destination areas.
- For sustainable ecotourism involvement of communities in planning and local management is essential.
- Since ecotourism brings many changes at the community level, therefore, community should have greater voice in development and conservation of their natural resources.

Ecotourism Linked Key Words

All the key words, which the stakeholders considered had some relevance for the community based ecotourism, were captured. The key words have been arranged in alphabetical order.

Accountability
Adventure tourism
Basic needs
Benefit sharing
Bringing people close to nature
Bringing tourists to tribal
Capacity building
Carrying capacity
Conservation (bio-diversity)
Conservers' society
Cooperatives
Cost benefit analysis
Cost effectiveness
Culture and traditions Eco-development
Eco-friendly
Ecological complexes–Ecozones
Employment
Empowerment
Environment education
Funding to initiate
Gene pool
Goodwill of local communities
Governance (at all levels)
Health tourism
Income opportunities
Indigenous technical knowledge
Infrastructure development
Linkages-co-ordination of various departments
Local industries
Losses
Maintenance of system
Manpower
Marketing Environment
Marketing Indigenous products
Monitoring Natural heritage
Networking (products)
Orientation
Peoples' participation
Privatization
Publicity
Recreation
Rehabilitation
Religious tourism
Replication model
Representation
Research
Revenue
Stakeholders
Sustainability
Sustainable development
Tribal development
Wildlife friendly
Women's participation

Further these key words were categorized into five groups, viz. Social, Economic, Ecological, Policy and Institution depending upon the perception of the stakeholders for further analysis. The social dimension constituted the maximum percentage.

(Percentage)

	Social	*Economic*	*Ecological*	*Policy*	*Institutional*
Overall	30	18	19	10	23
Multiple choice	42	25	27	13	38

Problems and Solutions

The details of the problems associated with the community based ecotourism and their solutions were identified as follows:

S.No.	*Problems*	*Solution*
1.	Infrastructure—bad roads	• Regular maintenance
2.	Funds	• Create autonomous body (cooperative society)
		• Government (state or central)
3.	Marketing of et	• Strengthening existing network
4.	Drinking water	• Efficient rain water harvesting
		• Rational use of water
		• Augmentation of drinking wate schemes
5.	Improper solid waste management	• Treatment plant
		• Recycling of waste
6.	Lack of trained man power for ecotourism	• Training in specialized institutions like IIFM
7.	Increase pollution—air, water	• Restriction on vehicle movement
		• Catchment area treatment
		• Wetland management
		• Using better operated vehicles in core areas
		• Harnessing/promoting non-conventional energy source
		• Using cable cart
		• Introducing bullock cart
8.	Exceeding carrying capacity	• Horse riding
		• Regulating number of tourists
		• Dispersing tourists by developing other spots
9.	Lack of health facility	• Creating health resorts to divert tourists
		• Emergency health services, rescue team
10.	Lack of proper security—law and order	• Introducing tourist police
		• Regular police patrolling including odd hours
		• Involving local people in law and order
11.	Exploitation of tourists	• Displaying rates of all commodities and services at a tourist spots

(*Contd.*

(*Contd.*)

12.	Exploitation of local inhabitants	• Monitoring by police party/local body having local representative • Local guides to be made more responsible • Awareness generation/ education
13.	Threat to ecology	• Regulatory control through properly laid out rules • Compulsory plantation by tourists • Banning shifting cultivation
14.	Soil erosion	• To provide free of cost planting material to local villagers • Effective soil conservation measures • Metalling of roads
15.	Forest fire initiated by tourists	• Awareness generation • Putting hoarding • Educating tourists • Interpretation of nature/ biodiversity by displaying boards • Putting up models and watchtowers • Creation of safari
16.	High expectations of tourists (disappointed by not seeing wildlife)	• Encouraging bird watching, butterfly • Habitat management (salt licks and waterholes) increasing forest cover • Establishing animal rescue centres • Creating mini zoo/aviary
17.	Shortage of tourists	• Organise special events/ festivals • Inviting educational institutions • Offering package tours • Establishing tourist information centres near bus stand, railways station • Promote incentive tourism • Create conducive environment

People—From Problem to Solution

In the background of the recognition of the role of the communities in the natural resource management through various policy measures like Joint Forest Management, Panchyati Raj Act etc., the perception of the stakeholders was captured about the change in the attitude of the people over the last one decade—people as problem in 1990 to people as solution in 2002.

People (1990)—Problem	*People (2002)—Solution*
Reluctance	Willingness
No legal framework	Framework existing
Lack of awareness	Awareness
Authoritative	Participative
Individual	Institutional
Restricted	Open
Resolution	Involvement
Lack of policy	Policy development
Low stake	High stake
Fun	Tool of management
Exploitation	Involvement

General Recommendations

Recommendations made by the stakeholders for the different aspects and issues can be seen on Pages 145-148.

Site Specific Recommendations

During the analysis, an adjoining area having the ecotourism was visited by the stakeholders. The following recommendations were made for the promotion of the ecotourism in the area.

1. Preparation of Ecotourism management plan.
2. Training of communities.
3. Infrastructure development, laying of trails, construction of day-shelter, transport arrangement, interpretation skills (guides etc.).
4. Local handicraft, cuisine, folksongs/music to be encouraged.
5. Bigger and open water bodies should be only be used for recreational purpose.

Aspect	*Issues*	*Plan of Action*	*Who will take Action*
Social	Proper planning: safeguarding local culture and tradition of the area as per the interest of the community	Education and awareness	All stakeholders/Concerned Institution/ Influential people in the village/Local community like Panchayat, youth club, Social worker/Government agencies/NGO/State department of Tourism/Forest Department.
	Promotion of local talent	Encouraging local cultural activities	
	Linking local festivals and rituals	Publicity	
	Conflict	Conflict resolution through some committee at the village level	
	Safety (social security of tourist)	Trust building through the committee	
	Promotion of local peoples' knowledge related with the integrity of the ecosystem	Application of PRA techniques, dissemination of knowledge through village level workshops	
Economic	Local employment generation	Economic survey	Govt. agencies/ Village level institutions / NGO / Implementing Agency/ Local bodies with support and capacity building by department and Institutions/ Training by Tourism department
		Assessment of resource potential	
		Develop backward and forward linkages	
		Capacity building	

(*Contd.*)

(*Contd.*)

	Use of local people, local products and cuisine	Capacity building through training, awareness programmes	
	Infrastructure development	Developing basic facilities by government agencies/local bodies/ funding agencies	
	Introduce local food/restaurant cuisine	Training	
Ecological	Promotion of indigenous species diversity	Enlisting and protection guidelines	Government Department/Forest Department/Research Institution/ Municipal Corporation/NGO/Eco-development committee/Pollution control/Panchayat
		Identify sensitive species	
	Conservation of biodiversity/ environment with particular emphasis to waste and waste water management	*In situ* and *ex situ* conservation measures	
		Involving and empowering FPC	
		Formation of task force	
	Waste disposal management	Installation of proper waste management facilities	
	Protecting conservation of local area and promoting better environment	Formation of "taskforce" at State/ district level	
		Formation of Ecotourism committees in nearby villages	

(*Contd.*)

(Contd.)

Policy	Empowerment of local communities	Develop suitable legislation through state and central level committees	Government Agencies/ Community/NGO/Con- cerned department
		Participatory policy formulation and its implementation	
	Integrated management	Development of integrated management plan	
		Involvement of institutions/experts	
	Participatory policy formulation and its implementation	Stakeholders participation	
	Mechanism to implement the policy	Integration of government, community, corporate, and NGO systems	
	Logistics	Roads/transports/lodging/boarding	
	Promotion	Marketing, proper media planning, package and theme based, networking	
	Liberalisation and reduction of restriction	Formulation of liberal legislation	
	Cross sectoral/inter departmental linkages	Establishment of linkages among all concerned institutions and departments/reduction conflict through coordination committee	

(Contd.)

(*Contd.*)

	Review and monitoring of tourism policy	Faster for better implementation mechanism	
	Monitoring and evaluation of ecotourism projects	Strict legislation with dos and donts, rules and act	
	Zonation of tourism activities I.E., virgin, reserved, undisturbed and restricted zone identification	Survey, observation, mapping and accordingly	
Institutional	Involvement of local communities	Institutionalise Ecotourism committees	Government/Tourism Depart-ment/ Local villagers/Local bodies
	Formation of State level monitoring committee	Follow up of action on policy (State tourism policy)	

6. Infrastructure development.
7. Basic facilities development.
8. Publicity and promotion of the destination.
9. Creation of local committees.
10. Securities and departmental co-ordination.
11. Promotion of local economic earning activities.
12. Long-term planning.

Strategy for Community-Based Ecotourism

The stakeholders made following recommendations for the short-term and long-term strategies for the community-based ecotourism:

Short-term	*Long-term*
• Identification of potential areas.	• Policy development
• Motivation of community	• Research and monitoring
• Infrastructure development	• Development of good practices
• Attractive income generation packages	• Empowering the people to self-sustain themselves
• Organising workshops for the local community	• Ensuring the ecological character of the area on a long-term basis
• Provide communication facilities	• Designing media campaigns and other effective outreach measures
• Formation of village (local) level committees	• Development of infrastructure
• Local fee (entry fee)	• Providing attractive and economic tourist packages
• Basic facilities development	• Infrastructure development
• Training for local youth and inform of Internship for tourism guides	• Market promotion ecotourism
• Awareness Programme	• Creating Task force
• Co-ordinating committee formation	• Periodical review of the policy
• Case study development	

11

Local Initiatives to Localise Ecotourism

An Exploratory Study in Kerwa Van Vihar National Park Catchments, Bhopal, India

A.K. Bhattacharya, Suchitra Banerjee and *Vandana Saxena*

Background

From Tourism to Ecotourism

Tourism in general has already become the most important civil industry in the world. According to World Travel and Tourism Council tourism is now the world's largest industry (Lindberg and Hawkins, 1999)[1]. Ecotourism is the most important segments of tourism (WTO 1984[2] and 1991[3]). The concept of ecotourism evolved in 1980s and this concept mainly developed in response to calls for sustainable forms of tourism. One of these was an increase in environmental consciousness, which had begun in late 1960s and gained momentum in 1980s (Figgis 1993[4], Wight 1993[5], Figgis 1994[6], Harris and Leiper 1995[7], Eagles 1996[8], Ceballos-Lascurian 1996[9]).

When the term *ecotourism* first appeared in the 1980s, it was loosely applied towards travel that included any aspect of nature. Most of these travel packages and destinations, however, were little more than mass tourism taken outdoors. Most imparted more impacts than benefits on local communities and habitats. Over the years, changing attitudes by travellers, tour operators, communities and conservationists have significantly narrowed the ecotourism focus.

Butler writes that whale watching that began in response to their depletion of the Humpback whale and Blue Whale in the 1960s (resulting in them being declared protected in 1967 and 1966 consecutively) marked the birth of the ecotourism (Cebalos-Lascurian 1996)[9].

As a result of the increased environmental awareness, eco-development arose as a model that sought to integrate cultural, social and ecological goals with development. It has been suggested that out of this popular school of thought, the concept of ecotourism was developed as an opportunity to put the ideas of eco-development into practice (Sagasti and Colby 1993[10], Nelson 1994[11]). Ecotourism, being promoted as a means of 'giving nature value' and hence of achieving sustainable tourism development could be a useful choice besides the other types of tourism (Kamra 2001)[12].

Ecotourism has been defined in different ways. Ecotourism was first defined by Ceballos-Lascurian as: "Travelling to relatively undisturbed or uncontaminated natural areas with the specific objective of studying, admiring, and enjoying the scenery and its wild plants, as well as any existing cultural manifestations (both past and present) found in these areas" (Commonwealth Department of Tourism 1994)[13].

The International Ecotourism Society defines Ecotourism as "Responsible travel to natural areas which conserves the environment improves the welfare of local people."

A general consensus has formed on the basic elements of ecotourism:

- Contributes to conservation of biodiversity.
- Sustains the well being of local people.
- Includes an interpretation/learning experience.
- Involves responsible action on the part of tourists and the tourism industry.
- Is delivered primarily to small groups by small-scale businesses.
- Requires the lowest possible consumption of non-renewable resources.
- Stresses local participation, ownership and business opportunities, particularly for rural people.

Community-Based Ecotourism (CBE)

CBE management refers to ecotourism programmes, which take place under the control and active participation of the local people who inhabit or own a natural attraction (Lindberg *et al.* 1999)[14]. CBE works closely with local people, provides incentive for conservation, and more and more often, works outside of parks in locally owned land. Communities are an integral part of ecotourism ecosystem. They can help sustain ecosystem. Communities are to be affected first by any influences of ecotourism destination areas. For sustainable ecotourism involvement of communities in planning and local management is essential (Coltman 1989[15], WTO 1993[16], Cater and Lowman 1994[17]).

Since ecotourism brings many changes at the community level, therefore, community should have greater voice in development and conservation of their natural resources. Ecotourism can play a significant role in livelihood security as integral parts of ecosystem. CBE implies that a community is taking care of their natural resources in order to gain income through operating a tourism enterprise and using that income to better their lives. It involves conservation, business enterprise and community development. Community managed tourism attempts to let communities decide what type of growth they would like to see and then help them implement their plans (Lindberg and Hawkins 1999)[1].

Planning for CBE

Ecotourism areas often include existing communities, specially of traditional people, and therefore, ecotourism plan must be holistic in approach and consider ways to conserve local cultural traditions and identities and how to bring benefits to these communities. Although the researches indicate that the CBE can be a powerful force for conservation for conservation by providing benefits to local people, but meeting such conservation objectives requires careful and proper planning and design (Lindberg and Hawkins 1999)[1].

Some planning principles for ecotourism are as follows (WTO 1998)[18]:

- Apply strict conservation measures to the natural areas to protect the flora, fauna and ecosystems and make existing archaeological of historic sites.
- Establish carrying capacity standards so that there is not over-development of the tourist facilities or over-use of the environment by the visitors.
- Develop small scale tourist facilities in environmentally suitable locations, with locally based design, use of local building materials, energy saving devices and proper disposal of waste

material. A visitor centre with exhibits about the site and local conservation techniques should be developed.

- Prepare and distribute ecotourism codes of conduct for the tourists and tour operators, and monitor applications of these codes.
- Provide well-trained tour guides who will give accurate information to tourists, educate tourists about biological diversity, conservation techniques and observe good conservation measures during tours.
- Integrate local communities into tourism development by providing them jobs and income from tourism, arrange village tours where appropriate and educate tourists about the local cultures including their economic activities and how to show respect for their cultural traditions.

The Study Area—Kerwa-Van Vihar National Park Catchments

Kerwa

India in general and Madhya Pradesh in particular having unlimited potential in Ecotourism should be geared up to reap the benefits in a new perspective. One of the essential ingredients of ecotourism is community participation and community benefits. Since the destination areas for ecotourism in Madhya Pradesh are already under tremendous biotic pressure, for sustainable ecotourism, it becomes essential that the local communities are actively involved in its management. So close to the concrete jungle of Bhopal the Kerwa jungle are endowed with scenic and aesthetic values, which attract around one lakh visitors annually. The interesting issue is that most of them are local tourists including villagers from adjoining villages.

The area is close to Van Vihar National Park and has tremendous tourism potential and has also an impact on the socio-economic condition of the villagers. The local communities can play a significant role in the development of ecologically and sustainable management of ecotourism module in the area, on the other hand the ecotourism in the area can play an effective role in the economic development of the villages in the ecotourism destination area. Total available area is 50 sq. km where Ecotourism impact area is 10 sq. km. The common salient feature of the area are:

- *Local*: It is hardly thirteen km from the main centres of the capital.
- *Forests + wildlife*: The areas is naturally endowed with forests and wildlife.
- *Waterbody*: There is a dam, perennial water stream and also some water bodies.
- *Nearest nature*: So close to the city the area provides a serene natural surroundings.
- *Topography*: The area has good landscape, suitable for trekking and aesthetic value
- *Wilderness*: The area also provides wilderness for nature lovers.

Van Vihar

Van Vihar, the smallest National Park of India and one of most unique in the world is situated in Bhopal, the capital of largest state of India. Bhopal lies in the eastern edge of Malwa plateau and situated in the central part of state bounded by Rajgarh district in the northeast, Guna district in the east and south-east and Sehore district in the west and south-west.

Nature has tremendous capacity to resurrect itself. Van Vihar National Park is a living example that reflects this capacity of Nature. Van Vihar National Park was established in 1983. A 445 hectares of degraded and denuded hillock adjoining the city and dotted with numerous villages on the other side was provided legal protection by declaring it a National Park on 26th January, 1983.

Both Van Vihar and Kerwa are situated closeby, the former being the zoo having the status of national park whereas the later falls in the revenue forests, part of which has been transferred to the Forest Department for protection.

The Study

Looking to the growing tourism in this area, a tourism potential survey was done to understand the perception of the various stakeholders for planning sustainable CBE in the area.

Objectives

With the hypothesis that the small towns and the rural areas have considerable potential for ecotourism (Gunn 1988), the specific objectives of the study were set to—

— Identify key emerging issues in CBE, specially the community benefits.
— Generate options to address the key issues.

The study was intended for the following outcomes—

— Approaches for short term and long term strategy for CBE.
— Suggested module for CBE for operationalisation—Kerwa as a pilot project.
— Recommendations for State ecotourism policy guidelines.

Approach

- The whole analysis was accomplished through participatory approach at all levels involving all the stakeholders ensuring active participation of the stakeholders.
- The stakeholders included officials of different levels from Forest, Tourism, Energy and Rural Developments, tourist operators, NGOs, Professionals, Scientists and Academicians, Community members, Village level institutions including forest committees and Gram Sabha members.
- The methodology included intensive and extensive discussions, focus group discussions, village and house survey through schedules, field visits by combined team of major stakeholders followed by brain storming sessions and presentations of results of the analysis and open house discussions in the workshop.

The study was conducted during the period of October 2000 to March 2002. The field surveys were carried out in all the seasons.

The responses were received from:

- 69 groups (approximately 500 individuals)—visitors invariably come in groups of 6 to 10.
- 40 households—households were selected through random selection.

However the questions were addressed to the heads of the groups/ households, but attempts were made to obtain the group response through participatory approach. Finally the findings were discussed in a workshop participated by the major stakeholders.

Tourists' Perception

Tourist Influx

Tourists influx data indicate that rainy season is the peak period when the number of visitors visiting Kerwa is the maximum, specially on Sundays and general holidays.

This is followed by winter and minimum influx is in summer:

Seasons	*Average Sunday (annual)*	*Per day other holidays (annual)*
Rains (July/August)	200 (16000)	1000 (8000)
Summer (May/June)	500 (4000)	125 (1000)
Rest of Months	1200 (38400)	600 (19600)
Average	1217	596
Total	*58400*	*28600*
Annual Total	*87000*	

Visitors Profile

Nearly 79 per cent of the people visiting Kerwa are local visitors coming from different parts of Bhopal while 21 per cent are outsiders.

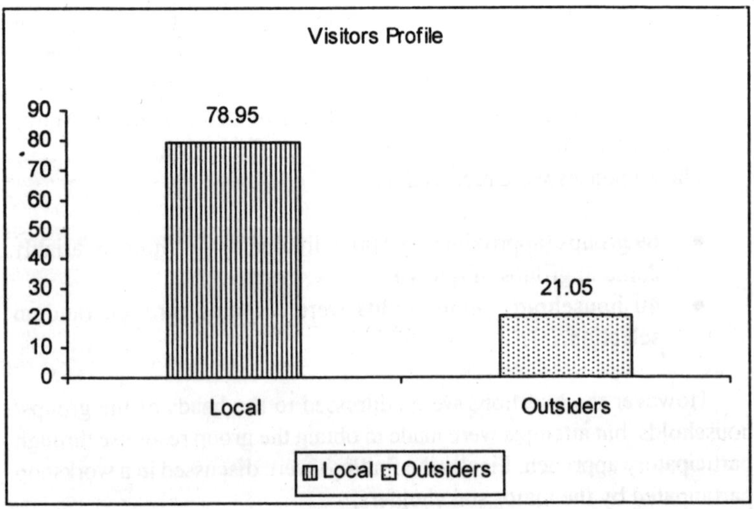

Education Level of Tourists

Most of the people visiting Kerwa are educated. The result of the survey indicates that 37.04 per cent group heads are postgraduates while the percentage of graduates is 33.33 per cent, followed by 18.52 per cent matriculates and 11.11 per cent technical.

Education	%
High school	18.52
Graduate	33.33
Post graduate	37.04
Technical	11.11

Occupation

The survey revealed that around 85 per cent group heads belonged to the service class, of which 15 per cent were from technical field while the rest were non-technical. Of the remaining fifteen per cent, 11 per cent are from business community and 4 per cent of visitors from the students community.

Income Groups

Most of the people visiting Kerwa belong to the middle class having the monthly income ranging between Rs. 3000 and Rs. 10,000, followed by 26 per cent between Rs. 10000 and Rs. 15000. The percentage of visitors whose income lies between Rs. 15001-20000 is 19 per cent, while that of between Rs. 1500-3000 is 15 per cent. A very small percentage of people visiting Kerwa is from the high income group.

Income group (Rs.)	%
1500-3000	14.81
3001-10000	37.03
10001-15000	25.92
15001-20000	18.24
> 20000	4.0

Purpose of Visit

The majority of visitors visit the area for recreation. The number of nature loving visitors is also quite good. However, from the discussion it emerged that the dividing line in the perception of the picnic, outing, and love for nature was quite thin.

Purpose of visit	*Percentage*
Close by picnic spot	57.8
Love for nature	17.5
Outing	15.7
Picnic spot and love for nature	3.5
Picnic spot and love for nature and outing	2.0
All purpose	3.5

Frequency of Visits per Month

About 91 per cent people visit Kerwa between 1-5 times a month. About 6 per cent people visit 6-10 times a month. Only a small percentage visit it for more than 15 times in a month.

Frequency	%
1-5	91.3
6-10	5.20
11-15	1.75
> 15	1.75

Other Adjoining Places being Visited

Besides Kerwa there are some other places also in and around Bhopal, which are of tourists attractions. These places are Kolar Dam, Bhojpur, Bhimbetika, Halali Dam, Lakeview, Kalia Sot Dam, Manbhavan ki Tekri (the highest peak of Bhopal), Delabadi (wildlife area), Lalghati, Chiklod, Raisen Fort and Indira Gandhi Dam. Percentage of the visitors who also visit these areas regularly area as follows:

Sl.No	*Name of Place*	*Percentage*
1.	Kolar Dam	40.7
2.	Bhojpur	33.3
3.	Bhimbetika	25.9
4.	Halali Dam	25.9
5.	Lake view	14.8
6.	Kalia Sot Dam	14.8
7.	Manbhawan Ki Tekri	7.4
8.	Delabadi	7.4
9.	Lalghati	3.7
10.	Chiklod	3.7

Willingness to Stay if Facilities Provided

Many of the visitors showed interest in staying in the area subject to the availability of proper amenities such as safe drinking water, better roads, trekking, Rock gliding, lodging, Park, Restaurant, Medical facilities, fuel, etc. The are ready to pay for these facilities.

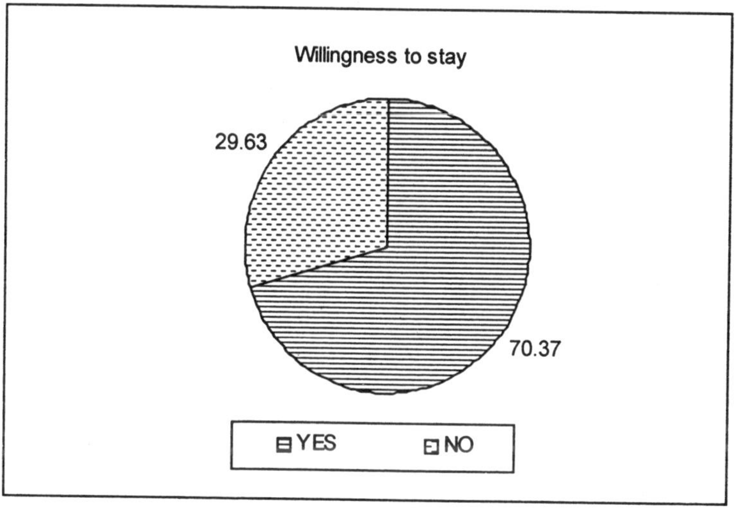

Number of Days Likely to be Spent

Majority (61%) of the tourists are willing to stay up to three days if adequate facilities are provided.

Days	*%*
1	14.11
2	33.33
3	13.11
4	0
5	0
>5	12.8
Did not respond	26.65

Facilities Required

The facilities desired mainly include safe drinking water and access.

Facilities	%
Water	18.5
Road	8.2
Trekking	7.0
Rock gliding	6.6
Lodging	9.4
Park	7.8
Restaurant	7.4
Medical facility	7.4
Fuel facility	3.7
Did not respond	24.0

Willingness to Pay for Stay

Majority of the tourists agree to pay up to Rs 200 per person per day for the necessary facilities.

Payment	%
50-100	22.22
101-200	14.81
201-300	0
301-400	7.41
401-500	0
>500	3.70
Did not respond	51.86

Problems and Difficulties

Visitors complained for non-availability of safe drinking water, poor public transport, and non-availability of proper shops of essential commodities. Also no trained guide is available for proper guidance about the area. There is no arrangement to check the visitors taking a dip or a swim in the dam for pleasure and fun, as a result there is always a fear of accident. Every year there are reports of few casualties. Besides, the visitors also face the problem of proper toilets and health care facilities.

Problems	Percentage
Water	64.9
Transport	22.8
Shop	8.7
Security	5.26
Guides	5.26
Shades	5.26
Others	5.26
Medical facility	3.5
Toilet facility	3.5

Willingness to Stay with Villagers

A large percentage of visitors (74%) are ready to stay with the villagers.

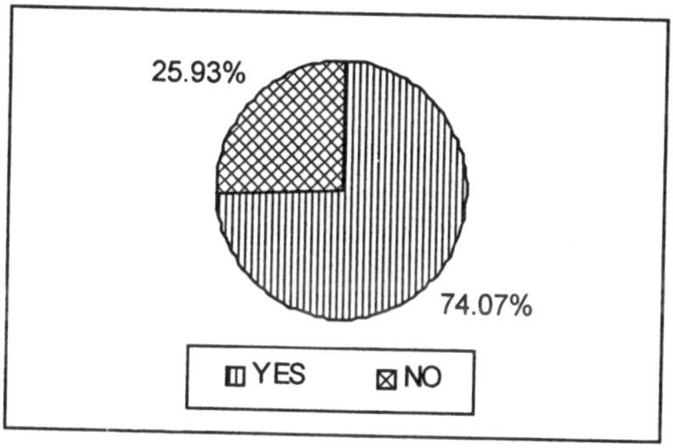

Interest in Local Culture and Cuisine

Visitors showed lot of interest in local culture including local folk songs and dances. Visitors were also found to be interested in local cuisine.

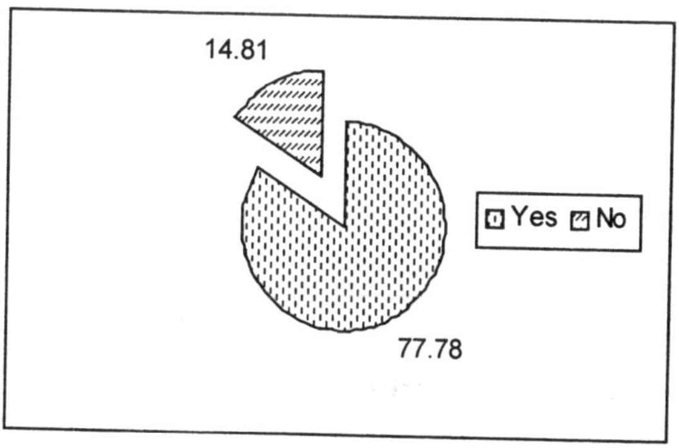

Interested in local cuisine

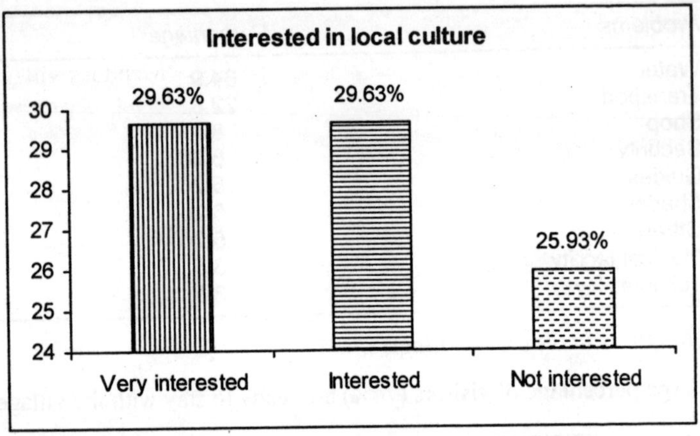

Villagers' Perception

Villagers' Profile

For the purpose of villagers' perception two adjoining villages named Mendora and Chichli were selected. These villages, being closest to the ecotourism destination area, were considered for the study.

Description	*Mendora*	*Chichli*
Number of families	64	32
Population	650	187
Total agriculture land (Acres)	NiL	77 A
Livestock	80	191
Distance from ecotourism destination area	2 Kms The village is on way to the area	3 Kms This village is situated on the other side of the area
Village area	8 acre	5 acre
Communities	3-ST, SC, General	2-ST, SC

Traditional Business of the Area

The traditional businesses of the area are agriculture (30%), Labour

(30%), animal husbandry (23.3%) and remaining people are involved in fishing (15.3%).

Traditional Business of the area	%
Agriculture	30.7
Labour	30.7
Animal husbandry	23.3
Fishing	15.3

Present Occupation

Most of the villagers of these two villages are occupied in 'animal husbandry' followed by nursery (plantation), agriculture, cowdung selling, Tendu leaves collection, fishing and others.

Options	%
Animal husbandry	72
Nursery	52
Agriculture	21
Cowdung selling	17
Tendu leaves collection	13
Fishing	11
Others	03

Handicraft and Articraft of the Area

Villagers feel that the '*broom', which is presently being made for* self-consumption, can have potential as handicraft. Besides this, some ladies are also involved in '*Rangoli'*.

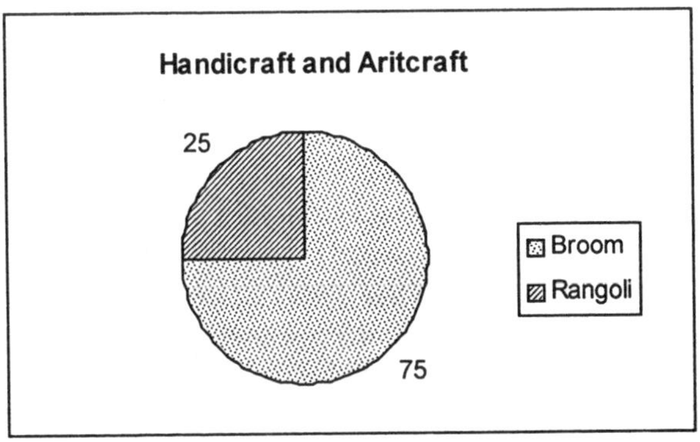

Conservation Activities by Committee

'Mendora Village Forest Committee' contributes to the conservation of forest. The committee is involved in plantation, fencing, checking illegal felling of trees, protection from fire, land filling, sanitation works etc.

Conservation Activities	%
Plantation	70.2
Fencing	15.7
Checking illegal felling of trees	8.5
Protection from fire	2.8
Land filling	1.4
Sanitation work	1.4

Awareness About Ecotourism

Most of the villagers are aware about purpose of the visit to the area by visitors, thus they are aware of ecotourism in their own perception. Nearly 76 per cent of them are aware of ecotourism and related activities.

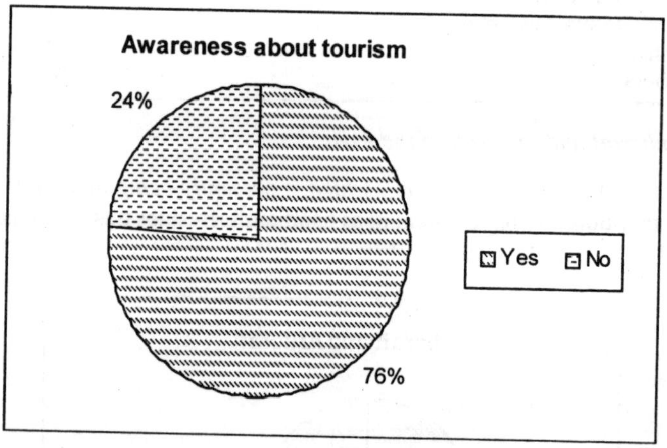

Reasons of Tourism

Through the perception of their interaction with the visitors, the villagers consider that people usually visit the place for picnic and recreation, to enjoy the nature, wildlife, fares and festivals, adventure sports and for religious purpose.

Reasons	*% Response*
Picnic and recreation	60
Natural beauty	48
Wildlife seeing	13
Fires and festivals	10
Adventure sports	03
Temple	02

Types of Tourists

As per the villagers' perception about 43.4 per cent people are local people who come from in and around Bhopal, while 42.2 per cent are outsiders, those come from different places. It is interesting to note that foreign tourists (14.4%) also visit the area as guests of the local people.

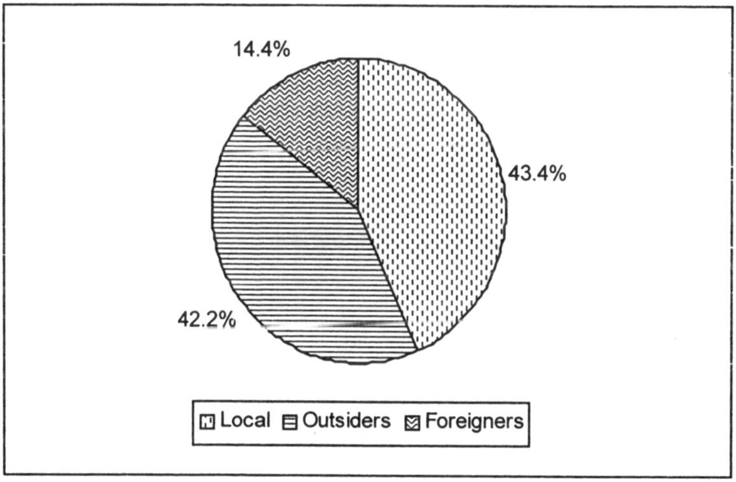

Age Class of the Tourists

According to the villagers most of the people who visit the area belong to the young generation (between 18-30 years), followed by middle class (31-50), children (below 18) and aged people (above 50).

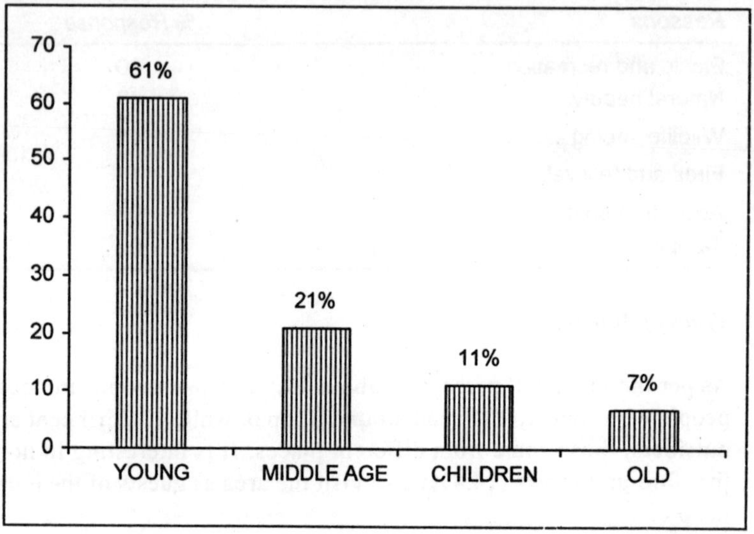

Benefits from Tourism

It has been observed that only a small percentage of local people are benefited from tourism. The main beneficiaries are petty shopkeepers, e.g. Grocery store, tea stall.

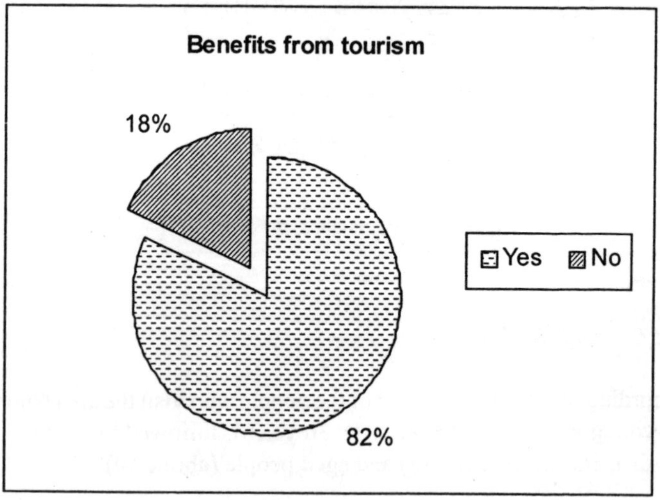

Willingness to Host Tourists

About 72 per cent people are in favour of hosting tourists, while 25 per cent are not in favour of tourism. 3 per cent percent did not respond.

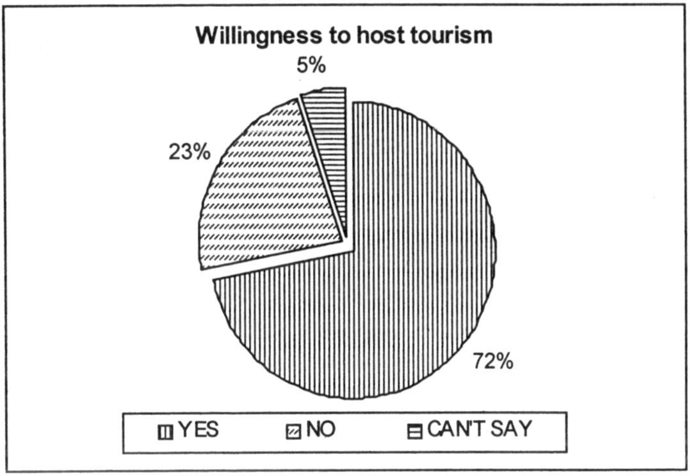

Willingness of Community to Control Tourism

Nearly 75 per cent of the villagers have been found to be interested in community control of the local tourism. They are of the opinion that village level forest committees could take over the activities of the tourism.

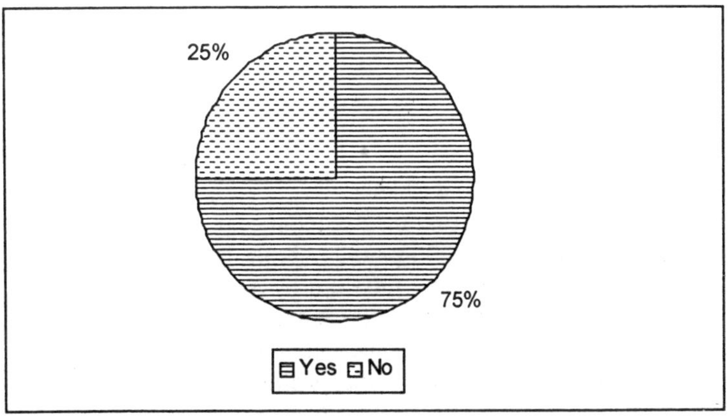

Interested to be Associated with Ecotourism

About 75 per cent of the villagers are willing to be associated with ecotourism and related activities. They are of the opinion that this may improve their economic status as they will get more opportunities to earn through tourism. Showing interest in various enterprises and earning activities, about 23 per cent people want to establish small shops, 23 per cent want to become tourist guides, 19 per cent are in favour of establishing hotels or restaurants, followed by trading (10.7%), general employment (8%), self-employment (5%), folklore and handicraft (2.2%), cottage industry (1.1) and adventure sports (1.1). Significantly, as against the opinion of men, the women appeared apprehensive that their children might be affected by undesirable behaviour and lifestyle of some of the tourists.

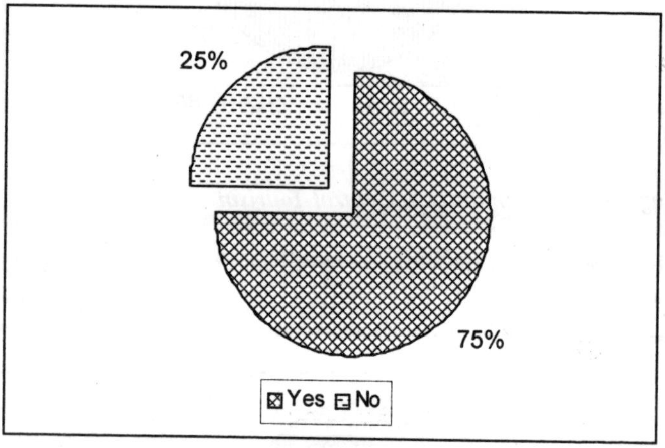

Infrastructure Need

The villagers of Mendora and Chichli villages reported that in order to adopt ecotourism for enhancing their economy, they need certain basic facilities for tourism development. About 18 per cent wanted infrastructuralstructural facility, followed by drinking water (12.5%), waste management and disposal facility (9.4%), electricity (9.4%), roads (9.4%), transportation (8.9%), communication (8.0%), Hoardings (6.7%), men and women toilets (6.7%), police station (4.9%), children park (3.1%), speed breakers (2.2%), medical facility (0.4%).

Community Based Implementation

Villagers were interested in ecotourism ventures if launched at the village level so as to enable their youths to get better job opportunities. Besides the male, women may also earn through home run centers of papad and pickles, tailoring and stitching, local cuisine.

Requirements	%
Building	17.9
Drinking water facility	12.5
Waste management facility	9.4
Electricity	9.4
Roads	9.4
Transportation	8.9
Communication facility (STD booth)	8.0
Boarding on the road	6.7
Basic toilet facility	6.7
Police station	4.9
Children park	3.1
Speed breakers	2.2
Medical facility	0.4

Villagers Expectations

Capacity Building

Villagers appeared quite clear that to take up ecotourism as an economic venture they require capacity building specially for—

- entrepreneurship development (self-employment ventures).
- tourism activities (guides/management).
- small scale industries like making of *papad, achar,* basket weaving and stitching.

Financial support: The villagers also consider that some initial financial is essential to in small scale industries, shops, dairy, livestock, raw material, starting business etc.

Infrastructure development: Villagers feel that some infrastructures development are required for operation of the whole tourism activities specially the sheds and other logistics.

Institutionalisation and regulations: The villagers firmly believe that in order to institutionalize the process, some regulatory measures are required. Institutional mechanism is considered necessary for providing logistics, support services, fuel supply (cowdung etc).

Reactions

Negative

- Villagers do not like tourists' behaviour to spread dirt and wastes anywhere on roads and other places thus spoiling the area by dumping empty bottles and waste food in water resulting in garbage and pungent smell.
- "*Gundagardi*" (hooliganism) is increasing because of tourism; sometimes youngsters come and shout under intoxication causing irritation to the people.
- Careless drinking/intoxication leads to dirty atmosphere.
- Few women complained that the tourists should not be allowed to enter the area as their growing up children may easily copy their bad habits.

Positive

- Some people running road side small shops of tea and snacks are benefited from tourists while others do not get much benefit.

Suggestions by Villagers

Hoardings at prime locations should be raised for drawing the attention of visitors towards tourists' code of conduct and better civic sense.

- Efforts should be made to educate tourists about proper disposal of waste in order to foster eco-friendly atmosphere and clean environment,
- A check post at the entry and near the dam be established to monitor anti-social, undesirable and accidental possibilities.
- Entry fee may be imposed.
- Fishing in the dam should be allowed so that the local community may be benefited.
- Publicity campaign be launched to enlighten the tourists as well

as the villagers about Environmental laws, rules and regulations framed by the government from time to time.

- Banners like "Make Your Trip Green and Clean", "Wandering one gathers honey", "Enjoy Nature—the Glory of God" be put up on all over the area.

Conclusions

1. The area has tremendous potential of ecotourism and with proper organized planning and implementation, ecotourism can make a breakthrough for the conservation and economic development of the area.
2. The CBE module can be implemented as the local people are aware and sensitive about the tourism ventures, thus they can be involved in planning and implementation of the ecotourism.
3. Most villagers strongly feel that ecotourism should be encouraged and more and more ecotours be promoted. While there is liking for tourists, there is strong disliking as well for the non-environmental tourist behaviours of contributing pollution.
4. There is need for capacity building for providing them opportunities for starting business at small-scale. The youths of the village are very much interested to be a part of ecotourism to act as tourist guides.
5. Women are also interested to take up the ecotourism related activities.
6. There is need to streamline the functioning of the village forest committee with a view to involve villagers in the development and management of the area.
7. An integrated holistic plan for the community-based ecotourism should be prepared through the involvement of all stakeholders and implemented through the local people and a strategy for the sustainable ecotourism management should also be evolved for all such areas.

NOTES

1. Lindberg Kreg & Deonal E. Hawkins (Eds) 1999. *Ecotourism: A guide for Planners and Managers*, Vol. I, North Bennington, VT, USA: The Ecotourism Society. Natraj Publishers, Dehradun, India

2. WTO, 1984. *Economic review of world tourism*, World Tourism Organisation, Madrid.
3. WTO, 1991. *Yearbook of Tourism Statistics*, World Tourism Organisation, Madrid.
4. Figgis, P., 1993. Ecotourism: Special interest of major direction? *Habitat*, 21 (1): 8–11.
5. Wight, P., 1993. Sustainable ecotourism: balancing economic, environmental and social goals within an ethical framework, *The Journal Tourism Studies*, vol. 4, no 2, pp. 2–10.
6. Figgis, P., 1994. Ecotourism: Panacea, Plague or Possibility. Eighth Richard Jones Memorial Lecture, October. University of Tasmania, Central for Environmental Studies, Hobart.
7. Harris, R. & Leiper, N. (Eds) 1995. *Sustainable Tourism: An Australian Perspective*, Butterworth-Heinemann, Chatswood, NSW.
8. Eagles, P.F.G., 1996. *Issues in tourism management in Parks: The experience in Australia. Australian Leisure*, vol. 7, no. 2, pp. 29–37.
9. Ceballos-Lascurian, Hector, 1996. Tourism, *Ecotourism and Protected Areas*, Cambridge, UK: International Union for Conservation of Nature and Natural Resources (IUCN).
10. Sagasti, F. & Colby, M., 1993. Eco-Development and Perspective's on Global Change from Developing Countries. In: *Global Accord*, Ed Chourci, N., MIT Press, London
11. Nelson, J.G., 1994. *The spread of ecotourism: some planning implications*, Environmental Conservation, 21 (3): 248-255.
12. Kamra, Krishnan A., 2001. *Economics of Ecotourism*, Kanishka Publishers, New Delhi.
13. Commonwealth Department of Tourism, 1994. *National Ecotourism Strategy*, Australian Government Service, Canberra, Australia.
14. Lindberg Kreg, Meganepler Wood & David Engeldrum (Eds) 1999. *Ecotourism: A guide for Planners and Managers*, Vol II, North Bennington, VT, USA: The Ecotourism Society. Natraj Publishers, Dehradun, India.
15. Coltman, Michael M., 1989. *Tourism Marketing*, Van Nostrand Reinhold, New York.
16. WTO, 1993. *Sustainable tourism development: Guide for local planners*, A Tourism and Environment Publication. World Tourism Organisation.
17. Cater, Erlet & Gwen Lowman (Eds) 1994. *Ecotourism: A Sustainable Option*, John Wiley & Sons, England.
18. WTO, 1998. *Guide for local authorities on developing sustainable tourism*, A Tourism and Environment Publication. World Tourism Organisation.

12

Economic Valuation of Ecotourism at Kerwa Region, Bhopal (M.P.)

Samita Vasudevan, Vanya Jha and *A.K. Bhattacharya*

Introduction

Ever since the second half of the twentieth century, concern about current and future use of our natural resources and environment has emerged at an increasing rate. This growing concern is accompanied by an increasing interest in so-called nature-based tourism or ecotourism, which can be defined as responsible travel to natural areas that conserves the environment and sustains the well being of local people. Presently, both benefits and threats have been observed resulting from the growing importance of tourism in environmentally sensitive areas.

India in general and Madhya Pradesh in particular is having unlimited potential in Ecotourism.

For realising this potential, two of the main aspects to be stressed upon are:

- — the needs of the domestic and local tourists
- — promoting tourism destinations outside forest areas.

The Kerwa Dam area near the city of Bhopal is one such site that has the potential to address both these aspects. Thus the present study has been conducted to obtain economic information about benefits that flow from Kerwa Dam Recreational Site near Bhopal, M.P.

The recreational value of an outdoor site is supposedly reflected in a visitor's willingness to pay for the visit. The present study adopts the

Travel Cost Method (TCM) to reveal this value. This method is based on conventional consumer demand theory where utility is maximised by the individual, and demand falls with rising prices. The method has been applied to estimate a value for the Kerwa Recreational Park situated near Bhopal. A standard assumption for interpreting travel costs as a valid proxy for the price of a trip is that the travel cost be incurred exclusively to visit that particular site and nothing else. For the Kerwa Recreational Park, this assumption often proves to be true and valid as people tend to go exclusively to the Dam site, due to its greater distance with any other recreational site in near proximity.

Background

The environment provides goods in term of resources (e.g. flora, fauna and minerals) and services (e.g. waste sink assimilation), a source of amenity services, use for recreational purposes and life support functions. While natural habitats in the past often seemed to have no reason for preservation, environmental goods are now-a-days increasingly being recognised as being valuable and limited in supply. Tropical rainforests for example protect watersheds, act as a habitat for potentially valuable species, and regulate climate. Furthermore, they provide a natural environment for recreational experiences and tourism, such as bird watching, nature photography, camping and outdoor education.

Attractions within the tourist sector are often dominated by the natural environment. Nature-based tourism and ecotourism are popular terms, associated with sustainable tourism. Driml (1997) developed a working definition of sustainable tourism in protected areas as follows:

> "sustainable tourism is tourism where natural environmental resources that are fundamental to preserve life support systems and the conservation of biodiversity are maintained, resulting in a positive and non-declining stream of net economic benefits, thereby retaining investment in adequate management."

Ecotourism thus is nature-based tourism that places an emphasis on conservation, education and interpretation of the natural environment.

There is a rising demand of nature-based recreational experiences, which is obvious; people want to spend more time and money in natural settings that are unspoiled as yet, as they become more aware of the value of pristine environments and their specific features. On the other hand, however, it can be argued that presence of more people in these areas will undoubtedly have impacts upon the environment, which may be negative. Nevertheless, the Ecotourism sector is growing rapidly. Visits to such areas bring along positive economic, social, and cultural effects as well as negative environmental impacts.

Economic benefits are gained through visitors who spend money, thereby creating economic activity and employment. Social benefits include opportunities for recreation, and the appreciation and enjoyment of nature. Cultural benefits are to be gained by promoting the understanding and revitalisation of the culture of the local community in the recreational site. However, the presence of people undertaking tourism and recreation activities in environmentally fragile areas also poses threat of adverse environmental impacts on the particular site.

Rationale of the Study

The present study has been undertaken to assess the economic value of the Kerwa region. This assigns a monetary value to the area incorporating only the use-value. The value so obtained can be utilized for assessing the feasibility of carrying out further developmental and conservation activities in the region. This is especially relevant in the context of the management plan that has been prepared for the region.

Objectives of the Study

It is quite evident that nature based tourism sites like the Kerwa Dam site has a potential to gain substantial benefits from tourism activities, provided that these are catered for in an environmentally sustainable way.

Furthermore, with rising high demands on resources, and a non-existent mechanism of 'user-pay' to obtain contributions towards management costs, there is an increasing need for more stability in funding arrangements. In most cases, natural environments have no entry fees or have only nominal fees imposed to cover some management costs. This is exactly the case in the site chosen for the purpose of our study, where no user fee is charged from the tourists, who visit the site. This does not necessarily reflect the "true" value to visitors of the opportunity to visit the site.

Thus, the specific objective of our study is to estimate consumer's demand for Kerwa Dam recreational site. By applying the Travel Cost Method (TCM), current visitor's demand will be assessed. In a second stage, demand will be predicted at various hypothetical entrance fees. This will provide an assessment of the benefits that are experienced by the visitors, i.e. an economic value is calculated for the area.

The explicit analysis of the benefits can be used to reveal what the site is "worth" to the public, and provide an economic justification for conservation, further development and management practices. Alternatively, the results the study could also be used by relevant authorities as a guideline for the introduction of entrance fees for the tourist site.

Methodology

Valuation Method

The method adopted for the present study is known as the Travel Cost Method (TCM), which is a commonly applied tool when valuing user benefits of environmental goods in monetary terms, such as in our case of a recreational park.

Harold Hotelling had originally suggested the Travel Cost Method (TCM) in 1947, when The US National Park Service had wanted to know how economic principles could be used to demonstrate economic values produced by national parks. Hotelling suggested that the travel costs an individual incurs to visit a recreation site could be used as an implicit price for that site's services. The basic premise of a TCM is that the number of trips to a recreation site will decrease with increases in distance travelled. Exploiting the empirical relationship between increased travel distances and associated declining visitor's rates should permit one to estimate a true demand relationship. If estimated empirically, this demand schedule could be used to compute the total benefits produced to park visitors, which were equal to any entry fees they paid plus their remaining unpriced benefits, referred to as consumer surplus.

TCM does not take any non-user benefits into account. This characteristic relates to the basic assumption of weak complementarity between the environmental asset and consumption expenditure. This concept was first discussed by Mäler (1974), stating that if the demand for a private good is zero, then the demand for the environmental service is also zero. For example, if the private good is swimming in a lake, and the environmental service is the quality of that lake, then if a person does not use this lake for recreation, he will most likely be indifferent to the quality of the water.

One of the drawbacks of the TCM is concerning the problem of multi-destination or multi-purpose trips. Since people often do not undertake the trip for one single purpose, or to visit one single site, assigning total travel cost to this one specific site would be incorrect and presumably overestimate demand. However, in this particular study of estimating the economic valuation of the Kerwa Recreational Site, we have assumed that all the visitors to the site are exclusive to the site, i.e. they undertake the trip for the sole purpose of visiting Kerwa Dam, and nowhere else. We have assumed this, because the main visitors to this site are locals from nearby Bhopal city, and not national and international tourists.

Sampling

Convenience sampling was used to select the respondents. A sample size of 20 respondents was taken and data was collected from the visitors to the area on a single day. Due to time constraints, a larger sample size could not be taken.

Tools

Questionnaires were used for data collection.

SPSS package was used for data analysis.

About Kerwa

Kerwa region is endowed with scenic and aesthetic values, which is in close proximity to the concrete jungle of Bhopal. The area is close to VanVihar National Park and has tremendous tourism potential and has also an impact on the socio-economic condition of the villagers. The local community can play a significant role in the development of ecologically and sustainable management of ecotourism module in the areas, on the other hand the ecotourism in the area can play an effective role in the economic development of the villages in the ecotourism destination area.

Location

Kerwa region is spread over an area of 50 sq. km. Limited within the north Latitude N 23° 18′ and Longitude E 77° 20′ Kerwa region lies with Mendora, Mendori, Sarotipura, Kekeriya, Ransudriya, Bhanpur,

Daulutpura and Chichilli villages.Bhopal station is 22 km and Habibganj railway station 18 km. The entry points in Kerwa region are from Bhadbadha Dam and Law Institute side and from Kolar Road, Rai Gas Agency side.

Terrain

The terrain is mostly undulating and hills of various heights and slopes are many in this tract. The attitude varies from 535 metres.

Forest

The forests found here are Tropical Dry Deciduous forest, with teak plantations.

Wildlife

Region holds a variety of wildlife. Different species of carnivorous and herbivores are reported from this area.

Water

The position of water availability is not good during pinch period. Many *nallas* flow from July to November. The river in region is Kerwa river which originates from Kerwa dam and flows to Mandideep. It holds water in pools in summer.

Prospects of Kerwa as an Ecotourist spot

Kerwa is an ideal site for promoting Ecotourism. The proximity to Bhopal city, easy approach roads, the extensive cover variety of fauna and flora contribute to prime potentially of ecotourism. However, lots of basic infrastructure amenities have to be provided for the visitors without jeopardizing ecological serenity of the ecosystem of Kerwa Region.

Tourism

Tourists, mostly from Bhopal visit the Kerwa Region mainly in daytime on Sundays and holidays. Wildlife visibility is not good but people see Peacocks, langurs etc. Also there are so many spots of historical, archeological and scenic interest. But at present they are neither preserved and the attempts to present it to the tourists have also been very poor. Hence all the tourists arriving are picnic-makers, not hardcore wildlife enthusiasts or adventure lovers.

Literature Review

Many different researchers have contributed to the now substantial body of empirical experience using Travel Cost Models for the Valuation of Environmental Goods, especially of tourist and recreational sites. To our knowledge, the use of economic valuation methods for non-market environmental goods and services in India has been extremely limited.

However, data from other countries may be examined to indicate the type of values identified. Menkhaus and Lober (1996)[1] used the Travel Cost Method to estimate the value that US ecotourists assign to Monteverde Cloud Reserve in Costa Rica. They arrived at a total annual US ecotourism value of US$ 4.5 million. Values such as this can be used to calculate revised (higher) entrance charges that more adequately reflect the ecotourism benefit for the area. This study arrived at an average entrance charge of US$ 40, which is considerably higher than the US$ 5-10 usually charged at national parks in Costa Rica (Lette & Boo, 2002)[2].

Similarly, Maille and Mendelsohn (1991)[3] found a consumer surplus for ecotourism demand in Madagascar of between US$ 276 and US$ 360 per person. Brown and Henry (1989) found that 265,000 to 300,000 tourists on safari in Kenya in 1989 received a consumer surplus of US$ 182-US$ 210 million, or around US$ 700 per tourist (Lette & Boo, 2002)[2].

In the Philippines, the value of watershed protection for marine tourism is estimated at US $14-19 million. In Costa Rica, the total value of their tropical forests (hydrological benefit, carbon sequestration, ecotourism, option for future pharmaceuticals, existence and option value) is estimated at US $102-213 per hectare per year. The net present value (at 8%) is calculated to be US $1,278-2,671 per hectare per year.

Fix and Loomis (1998)[4] compare non-market valuation techniques by applying a count data TCM and dichotomous choice CVM to a form of recreation for which it has not been previously applied: mountain biking. Due to mountain biking's increasing popularity these estimates of benefits may be useful in addressing conflicts. One of the most famous mountain biking sites in the U.S. (Moab, Utah) was chosen as the site for which to apply these two models. The benefits that were estimated for trips taken in the spring of 1996 are US$ 205 and US$ 235, for the TCM and CVM, respectively.

Greene *et al.* (1997)[5] estimated the demand for recreational fishing in Tampa Bay, Florida, to facilitate the environmental management of the bay. A nested random utility travel cost model is used to estimate access values. Results suggest that average annual values for the bay alone are $18.14 and $0.048 for participants and non-participants, respectively.

Research by Layman *et al.* (1996)[6] extends the standard travel cost method to develop estimates of the economic value of recreational chinook salmon fishing on the Gulkana River, Alaska, under existing and hypothetical fishery management conditions. Respondents were asked to state how the number of trips that they took to the study area would change if alternative fishery management practices were imposed. Three hypothetical management conditions were considered: a doubled 1992 sport fish harvest, a doubled daily bag limit, and a season bag limit of five. Each of the hypothetical fishery management conditions provides increased economic returns to anglers.

Study by Bowker *et al.* (1996)[1] examines per trip consumer surplus associated with guided whitewater rafting on two southern rivers in USA. First, household recreation demand functions are estimated based on the individual travel cost model using truncated count data regression methods and alternative price specifications. Findings show mean per trip consumer surplus point estimates between $89 and $286, depending on modelling assumptions and river quality. Magnitudes of these surpluses are very dependent on assumptions about the opportunity cost of time.

The analysis by Bell and Leeworthy (1990)[7] deals with tourists that come from significant distances to use principally beach resources. As earlier authors have argued, those that use the conventional travel cost methods (TCM) do not recognize its potential spatial limitations. One day trips as used by the TCM are certainly inapplicable to those coming from significant distances, such as tourists to Florida. The empirical data are consistent with the thesis that annual consumer demand by individual tourists for Florida beach days is positively related to travel cost per trip and inversely related to on-site cost per day. There are compelling reasons for treating recreational decision-making for tourists differently than for residents or those traveling relatively short distances. Employing the on-site cost demand curve for tourists using Florida's beaches, they found out the daily consumer surplus to be nearly $34.00.

The research by Wen (1998)[8] evaluating both tourism and tourist resources in China has considered only the positive contributions of tourism and ignoring the aspects of disparities between the east coast and the inland area, inbound tourism and domestic tourism, nature-based and city-based tourism and negative effects from tourism. On the micro-and site-specific level, tourist resources characterised by non-marketable and non-use values are not properly evaluated, exposing these fragile resources to the plight of exploitation. Commercialisation of tourist resources, including natural environment and ethnic cultures, threatens further development of tourism in China. Further, the limited applicability of the travel cost and the contingent valuation method in measuring the value of recreation sites in China is discussed.

Analysis and Results

For the analysis, the following steps were carried out:

1. Data was collected through the questionnaires on the following aspects:
   - number of visits to kerwa in the year
   - distance travelled to reach Kerwa (km)
   - roundtrip time taken (min)
   - mode of transportation
   - monthly income
2. The above data was used to calculate the following:
   - travel cost per km (cost of fuel × average mileage of the transport mode used)
   - cost of time per minute (derived from monthly income of the respondent)
   - total cost of roundtrip (roundtrip distance × travel cost per km)
   - total cost of roundtrip travel time (roundtrip time × cost of time per min)
   - total travel cost (total cost of roundtrip + total cost of roundtrip travel time)
3. The above data for each respondent was fed into SPSS data sheet and a regression equation taking number of visits per year as the Dependent variable and the total travel cost as the Independent variable was calculated (Annexure II). This regression equation gives us the demand function for the "average" visitor to the site.

   The equation obtained is:

   $$\text{Number of Visits} = 3.375 - 0.00992 \times \text{Travel Cost}$$

To maintain the simplest possible model, calculating the equation with only the travel cost and number of visits. If more independent variables (like gender and purpose of visit are taken, the regression model could be a better fit.

1. Using this equation the average number of visits for an "average" visitor were calculated taking the average travel cost (calculated by summing the individual travel cost and dividing by the number of respondents, i.e. 20). At this travel cost, the entry fee is taken to be zero.
2. Next, a demand function for visits to the site was constructed, using the results of the regression analysis. The first point on the demand curve is the total average visits to the site at current access costs (assuming there is no entry fee for the site), which in this example is 2.35 visits per year. The other points are found by estimating the number of visitors with different hypothetical entrance fees (assuming that an entrance fee is viewed in the same way as travel costs). The entry fees assumed for calculation were Rs.10, Rs.20, Rs.50 and Rs.100.

Entry Fee (Rs.)	*Average Number of Visits per year*
Nil	2.35
10	2.25
20	2.15
50	1.85
100	1.36

2. A Demand Curve was plotted using the above schedule. The area below this demand curve gives the average consumer surplus (calculated to be Rs. 49.6). This is multiplied by the total relevant population (the population of Bhopal as given in the Census of India, 2001, i.e. 14,33,875) to estimate the total consumer surplus for the site.

Thus, the value of the consumer surplus derived from the above step gives us the total economic value of Kerwa. This value comes out to be Rs.7,11,20,200, i.e. Rs. Seven Crore Eleven Lakh Twenty Thousand and Two Hundred.

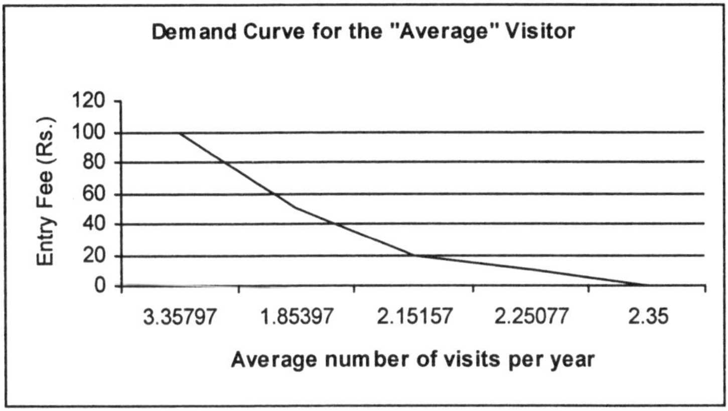

NOTES

1. Bowker, J.M., English, D.B.K. and Donovan, J.A., 1996. "Toward a Value for Guided Rafting on Southern Rivers", *Journal of Agricultural and Applied Economics*; 28(2).
2. Lette, H. and de Boo, H., 2002. *Economic Valuation of Forests and Nature: A support tool for effective decision-making*, Theme Studies Series 6, International Agricultural Centre (IAC), Wageningen, The Netherlands.
3. Tobias, D. and Mendelsohn, R., 1991. *"Valuing Ecotourism in a Tropical Rainforest Reserve"*, Ambio, 20:2,91-93.
4. Fix, P. & Loomis, J., 1998. "Comparing the Economic Value of Mountain Biking Estimated Using Revealed and Stated Preference," *Journal of Environmental Planning and Management*; 41(2).
5. Greene, G., Moss, C.B. & Spreen, T.H., 1997. "Demand for Recreational Fishing in Tampa Bay, Florida: A Random Utility Approach," *Marine Resource Economics*; 12(4).
6. Layman, R.C., Boyce, J.R. & Criddle, K.R., 1996. Economic Valuation of the Chinook Salmon Sport Fishery of the Gulkana River, Alaska, under Current and Alternate Management Plans, *Land Economics*; 72(1).
7. Bell, F.W. & Leeworthy, V.R., 1990. "Recreational Demand by Tourists for Saltwater Beach Days," *Journal of Environmental Economics and Management*, 18(3), May 1990.
8. Wen, J., 1998. "Evaluation of Tourism and Tourist Resources in China: Existing Methods and Their Limitations," *International Journal of Social Economics*; 25(2-3-4).

13

Assessment of Environmental Impacts of Tourism

Issues, Options and Tools

A.K. Bhattacharya, Ravi Sharma and *Suchitra Banerjee*

Introduction

Tourism is the fastest growing industry in the world and the world's largest civilian industry (Lindberg and Hawkins, 1993)[1]. According to a report by the World Travel and Tourism Council (WTTC, 2002)[2], India could generate 25 million additional jobs in the Tourism Sector by 2010. This is due to the reason that an increasing number of tourists now prefer to visit attractive natural environments or protected areas set aside for conservation. Undisturbed ecosystems, their plant and animal communities are critical in maintaining the clean air, clean water and healthy environments that are key tourism attractions in many destinations (Buckley, 1999)[3].

Located at the top of the environmental and industrial chain, tourism is extremely sensitive to environmental conditions and to the impacts others have on the system. In fact, the state of tourism itself may be a key indicator of system stability. Tourism, a multifaceted economic activity, interacts with the environment in the framework of a two-way process. On the one hand, environmental resources provide one of the basic "ingredients", a critical production factor, for the production of the tourist product: the natural and/or manmade setting for the tourist to enjoy, live in, and relax. On the other hand, tourism produces a variety of unwanted by-products, which are disposed, intentionally and unintentionally, to and modify the environment; the case of negative environmental externalities (Briassoulis, 1992)[4]. Moreover, economic activities besides tourism use up and modify environmental resources quality available for tourism purposes. Because of this linkage, the tourism sector needs increasingly to become a knowledge participant in the planning as the use of the environment and its impacts (Batta, 2000[5]; Manning, 1992[6]).

Impacts on biodiversity are particularly severe for large new tourism developments in relatively undisturbed areas, such as those in and around conservation reserves and other Protected Areas. These developments involve clearing vegetation, major disturbance to fauna through loss of habitat, noise, barriers etc. (Buckley 2001)[7], and a marine and freshwater pollution from discharges, which include nutrients, metals, pathogens, and toxic chemical compounds such as solvents (Warnken and Buckley 2000)[8]. Even without accommodation and large scale facilities, nature and adventure tours which are designed and managed for minimum impacts may have considerable impacts on biodiversity and other components of the natural environment. It is evident that not only has tourism grown rapidly worldwide, but also many of those concerned about the negative impacts of the destinations assume that mass forms of tourism are responsible for these problems (Cooper and Ozdil 1992)[9].

Tourism and Environmental Impacts

It is undeniable that tourism has enormous potentials for the environment conservation of environment. However, it must also be borne in mind that the balance between tourism and the environment is very fragile one. Many developing countries, anxious to reap the full benefits of tourism have, without undertaking a proper analysis of the potential impacts transformed their virgin areas into tourists' centers to cater for the tastes and desires of mass tourism. Such rapid development may lead to a complete transformation of an area, producing irreversible impacts to the natural environment.

Among the three impacts of tourism namely, the economic, socio-cultural and environmental, the economic impact has undoubtedly played a dominant role in tourism literature and policy making till 1960s. With the focus on the economic benefits obtained by the areas due to the development of tourism, the adverse non-economic, socio-cultural and environmental-impacts are totally ignored. Environmental costs continued to be neglected because of the prevalent belief of the nature being inexhaustible and renewable. This led to an indiscriminate and unplanned growth of tourism infrastructure in many countries and soon the negative effects in the form of social and environmental degradation started emerging. Specialists tours such as photographic safaris and wildlife watching, which can affect animals through noise, visual and scent disturbances, and by affecting predation and breeding behaviour. Similarly, wildflower tours can affect plant biodiversity if participants collect plants or fruit, introduce weeds or pathogens, or start fires (Buckley, 2002)[10].

In some destinations, tourism can produce a local economic boom leading to uncontrolled high-impact private development, high resource consumption, waste generation beyond the capacity of local waste treatment disposal systems, if any, and land clearance and harvesting with major impacts on biodiversity. In addition, infrastructure built for tourism may be used for illegal collection of endangered plant and animal species (Buckley, 2002) [10]. Small-scale operations may eventually turn into much larger and more destructive operations (Hunter and Green, 1995)[11].

The Existing Knowledge

The study of the Environment Impacts of Tourism is currently in a growing stage and more research is expected to appear. The first effort towards environmental Impacts Assessment was directed basically to Impacts of Leisure activities and especially outdoor recreation. The first studies concerning the environmental impacts of tourism appeared after the mid-seventies (Tangi, 1977[12], Bavd-Bovy and Lawson, 1977[13]) followed by more research activity in the 1980s. Useful Reviews on the subject have been made by Pearce (1985)[14], Farrel and McLellan (1987)[15], Farrel and Runyan (1991)[16].

The biological and ecological impacts of tourism have been studied in the case of specific environments—Islands, Coastal Zones, Alpine areas, National Parks etc. (Lindsay, 1986[17]; Nijkamp *et al.*, 1991[18]). Nijkamp (1980)[19]and Onyeanusi (1986)[20] developed the impact structure matrix combining environmental elements and the range of possible impacts of these elements from the development of tourism to a certain level (carrying capacity levels). To fill out the structure matrix a set of different tools are used (Social surveys, behavioural inquiries, multiple measurement techniques, ecological indicators etc.).

The study of the environmental impacts of tourism thus started basically after 1970s. The analysis of the environmental impacts of tourism has been predominantly qualitative and mostly descriptive. The type and intensity of the environmental impacts of tourism depends on the interaction between the type of tourism development, the socio-economic and other characteristics of tourists and the natural, socio-economic and institutional characteristics of the host area. The environment is being increasingly recognized as a key factor in tourism. In the last decade of the twentieth century, it was noted that tourism depends ultimately upon the environment, as it is a major tourism attraction itself, or is the content in which tourism activity takes place (Holden, 2000)[21].

The relationship between tourism and the environment is taking place on various levels. In addition to direct tourism impacts on the environment through, e.g. pollution, noise and disturbance, indirect, irreversible and long term consequences between tourism and environmental quality is characterized by dynamic feedback mechanisms (OECD, 1980 [22]& 1994[23]). It has been suggested that controlling the volume of tourism might alleviate the situation (Wheeler, 1994)[24], especially since the tourism is typically found in locations with fragile environments, such as mountains and coasts which are peripheral to the world economy (May, 1991)[25]. Ecotourism not only depends on mass tourism enterprises and infrastructure (air, travel, other forms of transports, tour operators, hotels etc.), but also involves the danger to eventually promote mass tourism itself. Unfortunately, the experience of alternative tourism in general shows that adventurous travellers have just served to open up destinations "of the beaten track " to large-scale tourism projects, accelerating the pace of social and environmental degradation of these areas. As a result of increasing globalization and liberalization, the competition within the international tourism industry and among tourist-receiving countries has become so fierce that there is hardly any margin left for social and ecological concerns. Many studies throughout the world have now documented the ecological impact of indigenous population on the environment (Bhattacharya, 2003a[26] & 2003b[27]).

Emerging Research Issues

In order to analyze the impacts of Tourism on the Environment as a necessary prerequisite for tourism planning and policy decisions, following major issues need to be addressed:

- Identification and analyses of various impacts of Tourism on environment in the target area.
- Assessment of the contribution of tourism in the observed or expected environmental modification in the destination area.
- Estimation of the demand for resources and the amounts of

residuals disposed to the regional environment.
- Analyses of the environmental impacts of tourism on the local communities.

Research Output

The expected research output will include:

- Better understanding of environmental impacts of Tourism in general and in the study area in particular.
- Scientifically validated report on the impacts of tourism in the study area:
  - — Direct Environmental Impact.
  - — Indirect Environmental Impacts through the changes in physical, demographic and other characteristics.
- Identification of tourism activities that are compatible/ incompatible with the area and suggesting mitigation measures.
- Guidelines and strategies for formulating effective policies, plan and programmes about sustainable tourism.

Models for Tourism Impact Research

Very few models have so far been validated and standardised for the tourism impact studies. Two models (Material Balance Model and Integrated Impact Analysis Model) are being discussed here. These models have potential to be used in the present or improvised form for the tourism impact analyses.

Material Balance Model for Tourism (*Source: Briassoulis, H. 1992*)[4]

The material Balance Model for Tourism (Kneese *et al.*, 1972[28]) is depicted as follows:

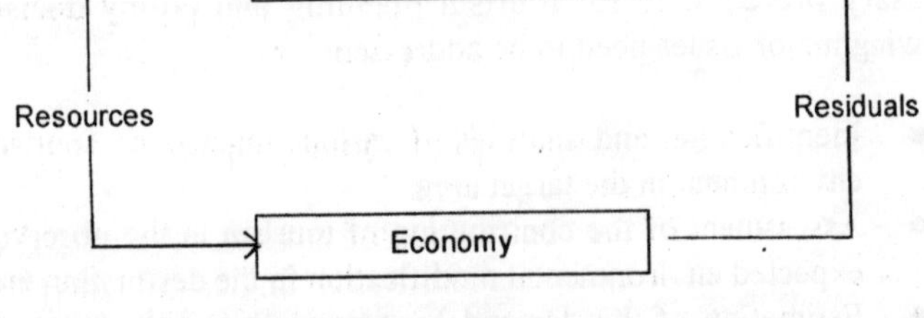

On the basis of the above framework, 4-major classes of

environmental impacts of tourism can be distinguished.

(a) Consumption-related impacts on resources.
(b) Production-related impacts on resources.
(c) Production-related residual impacts.
(d) Consumption-related residual impacts.

The resource impacts concern depletion and competition for resources between tourism and other economic activities while the residual impacts refer to pollution and environmental degradation of resources.

The Integrated Impact Analysis Model

(Redrawn from Briassoulis, 1992: Environmental impact of tourism—A framework for analysis and evaluation.)[4]

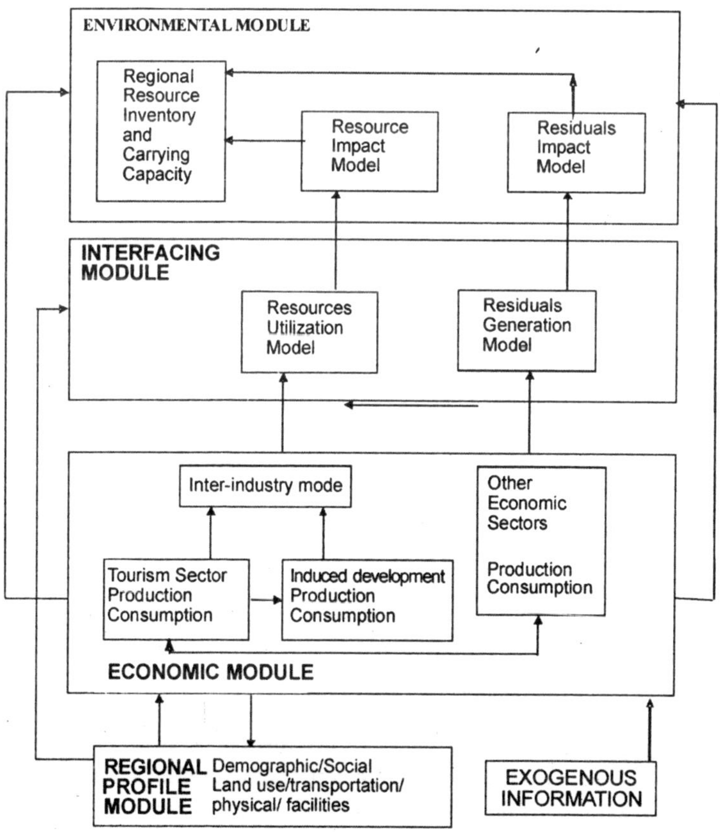

Impact Categorization

Environmental impacts can be categorized further on the basis of following criteria:

(a) Directedness of Impact
- Direct-impacts that directly emanating from tourism activities linked to tourism activities such as accommodation travel, sightseeing etc.
- Indirect-impacts that are caused by activities linked to tourism such as retail trade, entertainment etc.
- Induced-impacts due to non-tourist related activities, which are induced by tourism development of a region and would not otherwise occur.

(b) Spatial scale and extent: local, region, national as well as point and non-point.

(c) Temporal scale: short-, medium and long-term impacts.

Methodology

The methodology adopted to achieve the specific goal is consisted of both Experimental (Laboratory work, field work) as well as Non-Experimental (field studies, surveys, questionnaires). An Integrated Model for Tourism Impact Analysis is adopted to attain the specific goal. An Integrated Economic/Environmental model is suitable for the analysis of the Environmental Impacts of Tourism and Ecotourism. This Integrated model is best for *ex post* impact Assessment (for regions already developed touristically) and also for *ex ante* (for regions considering some forms of new or additional tourism development).

Three approaches can be adopted in the study

- Systematic information collection and analysis approach.
- Participatory approach.
- Consultative approach.

Framework of the Methodology

1. Gathering Information of the study Area.
2. Field and visitor survey for the purpose of determining Production and consumption characteristic of tourism and associated Sectors:

2.1 Production characteristic (Infrastructure and Superstructure):
- Transport Network (Travel, Roads and Transportation).
- Water supply and sewage and waste disposal.
- Utilities like electricity, telephones etc.
- Man-made attractions—monuments, buildings, facilities etc.
- Specialized facilities (like trekking, picnicking etc.).
- Accommodation (Hotels, campings etc.) (No. of beds).
- Shops etc. (foods, souvenirs etc.)

2.2 For consumption characteristic:
- Number of tourists visiting annually.
- Types of tourists.
- Reason for visitation.
- Frequency of visits.
- Willingness to pay (WTP).
- Duration of stay/length of tourists' period (Tourist night).
- Average consumption of water, energy etc. per tourist per day.
- Awareness about Ecotourism and environmental issues.
- Environmental Attractions.
- Average consumption expenditure by type of good and service.

3. Data collection for Regional Profile:
   - Demographic
   - Social
   - Physical structure of the Region.
4. Determination of Categories of Environmental resources. Identification of the scope of Impacts:

   A. Identification of impacts.

   B. Categorization of Impacts
   - Negative and Positive impacts.
   - Primary and Secondary Impacts.
   - Short-term and Long-term impacts.
   - Individual and Cumulative impacts.
   - Socio-cultural and Economic impact (Interfacing with ecological impacts)

   C. Study of following Impacts:

S.No.	*Elements*	*Parameters*	*Indicators*	*Measures*
1.	Flora	Distribution, Floral diversity composition		Quadrate study (IVI)
	Human	Human threats. (Noise pollution due to tourists and vehicles)	Average No. of days in which pollution standards are exceeded per year	Measuring decibel levels at key locations.
		Human encroachments	Human population in the surrounding areas	Nos. of peoples within adjacent to the tourism area in study area.
2.	Agriculture	Land use pattern		Determining the change in land use pattern by comparing with secondary data collection.
3.	Land	Characteristics (soil)	Solid Waste and Visual effect	Visual spectrum method.
		Land use Pattern	Changes in land use pattern.	Determining changes in land use pattern by comparing with secondary data pertaining to the change in land under use and under tourism activities.

(Contd.)

(*Contd.*)

			Erosion.	Field determination for measurement of erosion potential.
			Per cent of land abandonment in last decade	Secondary data collection pertaining to land and change in land pattern.
4.	Roads / transport	Clearing of vegetation		Site evaluation and interviewing forest department Staff, drivers, guides and tourists.
		Road erosion and road deg-radeation due to vehicular traffic		
		Dust and smoke creation due traffic on unsealed roads		
		Loss of access to key sites		Questionnaire schedule
	Buildings/ constructions	Changes in land use pattern due to construction.		Determining land use pattern by comparing with secondary data collection of change in land use during last decade.
		Visual effects		Site evaluation and

(*Contd.*)

(*Contd.*)

		Rate of growth of tourism in area leading to more facilities being constructed		collecting the data regarding the change in tourist numbers and tourism development during the past few years.
	Villages	Socio-cultural and economic impacts: (a) Change in cultural pattern. (b) Problems due to tourists. (c) Employment generation (alternative livelihoods). (d) Income generation (e) Objection from the local people against the facility	— Nos. of retail establishments/No of establishments serving local needs (as opposed to tourists) — Per cent of local establishments open year around. — Per cent of locals involved. — Per cent of jobs supported by tourism.	By questionnaire survey and field survey to collect the information regarding the social, cultural and economic activities.

(*Contd.*)

(Contd.)

		Social issue created by the tourism	(a) Alien culture (b) Nuisance created to the local people. (c) Eve-teasing of local women. (d) Socio-inequalities created within the community. (e) Problems from local people to tourists.	By questionnaire schedule.
	Visitors experience	(a) Pollution in tourism area. (b) Dust and smoke generated by vehicles. (c) Information imparted about nature through interpretation facilities (if any) (d) Willingness of the tourist for another visit.		By questionnaire schedule.
5.	Economic	(a) Increased employment or increased dependence of local people on tourism for their livelihood. (b) Increased cost of living in and around the tourism area (c) Increased earning from tourism		By field survey and questionnaire.

(Contd.)

5. Synthesis of data, interpretation and Impact Analysis with the Integrated Model and Assessment.
   5.1. Assessing for preparing economic module:
   - Description of Economic structure and tourism's position.
   - Direct, indirect and induced impacts of changes in production and consumption characteristic of Tourism.
   - Find Tourism induced development effects.

   5.2. Assessing for preparing regional profile module:
   - Changes in Regional environment according to tourism requirements.
   - Tourists effects on Regional profile are assessed such as population growth caused by tourism-related employment, Land-use etc.

   5.3. Assessing for preparing interface module:
   - Assess the demand of tourist and related physical changes.
   - Types and amounts of residual generated.

   5.4. Assessing for preparing environmental module:
   - Impacts on the Quality and Quantity of regional resources.
   - Linking various modules.
   - Impact Analysis to assess the contribution of tourism on the environment:
     - — Environmental Impact directly during the production and consumption of the tourist product.
     - — Environmental Impact indirectly, through the changes in economic, demographic, physical and other characteristics of the region.
6. Use of simple statistical tools for Analysis of Impacts.
7. Impact evaluation of tourism to address main issues-
   - Relative importance of tourism direct, indirect and induced impacts on environmental resources.
   - Distinguishing the impacts as beneficial or negative on the environment.

Objectivewise Parameters for Impact Assessment

S. No.	*Objective*		*Parameters*
1.	To identify and analyze the various impacts of tourism on Environment in the study area.	(a)	Field survey (Questionnaire schedules), Secondary data collection regarding tourist annual flow, demographic pattern, change in land use pattern during last years etc.
		(b)	Identifying parameters and indicators of
		(c)	Environmental elements (physical ecological, built etc.).
		(d)	Measuring the parameters by using techniques /tools/ instru-ments etc.
		(e)	Assess the impacts on environ-ment due to tourism.
2.	Assess the contribution of tourism in the observed or expected environmental modification in the study area.	(a)	Analyzing different environmental parameters and its impacts due to tourism.
		(b)	Synthesis of data and interpret-tation by linking different modules and impact analysis with integratec model and assessment.
		(c)	Use of simple statistical tool for analyzing the impacts.
		(d)	Enlist the different environmental impact identified due to tourism.
3.	To estimate demand for resources and the amount of residuals disposed of to the regional environment.	(a)	By questionnaire /schedule, collect data, which will specify the tourists demand and facilities provided, pressure indicator.
		(b)	Secondary data collection of the standards fixed by the agencies.
		(c)	Compiling the data of indicators oi environmental elements.
		(d)	Comparing the calculated residuals disposed with the standards and its related environmental consequences and health hazards.

(*Contd.*)

(Contd.)

S. No.	*Objective*	*Parameters*
4.	To analyse the effects of environmental impacts of tourism on the local communities.	(a) Collection of primary data through focus and household through structured questionnaires and secondary data collection like tourists flow, land under use etc. (b) Identifying different environmental parameters and measuring the effects of the parameters. (c) Logically analyzing and evaluation and linking different determinants. (d) Statistical analysis of the effects of the impacts.

NOTES

1. Lindberg, K. & Donald, E. Hawkins, 1993. *Ecotourism—A guide for local planners and managers*, North Bennington, VT, USA: The Ecotourism Society.
2. Warnken, J. & Buckley, R., 2000. Monitoring diffuse impacts: Australian Tourism developments, *Environmental Management* 25 (4), 453-461.
3. Buckley, R.C., 1999. Tourism and biodiversity: land-use, planning and impact assessment, *Journal of Tourism Studies*, 10 (2): 47-56.
4. Briassoulis. H., 1992. Environmental impact of tourism: A framework for Analysis and Evaluation, In: Briassoulis, H. and Straaten, Jan Van der (Eds.). *Tourism and the Environment: Regional, Economic and policy Issues*. Kluwer Academic Publishers: 11-22.
5. Batta, R.N., 2000. *Tourism and the Environment—A quest for sustainability*, Indus Publishing Company, New Delhi.
6. Manning, Edward, 1992. Managing Sustainable Tourism—Indicators for Better Decisions: *Proceedings of the 1992 World Congress on Adventure Travel and Ecotourism*.
7. Buckley, R.C., 2001. Environmental Impact of Ecotourism, In D. Weaver (Ed)., *The Encyclopedia of Ecotourism*, CABI, Wallingford: 379-394.
8. WTTC 2002. *Corporate Social Leadership in Travel & Tourism*. London U.K., World Travel and Tourism Council.
9. Cooper, C.P., and Odzil, I. 1992. 'From mass to responsible tourism—the Turkish Experience', *Tourism Management*, December 13 (4): 377-386.
10. Bukley, Ralf, 2002. Tourism and Biodiversity in North and South, *Tourism Recreation Research* Vol. 27(1), 2002: 43-51.

11. Hunter, C. & Green, H., 1995. *Tourism and the Environment: A sustainable relationship*. Routledge, London.
12. Tangi, M., 1977. Tourism and the Environment, *Ambio*, Vol. 6, pp. 336-346.
13. Baud-Bovy, M. & Lawson, F., 1977. *Tourism and recreation development*, London: The Architectural Press.
14. Pearce, D.G., 1985. 'Tourism and environment research: A review', *International Journal of Environmental Studies* 25: 241-255.
15. Farrell, B.H. and McLellan, R.W., 1987, 'Tourism and Physical Environment research', *Annals of Tourism Research* 14(1): 1-16
16. Farrell, B.H. and Runyan, D., 1991. Ecology and Tourism, *Annals of Tourism Reserach*, 18 (1): 26-40.
17. Lindsay, J., 1986. *Carrying capacity for tourism development in national parks of the United States*, UNEP Industry and Environment. Jan/Feb/March. 8(1): 17-18.
18. Nijkamp, P. and Bergh, J.C.J.M. van den, 1991. Operationalizing sustainable development-dynamic ecological—economic models. *Ecological Economics* 4, 11-33.
19. Nijkamp, P. 1980. *Environmental policy analysis*, New York: John Wiley and Sons.
20. Onyeanusi, A.E. 1986. Measurements of impacts of tourist off-road driving in grassland in Masai Mara National Reserve, Kenya: a simulation approach. *Environmental Conservation* 13 (4), 325-329.
21. Holden, A., 2000. *Future of Tourism's Relationship with Environment, Environment and Tourism*. London, Routledge.
22. OECD, 1980. *The impacts of tourism on the environment*, Paris: Organisation for Economic Cooperation and Development.
23. OECD, 1994. *Environmental Indicators—OECD*, Core Set, Paris.
24. Wheeler, B., 1994. Ecotourism: A reuse by any other name. In: Cooper, C. and Lockwood, A. (eds). *Progress in Tourism Recreation and Hospitality Managemen*. Vol. 6, Chichester: Wiley, pp. 3-11.
25. May, V., 1991. "Tourism, environment and development: Values, sustainability and stewardship", *Tourism Management*, 12 (2): 112-118.
26. Bhattacharya, A.K., Banerjee, S. and Saksena, V., 2003a. "Local initiatives to localize Ecotourism—An Exploratory study in Kerwa Van Vihar National Park Catchments, *Journal of Tourism Recreation Research*, Vol 28(1) 2003.
27. Bhattacharya, A.K., Banerjee, S. & Saksena, V., (2003b). Community based Ecotourism—Identification of issues, options and strategy through Stakeholders Analysis. Paper submitted for publication.
28. Kneese, A.V., Ayres, R.U. & D'Arge, R.C., 1972. *Economics and the Environment. A Materials Balance Approach*. Baltimore, Md.: Johns Hopkins University Press.

14

Relevance of Carrying Capacity and Eco-Development Linkages for Sustainable Ecotourism

A.K. Bhattacharya and *Suchitra Banerjee*

Introduction

The understanding of the concept of carrying capacity (CC) as a determinant of sustainable ecotourism (SET) raises many issues, which suggest further analysis of the existing concept of CC and exploring different viable models for using CC as an effective tool for SET. CC has been defined differently in different perspectives but because of the complexities of variables associated with CC it has not been possible to arrive at any simple holistic definition of CC, which could be applied and used with respect to SET. Most of the definitions of CC mainly relate to the visual impacts of the total load of the number of tourists on the physical environment along with some notional impact on social, biological and psychological environments. These impacts endure the limitations of being notional, unexplained and non-measurable. The absence of certain well defined verifiable indicators and quantitative variables of CC makes the whole phenomenon more complex and also makes it difficult to establish linkages between CC and SET.

Some of the researchers (WTO, 1984[1]; O'Reilly, 1986[2]; Pearce, 1989[3]) have reservations about the usefulness of CC, as they consider that CC has very limited use in SET, whereas others (Romeril, 1989[4]; Martin & Uysal, 1990[5]; Hunter & Green, 1995[6]) appreciate that CC is an essential measurement for SET and it has further been accepted as an effective tool for tourism managers by many researchers (Wall, 1983[7]; Williams & Gill, 1994[8]; Hunter & Green, 1995[6]). Therefore, the premonition that CC cannot be measured quantitatively and hence has no practical value is a matter of concern and demands further research works, because the quantitative measurement of CC has not been possible mainly owing to lack of appropriate research data (Romeril, 1989[4]; Buckley, 1995[8]) and also adherence to the traditional approaches (Williams & Gill, 1994)[9]. Matching our requirements a system can be evolved to assign empirical values for quantitative estimation of CC. Statistical and mathematical correlationships for the quantitative measurement of CC can be developed by using Environment Impact Assessment (EIA) techniques and matrices (Tivy, 1973)[10]. Impacts can be quantified in many cases and gauged in others, which means that they may be measured and quantified in discreet ways. Empirical models for determination of quantitative measurement of CC (Canestrelli & Costa, 1991)[11] need to be further verified and modified with more field research data and experiments.

Significantly more investigations and information are required regarding the cumulative impacts associated with ecotourism (Stabler, 1994)[12] as the understanding of the cumulative impacts is very essential to analyse the total CC for SET. By adopting an alternative approach through identifying and establishing the appropriate factors influencing CC, it is possible to use the CC management technique for SET (Williams & Gill, 1994[8]). In India, as in the case of most of the developing countries, ecotourism is on taking off stage and growing rapidly. On one hand, as a result of the gradual phenomenal increase in the population of human and livestock, there is constant consequential increase in biotic pressure on the natural resources including the forest lands, whereas on other, concomitantly, the ecotourism is assuming a larger dimension, which might result into a conflict in sustainability of development. Therefore, it is appropriate and high time to devise and standardise tools which might help to derive a balance between these two challenges and evolve a realistic and pragmatic package of solutions for managing these challenges. With the existing knowledge, it appears that the technique of CC can be used as a tool for meeting these long term challenges for better management of natural resources for SET.

Carrying Capacity for Set

The concept of CC with respect to SET needs to be perceived in a different perspective and requires the visualization of the whole process with a more comprehensive viewpoint. The concept of CC is very old in wildlife management, and was used for the first time by Dasmann in 1945 (Wall, 1983)[7] for assessing the capacity of the forests for grazing by animals. The technique of CC is very pertinent and has special relevance for the protected areas (PAs) for analyzing the interactions of the biotic pressure, ecotourism and PAs ecosystem to maintain the natural equilibrium, and the judicious and scientific determination of CC can serve to be an effective tool for the management of PAs and SET. Since the primary objective of the PAs is the management of wildlife, therefore, biodiversity conservation becomes the priority area for the CC consideration in PAs. The definition of CC as propounded and applied by Lindsay (1980)[13] for PAs emphasises all aspects concerning SET but apparently does not take into consideration the probable long term impacts of ecotourism on biodiversity. The qualitative and quantitative implications of ecotourism on wildlife, viz. species diversification, health, and even breeding need to be essentially considered for the determination of CC for SET in PAs. Thus the quantitative measurement of CC for certain variables also becomes a prerequisite for assessing the scope for SET in PAs. The example of occasional discontinuation of the visit of the migratory Siberian Cranes indicates that sometimes the attitude and behaviour of the ecotourists plays a significant role as a variable of CC towards SET in PAs. In India and other developing countries the displacement of the communities from the National Parks (NPs) and their rehabilitation may act as a limiting factor for CC towards SET. Thus the concept of destination area as extended by Hovinen (1982)[14] and Martin and Uysal (1990)[5] needs to be redefined or modified for the Indian NPs, specially when the communities have been displaced outside the NPs areas. Because of the ban on prevailing community rights in NPs and Wildlife Sanctuaries (WLSs), the impact of ecotourism in such cases goes beyond the destination areas through the ecotourists and NPs based community linkages. Thus in order to determine CC for SET in PAs it becomes essential to include the impact of ecotourism even outside the destination area, taking into consideration the interrelationship of NPs based community and ecotourists visualising both human and non human factors. In reference to SET in PAs the level of tolerance of the ecosystem is more important variable *vis-à-vis* the load of number of ecotourists.

Unscientific management of ecotourism in PAs may cause more environmental damage than that caused to any other destination areas (Ashton, 1992)[15]. For PAs the ecotourism may be designated as 'Environmental Tourism', which not only covers all the parameters of ecotourism but also provides a wider horizon to include all environmental factors of the destination area—both micro and macro—viz. ecological, social, cultural, psychological and anthropological etc.

Eco-Development (ED)

The ED model, originally advanced by UNEP (Dasmann, 1985)[16], aims towards an ecologically sustainable form of development through the economic sustainable development of local communities dependent on the environment for their basic needs. Thus ED attempts at overall development through which the local communities become self-reliant to achieve both their basic needs and help them in sustaining biodiversity. Some researchers prefer to use the concept of ED for both ecological and economic development, and thus consider that 'eco' stands for both 'ecological' and 'economic' (Gupta, 1988)[17]. The concept of ED has been pursued through a series of projects in management practice of PAs (Collins, 1994[18]; Rao & Geisler, 1990[19]). IUCN (1984)[20] has reported human encroachment as the second of the top ten kinds of threat facing 43 of world's most threatened PAs, which manifests the significance of ED in the management of the PAs.

The term 'ED' has been adopted in India since 1980s and a number of definitions have been proposed (World Bank, 1995)[21]. Most commonly it has been applied in connection with PAs such as NPs and WLSs to define a strategy for protecting ecologically valuable areas from unsustainable or otherwise unacceptable pressures resulting from the needs and activities of the people living in and around such areas. It is based on the assumption that if alternative natural and social resources are created in areas peripheral to the PAs, exploitation of resources with in the PAs will be reduced. There is also an assumption that the involvement of the PAs management in the creation of employment opportunities in fringe areas resources, which may include fodder, veterinary services, water services, and other village infrastructures, will motivate communities to collaborate in the protection of forest areas. ED involves participatory management of resources but communities do not receive direct benefits of forest produce from the PAs. In other forest areas managed through Joint Forest Management, the communities dependent on them derive usufruct benefits by sharing of products. This anomaly in usufruct sharing mechanism in PAs and adjoining non PAs may lead to a conflict in sustainability of PAs and this issue needs to be examined from the viewpoint of management.

In many of Indian states including Madhya Pradesh, Gujarat, West Bengal and others, the term ED has been used in a more general way to denote ED of villages close to forest lands. The more general usage is referred to as 'Village Resource Development Programme', which is similar to ED in trying to reduce pressure on forest land by providing alternative sources of income. The proposed strategy for biodiversity conservation would be to focus on improved PAs management of high priority areas of biological importance, in which human impacts are readily manageable. Also the management of lower priority would be maintained through fewer resources. PAs management would be complemented by ED in areas peripheral to PAs. The PAs network, in particular, linkages between existing PAs, would be developed through identification of relatively undisturbed satellite areas and for the intervening zone, the preparation of management plans that are compatible with biodiversity conversation. Thus the issues related to ED with respect to new strategy of management of PAs and its impact on SET need to be thoroughly examined.

Carrying Capacity and Eco-Development Linkages for Set

The preceding paras clearly suggest that in the management of PAs the ED approach is going to have great impact on CC with respect to SET. The impact of communities living in and around the PAs on the CC of the destination area needs to be critically examined for the determination of total CC. The different indices of interrelationship and interactions between the PAs and the dependent communities have to be linked with other factors/variables of CC. Whether the local communities along with the ecotourists can be used as contrivance for conservation of biodiversity and whether ecotourism can be practised as an incentive for the local people for the conservation leading to SET are challenging issues to be addressed by the managers of PAs, and the success of SET in PAs will largely depend on this complex but natural relationship. The knowledge of CC–ED linkages should prove to be an effective tool for evolving suitable strategy for both integrated management of PAs and also environmentally compatible ecotourism. Participation of local people in managing PAs becomes essential for a fundamental shift away from the old conventional approach to conservation (Williams & Gill, 1994)[8]. The necessity for incorporating the needs and perceptions of local people in the establishment and management of the resources was first emphasized by Man and Biosphere Programme in 1979 (UNESCO, 2002)[22].

Conservation of biodiversity and integration of protected area management with the needs of local people and their development has emerged as a priority for the management and sustainability of NPs and SET (Wells & Brandon, 1995[23]; Buckley, 1995[9]). Conservation, thus, requires a perspective which stretches beyond the NPs boundaries. Many case studies have indicated that the community involvement in the SET of PAs has not been very effective mainly because of two reasons. First, the participation has not been beneficiary oriented and, therefore, has not been sustainable and secondly, methods of participation have not been matched with requirements and thinking of the local people. Although community involvement has widely been accepted as an essential key element and prerequisite for PAs management and SET (Sadler, 1990[24]; Burchett, 1991[25]; Wight, 1993[26]; Lay, 1996[27]), but no pragmatic and effective devices for active involvement have so far been evolved. Thus it becomes imperative that specific models for community involvement are required to be evolved for better cooperative relationship between PAs and local people (Wells & Brandon, 1995)[23]. Long term success of SET ultimately depends upon community perception and involvement in planning and management of SET (Craik, 1991)[28].

World Bank and Future Strategy for Set

In countries, where World Bank sponsored forestry projects are being implemented, the SET will largely depend upon the strategy being adopted for PAs management under the project. In India the World Bank Forestry Project encompasses the component of Biodiversity Conservation for PAs, which aims at improved management of high priority PAs through provisions of support for development of management plans, staff training, equipments, and other facilities. There is special provision for funds to support the development of alternative resources of income for communities resident within PAs and in areas peripheral to them. A low level of project funding would be made available for low priority PAs whilst the identification and planning of linkages to complete the PAs network would be finalized. It would be interesting to examine the impact of the provisions of the project on the management of the PAs with respect to the SET.

Scope of Research

Based on the analysis of the existing knowledge of the subject, as discussed in the preceding paragraphs above, and the evaluation of the strength and gaps in the knowledge, the following issues need to be taken into consideration:

1. With the growing environmental awareness, the ecotourism is going to assume very large dimension in the recent future, more widely in the developing countries, and the PAs will be the main destination areas for the majority of ecotourism.
2. CC can serve as one of the most effective tools for the management of PAs with respect to promotion of SET. The lack of adequate field data and research work act as an impediment to the use of CC as a practical management supplement for SET. Relevant techniques are required to be generated to analyse the cumulative impacts affecting the total CC.
3. It is possible to evolve certain correlationship/matrices for the quantitative measurement of CC for SET in PAs. With the help of some relevant research work and field data practical models for quantitative measurement of CC may be developed.
4. Because of lack of certain well defined objectively verifiable indicators the effectiveness of CC technique for SET in PAs has been impaired. Based on practical research and field data some simple responsive applicable indicators may be formalized for the use and monitoring of CC techniques for SET in PAs.
5. The process of ED influences CC largely in PAs, and the CC in PAs is regulated by the factors involved in the ecologically oriented economic development of the communities dependent on the PAs. The analyses of the synergistic impact of the ecotourism and the communities on the PAs by scrutinising the CC-ED linkages may provide some pragmatic solutions for the challenges likely to be met for SET in PAs.
6. Involvement of communities living in and around PAs is an essential element for SET. Realistic and pragmatic models ensuring active involvement of communities may be evolved by examining and extending CC-ED linkages for SET. This will in turn facilitate SET and also protect PAs which will contribute to both management of PAs and SET.
7. Under the new forestry projects being assisted by World Bank there is a distinct shift in strategy of management of PAs, from

the traditional top bottom approach to bottom top approach, the emphasis being on community participation in forest management. The implementation will have many challenges mainly because the existing models of people's participation have not been found to be effective.

8. In spite of the obvious significance and relevance of CC for SET, no organised study on the CC with respect to SET in PAs in India has so far been pursued. A detailed pilot study should be fruitful to develop a sound strategy for SET in many developing countries.

Parameters of further Investigations

In the light of the above mentioned background it clearly emerges that some more intensive and integrated research works are necessary to

(a) redefine the CC for PAs, with respect to SET, specially in developing countries,
(b) evolve certain applicable models for quantitative measurement of CC,
(c) determine ED and CC linkages for SET, and
(d) design appropriate management modules suitable for SET in PAs.

In order to bridge the gaps in the existing knowledge on the subject the research should aim at the following objectives:

1. To critically analyse the process and frameworks of CC in PAs with respect to SET, by identifying the various factors/ parameters linked with and/or influencing CC and their interactions and cumulative impacts.
2. To develop certain empirical/statistical correlationships/matrices for the determination of quantitative measurement of CC with respect to SET in PAs.
3. To study the linkages between the CC and ED in PAs by evaluating the qualitative and quantitative impacts of the communities on the CC—both individual and synergistic.
4. To identify and standardize certain objectively verifiable and applicable indicators for CC and ED linkages and SET based on the field standards.
5. To evolve certain realistic models/package for CC and ED linkages for the management of PAs and SET.

6. To critically evaluate the cumulative and synergistic impacts of the World Bank Forestry Project on the management of the PAs and SET.
7. To develop an appropriate package for training and education for SET for all levels involved in ecotourism including the local communities.

Design and Methodology for Research

The methodology of the study should, however, be decided on the basis of the available literature and preliminary reconnaissance survey of the area, and nevertheless, and following criteria may be taken into consideration while finalising the methodology:

1. Review of the available literature/case studies and collection of data available on the subject.
2. Critical examination and analysis of the existing management practices and Working/Management Plans being implemented and their impacts in PAs with respect to SET and scrutiny of the mechanisms of various operations.
3. Use of participatory rural appraisal/rapid rural appraisal techniques for the study of the correlationships of communities and PAs with respect to their dependence on PAs and their involvement in management practices.
4. Development of empirical correlationships for quantitative measurement of CC on the basis the existing EIA techniques.
5. The study area (PAs) may be selected on the basis of following norms:
    - The National Park attracting the maximum number of ecotourists.
    - The National Park receiving the minimum number of ecotourists.
    - The largest National Park (in terms of area).
    - The smallest National Park.
    - The National Park where ED and community participation are successful.
    - The National Park where ED and community participation are reportedly failure
    - Comparison of the observations with those of a NPs of some other developing countries and also with those of developed

countries, where ecotourism is rapidly growing as an industry.

Conclusion

There is considerable scope for research on the concept of CC and its use as a management tool for ecotourism in PAs, specially in developing countries, where the ecotourism may add to the cumulative impact of the already existing biotic pressure on the PAs. Exploratory research is essential to evolve some viable mechanisms and models to involve the local communities in the management of ecotourism in these areas. Sustainability of ecotourism will largely depend on the active involvement of the local communities. Ecotourism and ED linkages may probably be used as tool for the conservation of the natural resources and sustainable development.

NOTES

1. WTO, 1984. *Economic review of world tourism*, World Tourism Organisation, Madrid.
2. O' Reilly, A.M., 1986. Tourism carrying capacity, *Tourism Management*, 7 (4), 254–258
3. Pearce, Douglas, 1989. *Tourist Development*, II ed. Longman Scientific.
4. Romeril, M., 1989. Tourism and the environment-accord or discord? *Tourism Management*, 10 (3): 204-208.
5. Martin, B. S. & Uysal, M., 1990. An examination of the relationship between carrying capacity and tourism lifecycle: management and policy implications. *Jour. Env. Manag*, 31: 327-333.
6. Hunter, C. & Green, H., 1995. *Tourism and the Environment: a sustainable relationship*, Routledge, London.
7. Wall, G., 1983. Cycles and Capacity: A Contradiction in Terms ! *Annals of Tourism Research*, 268-269.
8. Buckley, R.C., 1995. *Where Tourism and Ecology Meet*, International Conference: Ecotourism: Concept, Design and Strategy, February 6-8, Bangkok, Institute of Ecotourism, Srinakharinwirot University.
9. Williams, P.W. and Gill, A., 1994. Tourism CC management issues. In: *Global Tourism: The next decade*, Ed Theobold W., Butterworth Heinemann, Oxford.
10. Tivy Joy, 1973. *The concept of determination of CC of recreational land in the USA.* Countryside Communication, Perth, Scotland, Occasional paper No. 3, Perth, Scotland.
11. Canestrelli, E. & Costa, P., 1991. Tourist CC A Fuzzy Approach, *Annals of Tourism Research*, 18: 295-311.

12. Stabler, M., 1994. An economic critique of tourism environmental auditing. Tourism Research and Education in Australia: *Proceedings from the Tourism Research and Education Conferences*, Eds Faulkner, B., Fagence, M., Davidson, M. & Craig, Smith S. Bureau of Tourism Research, Canberra.
13. Lindsay, J., 1980. Compatibility planning for different types of outdoor recreation and resources. In: *Tourism Planning and Development Issues* (D. Hawkins, *et al.*, eds) pp. 138–47, Washington DC, George Washington University.
14. Hovinen, G.R., 1982. Visitors Cycle: Outlook for Tourism in Lancaster Country, *Annals of Tourism Research*, 9: 565-583.
15. Ashton, R., 1992. World Tourism Trends in Tourism and Conservation (Ecotourism). In: *Proceedings of the 1992 World Congress on Adventure Travel and Ecotourism*, The Adventure Travel Society Inc., Englewood, Columbia. pp. 67-73.
16. Dasmann, R.F., 1985. Achieving the the sustainable use of species and ecosystems. *Landscape Plan*, 12 (3): 211-219.
17. Gupta, A., 1988. *Ecology and development in the third world*, Routledge, USA.
18. Collins, C., 1994. Conservationists call on tourism industry to join green movement, *The Australian*, 14 April, p. 6.
19. Rao, K. and Geisler, C., 1990. The social consequences of protected areas development for resident populations, *Society and Natural Resources*, 3:19-32. (www.gu.edu.au/centre/capsm/Corbett.htm—101k).
20. IUCN, 1994. *Ecotourism Consultancy Programme*, IUCN, Mexico.
21. World Bank, 1995. *The India Eco-development Project: Project Document*, The World Bank–GEF, World Bank, World Bank India.
22. UNESCO, 2002. UNESCO, *Man and the Biosphere Programme* (WWW.UNESCO.ORG/MAB/, 3 AUG, 2002).
23. Wells, M. & Brandon, K., 1995. *People and Park: Linking protected area management with local communities*. The World Bank, World Wildlife Fund, USAID, Washington DC.
24. Sadler, B., 1990. Sustainable development, northern realities and the design and implementation of regional conservation strategies. In: *Achieving sustainable development through northern conservation strategies*, pp. 3-31, University of Calgary Press, Calgary.
25. Burchett, C., 1991. *Ecologically sustainable development and its relationship to Aboriginal Tourism in the Northern Territory*, Ecotourism: Incorporating the Global Classroom, Canberra, Bureau of Tourism Research.
26. Wight, P.A., 1993. Ecotourism: Ethics or Eco-Sell? *Journal of Travel Research*, 31 (3): 3-9.
27. Lay, B., 1996. Community Involvement in the Establishment of an Access Network to Scenic Areas. Paper presented at the Best Practice Ecotourism Conference, Coolangatta, Qld.
28. Craik, J., 1991. *Resorting to Tourism: Cultural Policies for Tourist Development in Australia*, Allen and Unwin, North Sydney.

15

Estimating the Total Carrying Capacity of Protected Areas with Respect to Tourism Activities

A Case Study of Bandhavgarh National Park, Madhya Pradesh, India

A.K. Bhattacharya and *Thriveen Shanker*

Background

Tourism is the largest growing civilian industry that employs the maximum number of persons, both directly and indirectly (WTO, 1999)[1]. It is also the sector which has a high multiplier effect of 3.2—i.e. for every rupee invested in building the infrastructure, the revenue generated is more than three times (Singh, 1996)[2]. But these meritorious attributes have not manifested into ground reality, as tourism is yet to become a powerful engine of socio-economic upliftment for the local communities or to be utilised as a tool for natural resource management in India. On the contrary, the uncontrolled growth of tourists and tourism activities in the areas of natural beauty and historical significance is exhausting the very resources that transform an area into a tourist destination.

A basic problem which arises while managing the impacts of tourism on environment is the difficulty in quantifying the magnitude of the ecological resources and determining what proportion of its potential should be committed to the baseline demands imposed within the natural environment itself. It is important to specify these capacities so that human activities can be reoriented to operate within these natural constraints. It was at this juncture that carrying capacity concept was recognised as a tool for establishing management standards for natural resource managers (Wall, 1983[3]; Williams & Gill, 1994[4]; Hunter & Green, 1995[5]).

Carrying Capacity (CC) Concept in Tourism Management

Even though CC concept was used in many other fields of natural resource management, it was adopted by the tourism sector only in 1960s (Lucas, 1964[6]; Wagar, 1964[7]). In the context of tourism, World Tourism Organisation (1993)[8] defines CC as, the level of visitor use an area can accommodate with high levels of satisfaction of visitors and few impacts on resource.

But the estimation of the comprehensive CC of a destination area has always been a factious proposition, with arguments upwelling for and against its utility both as a criterion for impact appraisal and as a management device. As it has a multi-dimensional existence, quantification of CC is not easy and thus many researchers have doubts about its usefulness (WTO, 1984[9]; O'Reilly, 1986[10]; Pearce, 1989[11]). But another section of thinkers believe that the fault lays not with the concept itself, but due to lack of appropriate research data (Romeril, 1989[12]; Buckley, 1995[13]) and also adherence to the traditional approaches (Williams & Gill, 1994)[4]. This study pursues a new methodology to estimate the comprehensive CC of a tourism destination by estimating the impacts on the indicators identified and then extrapolating it to the individual components of CC.

The site selected for the study was Bandhavgarh National Park in Madhya Pradesh, which is one of the Tiger Reserves of India. Bandhavgarh is an ideal site for such a study as it is a tourism destination of international reputation with relatively high visitor influx and one in which the tourism infrastructure has been in existence for an extensive duration.

The Broad Objectives of this Study were as under:

- To collect data on the number of visitors and vehicles arriving at the protected area each year and the amount of facilities available for the visitors outside the protected area.
- To collect information on the impacts of tourism on the different components of carrying capacity of the tourism destination from the different stakeholders using questionnaire surveys and interview schedules.

- To calculate the percentage of total carrying capacity left for utilisation by the tourism activity from the estimated impacts on indicators and components.

Methodology

The methodology was selected on the assumption that a tourist destination has a CC of 100 per cent before the tourism activity was initiated. The adverse impact of these activities reduces its CC, while management initiatives can augment it.

The total CC is divided into five integral components, viz. Ecological, Facility (or Infrastructure), Social, Economic and Visitors' experience-based carrying capacities. After field visits and consultations with the experts, appropriate indicators were identified for these components which can predict the impact due to tourism on the respective component.

The methodology consists of a two-tier impact analysis, where the impacts of tourism on the indicators are at first evaluated Indicator Quality Unit or (IQU) and multiplied by the relative importance of each indicator in predicting the impact Parametric Importance Unit (PIU). The stakeholders are asked to examine whether the indicators selected are showing any kind of deviation from the *status quo* and on the basis of the percentage of people agreeing to the existence of impacts is taken as IQU. This is based on the assumption that the number of people agreeing to the impact statement is directly proportional to the severity of impact. The PIU is calculated from the arithmetic mean of scores given by experts based on the ability of the indicators to accurately predict the impact on that component.

The IQUs and PIUs of all the indicators of a component are added up to obtain the total impact of tourism on that component Carrying Capacity Impact Unit (CCIU). Then the relative importance of each component in determining the total CC Component Importance Value (CIV) is multiplied to the individual CCIU of each component to get the specific CC percentage left. The summation of all the CC percentages of the components will give the Total CC remaining in the destination area with respect to tourism activity.

Results

The results obtained by the step-by-step method to obtain the total CC of Bandhavgarh National Park is given in Tables 15.1 to 15.7.

Table 15.1: IQU of Ecological Carrying Capacity

	Indicator	*% of Agree-ment*	*IQU*	*Final PIU*	*CCIU*
(a)	Less animal sighting during the trip inside the PA (N)	10%	0.9	6.797	6.117
(b)	Crowding of vehicles around the animals, especially tiger (N)	80%	0.2	13.594	2.719
(c)	Aggressive behaviour of animals towards the vehicles (N)	50%	0.5	13.594	6.797
(d)	Disturbance caused to the tigers due to elephants (N)	40%	0.6	10.681	6.409
(e)	Road erosion and degradation of roads due to vehicular traffic (N)	40%	0.6	5.826	3.496
(f)	Dust and smoke generated by the vehicles inside the PA (N)	70%	0.3	12.623	3.787
(g)	Reduction of ocular forest quality in and around the tourism zone (N)	10%	0.9	7.768	6.991
(h)	Pollution in the tourism zone—air pollution or solid waste pollution (N)	0%	1	11.652	11.652
(i)	Awareness generated about the environmental protection in visitors (P)	70%	0.7	9.71	6.797
(j)	Importance given to wild animals by the local people (P)	93.3%	0.933	7.755	7.235

Total carrying capacity impact unit for ecological aspects = 62 per cent or 0.62

Table 15.2: IQU of Facility carrying capacity

Indicator	*% of Agree-ment*	*IQU*	*Final PIU*	*CCIU*
(a) Legal restrictions present during construction of facility (P)	0%	0	19.98	0
(b) Rate of growth of number of tourists in the area, leading to more facilities being constructed (N)	33.3%	0.333	31.18	10.383
(c) Water or natural resource shortage faced by the facilities (N)	0%	1	13.32	13.32
(d) Objections from the local people against the facility (N)	0%	1	15.54	15.54
(e) Willingness of the local people to sell their land for facility construction purposes (P)	76.6%	0.766	19.98	13.305

Total carrying capacity impact unit for facility aspect = 52.548 per cent or 0.525.

Table 15.3: IQU of Social Carrying Capacity

Indicator	*% of Agree-ment*	*IQU*	*Final PIU*	*CCIU*
(a) A social issue created by the tourism—Alien culture (N)	20%	0.8	18.24	14.59
(b) A social issue created by the tourism—Nuisance created to the local people (N)	0	1	10.64	10.64
(c) A social issue created by the tourism—Drinking culture (N)	3.3%	0.967	7.6	7.349
(d) A social issue created by the tourism—Eve-teasing of local women (N)	0	1	9.12	9.12

(Contd.)

(Contd.)

	Indicator	*% of Agree-ment*	*IQU*	*Final PIU*	*CCIU*
(e)	A social issue created by the tourism—Social inequalities created within the community (N)	20%	0.8	18.24	14.592
(f)	Preservation of local art and culture due to tourism (P)	23.3%	0.233	10.64	2.479
(g)	Improvement in the functioning of local institutions like Panchayat due to the growth of tourism (P)	26.6%	0.266	16.4	4.362
(h)	Problems from local people to tourists (N)	0%	1	9.12	9.12

Total carrying capacity impact unit for social aspects = 72.252 or 0.723 per cent

Table 15.4: IQU of Economic Carrying Capacity

	Indicator	*% of Agree-ment*	*IQU*	*Total PIU*	*CCIU*
(a)	Increased employment or increased dependence of local people on tourism for their livelihood (P)	63.3%	0.633	68.75	43.519
(b)	Increased cost of living in and around the tourism zone (N)	10%	0.9	31.25	28.125

Total carrying capacity impact unit for economic aspect = 71.644 per cent or 0.716

Table 15.5: IQU of Visitors' Experience based CC

Indicator	% of Agree-ment	IQU	Total PIU	CCIU
(a) Pollution in the tourism zone (N)	0%	1	12.21	12.21
(b) Overcrowding inside or outside the PA (N)	14%	0.6	16.75	10.05
(c) Dust and smoke generated inside the PA, disabling the enjoyment of the wilderness (N)	33.3%	0.667	15.54	10.365
(d) Quality of guides and the interest generated by them in nature (P)	50%	0.5	14.43	7.215
(e) Information imparted about nature through interpretation facilities (P)	33.3%	0.333	15.54	5.175
(f) Feeling of worthiness of the visit (P)	90%	0.9	13.32	11.988
(g) Willingness of the tourist for another visit (P)	96.6%	0.966	12.21	11.795

Total carrying capacity impact unit for visitors' experience aspect = 68.798 per cent or 0.688

Table 15.6: CIV for Bandhavgarh National Park

Component	Values given by the experts					Final
	1	2	3	4	5	
Ecological	55%	50%	60%	60%	60%	55.884
Facility	10%	15%	5%	10%	15%	10.784
Social	5%	5%	5%	0%	5%	3.924
Economic	15%	10%	15%	10%	10%	11.764
Visitors' experience	15%	20%	15%	20%	10%	17.648

Table 15.7: Percentage of CC left in Bandhavgarh National Park

Component	CCIU	CIV	% of CC left for tourism
Ecological CC	0.62	55.884	34.648
Facility CC	0.525	10.784	5.661
Social CC	0.723	3.924	2.837
Economic CC	0.716	11.764	8.423
Visitors' Experience based CC	0.688	17.648	12.142

Thus the total carrying capacity left in Bandhavgarh National Park for tourism activity = 63.711 per cent.

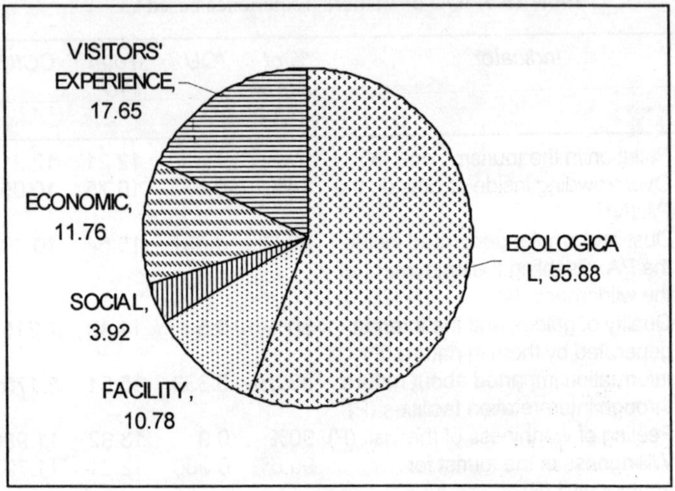

Fig. 15.1: Component Importance Value for Bandhavgarh National Park

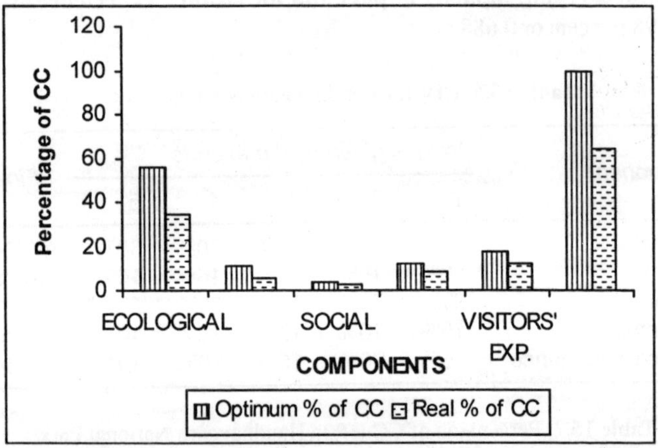

Fig. 15.2: Percentage of Carrying Capacity left versus Optimum Carrying Capacity in Bandhavgarh National Park

Discussion

The CCIU values of different components indicate that Facility or Infrastructure CC and Ecological CC are most affected by the tourism, as the CC of these components are reduced by 50 per cent and 40 per cent respectively. These values are within or near the moderate impact zone and should be a serious concern for the authorities. The important indicators for the ecological impact were the crowding of vehicles around the animals (especially tiger), dust and smoke generated by the vehicles inside the PA and aggressive behaviour of animals towards the vehicles. The facility CC was reduced mainly due to problems such as lack of any legal restrictions on the construction of facility around PA and high rate of growth in the number of tourists in the area, leading to the construction of more facilities. The other components are in a relatively better condition, with a CCIU near 70 per cent.

The CIU scores indicate the predominance of ecological CC in determining the total CC. But a large number of the tourists who visit Bandhavgarh are hardcore nature-lovers and thus the Visitor's experience-based CC is also very important. The factors that are leading to a reduction in this component are the poor quality of guides and the interest generated by them in nature, information imparted about nature through interpretation facilities, dust and smoke generated inside the PA, disabling the enjoyment of the wilderness and to a lesser extent the overcrowding inside the PA especially during the tiger sightings.

The overall impact produced by tourism on the total carrying capacity of Bandhavgarh National Park belongs to the 'Low' impact category (61-80 %), but the Facility CC has reached the 'Moderate' impact zone.

Conclusion

The tourism activity has reduced the total CC of the area by about 36 per cent from its original state. Though the impact still remains within the low impact category, it is perilously close to the MODERATE impact level (41-60).

Management initiatives are required to mitigate the adverse impacts arising from the tourism activities and special emphasis should be laid upon the sensitive components as Ecological and Facility CC in this case. There should be the formulation of a state level authority to control the construction of facilities around the PAs and regulate the flow of tourists within the CC limits of the destination area. The initiatives such as proper training of guides and including the local communities as a stakeholder in the tourism activity can alleviate the impacts on Visitor's experience-based and Social CC respectively. Calculation of Total CC is important from the perspective of tourism management, as it gives a holistic view of impacts due to tourism. The reductionism exhibited in management of tourism can only lead to pervert mitigation measures. So total CC is a better tool than the individual capacities for controlling and coordinating the tourism activities carried out by different stakeholders.

NOTES

1. WTO, 1999. *World Tourism Statistics*, World Tourism Organisation, URL:www.world-tourism.org
2. Singh, Ratandeep, 1996. *Infrastructure of Tourism in India*, Kanishka Publishers, New Delhi, p. viii.
3. Wall, G., 1983. 'Cycles and Capacity: A Contradiction in Terms!' *Annals of Tourism Research*, pp. 268-269.
4. Williams, P.W. and Gill, A., 1994. 'Tourism CC Management Issues'. In: *Global Tourism: The Next Decade*, Ed Theobold, W., Butterworth, Heinemann, Oxford.
5. Hunter, C. and Green, H., 1995. *Tourism and the Environment: a Sustainable Relationship*. Routledge, London.
6. Lucas, R.C., 1964. 'Wilderness perception and use: the example of boundary water canoe area', *Natural Resources Journal* 3(3): 394-411.
7. Wagar, J., 1964. 'The carrying capacity of wildlands for recreation', Forest Service Monograph, *Society of American Foresters*, 7: 23.
8. WTO, 1993. *Sustainable Tourism Development: Guide for Local Planners*, World Tourism Organisation, Madrid.
9. WTO, 1984. *Economic review of world tourism*, World Tourism Organisation, Madrid.
10. O'Reilly, A.M., 1986. 'Tourism carrying capacity', *Tourism Management*, 7 (4), 254–258.
11. Pearce Douglas, 1989. *Tourist Development*, II ed. Longman Scientific.
12. Romeril, M., 1989. Tourism and the environment—accord or discord? *Tourism Management*, 10 (3): 204-208.
13. Buckley, R.C., 1995. *'Where Tourism and Ecology Meet'*, International Conference: Ecotourism: Concept, Design and Strategy, February 6-8, Bangkok: Institute of Ecotourism, Srinakharinwirot University.

16

A New Methodology for Estimating the Total Carrying Capacity of Tourism Destinations

A Case Study of Pench National Park, Madhya Pradesh, India

A.K. Bhattacharya and *Thriveen Shanker*

Background

The industrial growth of the 20th century has instigated a proportional increment in the field of mass tourism. The apparent progenies of these altered economic conditions like the modified work culture with paid holidays (Urry, 1990)[1], demographic and social changes (e.g. dual income households and a growing proportion of single adults), relaxation of immigration restrictions, increased paid leave and more flexible working time, improved education and awareness of travel possibilities, and technological improvements (Ceballos-Lascurian, 1992)[2] have all played a part in this progress. But this evolution of tourism facilities was not in tune with the resource limits of the destination and resulted in skewed development of infrastructure as well as subsequent social and environmental impacts.

But the notion that mass tourism is a bad thing is rather simplistic and it has been suggested that controlling the volume of tourism might alleviate the situation, especially since the tourism is typically found in locations with fragile environments, such as mountains and coasts which are peripheral to the world economy (May, 1991)[3]. Carrying capacity is often cited as a framework in which the objective of regulating the scope of tourism in a destination can be achieved (Hunter and Green, 1995[4]; Inskeep, 1991[5]; McIntyre, 1993[6]; O'Reilley, 1986[7]; WTO, 1993[8]).

Carrying Capacity (CC) of Tourism Destinations

The concept of carrying capacity is very old in wildlife management, and was used for the first time by Dasmann in 1945 for assessing the capacity of the forests for grazing by animals (Wall, 1983)[9]. It was from early 1960s that this concept was applied in recreational research for the purpose of determining the ecological disturbance from use (Lucas, 1964[10]; Wagar, 1964[11]).

World Tourism Organisation (1993)[8] defines CC as, the level of visitors use an area can accommodate with high levels of satisfaction of visitors and few impacts on resource. Mathieson and Wall (1982)[12] defines it as the maximum number of people who can use a site without an unacceptable decline in the quality of experience gained by visitors.

The concept of carrying capacity of tourism activities can be explained on the basis of following interrelated elements: (1) The amount of use of a given kind, (2) a particular environment can endure, (3) over time, (4) without degradation of its sustainability for that use (Fennel, 1999)[13]. The technique of CC is very pertinent and has special relevance for the protected areas (PAs) for analyzing the interactions of the biotic pressure, ecotourism and ecosystem to maintain the natural equilibrium, and the judicious and scientific determination of CC can serve to be an effective tool for the management of PAs and sustainable ecotourism (Bhattacharya and Banerjee, 2003)[14].

But the methodology to be adopted for estimating a comprehensive carrying capacity for tourism destination areas is still a bone of contention. It may be due to the following reasons:

- It is still uncertain whether the basis of fixing the carrying capacity should be in proportion to the number of tourists allowed or the impacts caused due to them.
- Many of the factors used for the estimation of CC are qualitative and their interdependencies unknown. Thus their estimation becomes subjective.
- The different sub-components of Total CC like ecological CC or social CC may have different scales and units and thus it becomes difficult to estimate comprehensive CC by integrating all these elements.

Thus it was necessary that an integrated method be adopted for calculating the Total CC and it acts as the background for this attempt.

Methodology Used for the Study

The methodology is adopted from Batelle Environmental Evaluation System (BEES) used in Environmental Impact Assessment (EIA) studies and Limits of Acceptable Change (LAC) framework from the field of tourism research. The method is a two-tier system, where the impacts are at first estimated for the individual indicators of components and then for the components itself.

The different steps in calculating the total CC are as follows:

1. The total CC is divided into five components, i.e. Ecological CC, Facility (or Infrastructure) CC, Social CC, Economic CC and Visitor's experience based CC. Fix the relative importance of these components as percentages of Total CC. This is calculated from the ratings given by the experts in the subject, who are familiar with the study area. Then arithmetic means of the different ratings are taken to be the Component Importance Value (CIV) of respective components of CC.
2. Select local level indicators for each of the components to identify the impacts of tourism on the five components of carrying capacity. The indicators were selected after discussions with experts of ecotourism and field visits.
3. Make a Value Function Curve (VFC) for each indicator on the basis of percentage of people who recognise the impact of tourism on that specific indicator and respective quality of the indicator. So greater the number of people who recognises the impact, greater is the impact.

There will be two types of VFC, for negative and positive indicators. Negative indicators are those indicators, which are given a higher value when lesser number of people states that there is an impact on it. The positive indicators will be given higher value as the numbers of people agreeing with the statement of impact on it increases. Thus the indicators selected are classified as positive and negative indicators.

The value function curve for a negative indicator is given in Fig. 16.1.

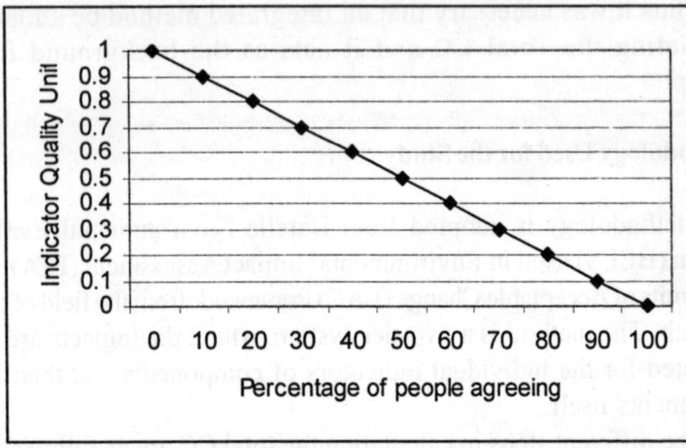

Fig.16.1: Value Function Curve of Negative Indicators

The positive indicators will have the following kind of VFC is given in Fig. 16.2.

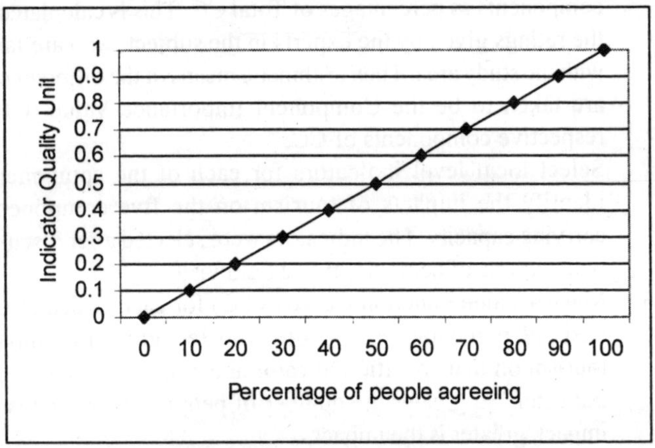

Fig. 16.2: Value Function Curve of Positive Indicators

If no one among those surveyed says there is an impact on the positive indicator it will be given an Indicator Quality Unit (IQU) of 1. If 1-10 per cent of people consider there is an impact then the value given will be 0.9, 11-20 per cent will be 0.8, 21-30 per cent will be 0.7 etc. and 91-100 per cent will get an IQU of zero. In the case of the negative indicators the valuation will be *vice versa*. It is based on the assumption that the impact is directly proportional to the number of people who recognise that there is an impact.

1. Fix a Parametric Importance Unit (PIU), for each of the indicator. Delphi technique is used in BEES to fix this value. But due to time constraint, another technique was adopted. In this method, a set of experts was asked to give ratings for the indicators on the basis of its ability to accurately predict the impact. The experts were given the guidelines for rating, which is as follows:
   - *High importance*: These are the indicators that directly indicate the impact as well as the chances that the occurrence is only due to tourism activity is also high. The impacts on these indicators are directly observable and the cause-effect relationship can be easily produced. So it should be an important indicator of impact on that component.
   - *Medium importance*: These are also indicators that directly indicate the impact, but the chance of tourism activity being the lone causative factor is doubtful. Thus the cause-effect chain is not clearly discernable. Thus these indicators should be of medium importance.
   - *Low importance*: These indicators are indirect signs of an impact. They are not directly observable or quantifiable. The possibility that the reason for the impact on these indicators is due to tourism is very difficult to establish. Due to this reason, these indicators should be assigned only a low importance.

Once the data on ratings is collected, then a value was assigned to these ratings, such 'x' for Low importance, '2x' for medium importance and '3x' for high importance indicators. A summation of the rating of all the experts were done and the arithmetic mean is calculated. This will be the Parametric Importance Unit of that indicator.

2. Multiply the Indicator Quality Unit (IQU) with Parametric Importance Unit (PIU) of each indicator to get the Carrying Capacity Impact Unit (CCIU) for that indicator.
3. Sum up the CCIU for all indicators to get the Total Carrying Capacity Impact Unit for that component of CC.
4. Multiply the Total CIU for the component with its respective Component Importance Value (CIV) as calculated in step no.1. For example, the Total CIU of Social Carrying Capacity is obtained as 2 out of a maximum possible 3, and its Component

Importance Value (CIV) is 15 per cent, then for that area the percentage of Social Carrying Capacity's quality is (2/3) × 15 = 10 per cent out of a maximum possible 15 per cent.

5. Sum up all the carrying capacity percentages for all the components to get the total carrying capacity value in percentage out of a total possible 100 per cent. This indicates the level of impacts caused due to tourism activities. Lower is the value, greater is the impact caused.
6. Set standards for the total impact and compare it with obtained percentage. The standards was set as follows:

   0-20% = Very High Impact on Carrying Capacity.
   21-40% = High Impact on Carrying Capacity
   41-60% = Moderate Impact on Carrying Capacity
   61-80% = Low Impact on Carrying Capacity
   81-100% = Very Low Impact on Carrying Capacity.

7. The percentage of CC and the standards obtained from the study can form the basis of formulating the management plans for individual PAs. Additional emphasis of mitigation can be given on those components or indicators that are shown to be impacted heavily by the tourism activities.

Results

A total of 33 indicators were identified for the 5 components of CC. These indicators were rated as those of high, medium and low importance by the experts and the stakeholders were asked for their perception of impact that has occurred on these indicators due to tourism. From this data IQU and PIU were calculated for each of the indicators. Then CCIUs for the components were found out by multiplying the IQUs and PIUs and it was as follows:

Table 16.1: Total Carrying Capacity Impact Unit of Components of CC

Component	*CCIU*
Ecological CC	84.518 % or 0.845
Facility CC	80.02 % or 0.8
Social CC	63.48 % or 0.635
Economic CC	66.09 % or 0.661
Visitor's experience based CC	78.894 % or 0.789

Table 16.1 shows that the component most affected by the growth of the tourism activity in Pench National Park is the Social CC, which has been reduced from the original level of 100 per cent to about 64 per cent. The least affected is the Ecological CC, which is reduced by only 15 per cent from the original stage.

Table 16.2: Component Importance Values of Components of Total CC

(In per cent)

Component	*Values given by the experts*					*Final CIV*
	1	2	3	4	5	
Ecological	75	80	65	70	75	73.27
Facility	5	5	10	5	5	5.94
Social	5	0	5	0	5	2.97
Economic	10	10	10	10	5	8.91
Visitors' experience	5	5	10	15	10	8.91

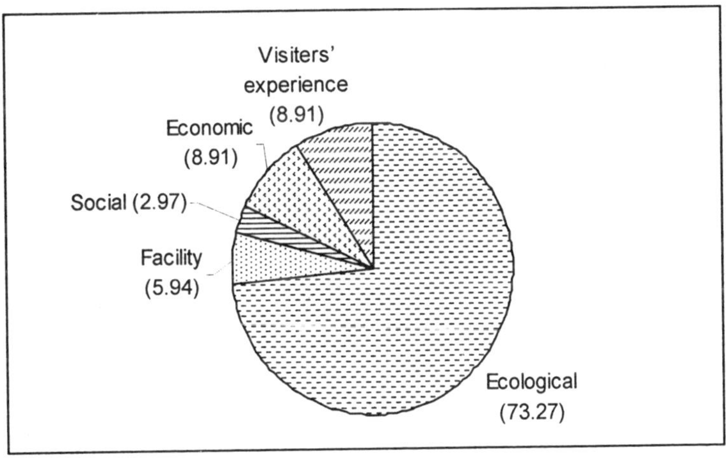

Fig 16.3: Component Importance Value for Pench National Park

The table indicates that the experts apprehend a dominant role of ecological aspect in determining the carrying capacity of Pench National Park *vis-à-vis* other components.

Table 16.3: Percentage of CC left for Tourism Activity in Pench National Park

Component	*CCIU*	*CIV*	*% of carrying capacity left for tourism*	*% of CC reduced from the original level*
Ecological CC	0.845	73.27	61.913	15.5
Facility CC	0.80	5.94	4.752	20.0
Social CC	0.635	2.97	1.886	36.5
Economic CC	0.661	8.91	5.89	33.9
Visitors' Experience based CC	0.789	8.91	7.03	21.1
Total	—	—	81.47	18.53

Thus the total carrying capacity left in Pench National Park for tourism activity is 81.47 per cent.

According to the standard table ratings, the impact is very low (81-100 %), but is perilously close to enter the low impact level (61-80 %).

Discussion

The fact that the number of tourists has increased enormously only in the last few years may be the reason for ineffectual impacts on ecological or infrastructural aspects of the area. As the benefits from the tourism activity becomes more pronounced, social CC may be superceded by other components in its CCIU. The impact on social aspect is mainly due to the apprehension towards the creation of social inequalities and lack of any apparent social benefits from the growth of tourism. But the relative importance of components indicates that the ecological aspect of the area is most prominent in deciding the total CC of the destination area. It may be due to the level of growth of tourism at present in and around Pench National Park, where the infrastructures such as tourist resorts are underdeveloped.

The percentage of CC reduced from the original value indicates a fall of around 35 per cent for both social and economic CC, but the increased benefits from tourism can act as the compensatory mechanism and these values may increase in the future. But a decline of 20 per cent in the visitor's experience CC is a more serious concern. The factors such as quality of guides, their ability to generate interest in tourists about nature, and information imparted through the interpretation facilities

(Interpretation Centres, Natural museums, Amphitheatres etc.) will have to given more attention for the mitigation of this impact.

Conclusion

The total CC of Pench National Park has declined by 18.53 per cent due to the tourism activities. This still belongs to a 'Very Low' impact category as the area has retained more than 80 per cent of its CC, but it is perilously close to entering into the 'Low' impact zone. Though there is no large-scale deterioration of standards in tourism but the aspects such as reduction in visitor's satisfaction or alienation of local people from the tourism can become the Achilles' heel if not managed properly. Management initiatives such as superior training of guides, creation of interpretation centres and involving the local people in tourism activities by programmes like Community-based ecotourism can ensure the growth of sustainable and equitable tourism.

The new method adopted was able to predict the total CC of the destination areas of tourism, including the distinct percentage of reduction in each indicator and component. This can be a useful tool in the hands of the policy-makers and managers, as it can pinpoint to the exact element that will have to be managed intensively. It is also useful for monitoring and evaluation where the CC estimated can be the baseline data for periodic examination. It may also be utilised as the basis for conducting Environmental Auditing on tourism sector.

NOTES

1. Urry, J., 1990. *The Tourist Gaze: Leisure and Travel in Contemporary Societies*, Sage Publications, London.
2. Ceballos-Lascurian, H., 1992. *Overview on ecotourism around the world: IUCN's Ecotourism Programme*, IUCN Ecotourism Consultancy Programme, Mexico.
3. May, V., 1991. "Tourism, environment and development: Values, sustainability and stewardship", *Tourism Management*, 12 (2): 112-118.
4. Hunter, C. and Green, H., 1995. *Tourism and Environment: a sustainable relationship*, Routledge, London, pp. 10-121.
5. Inskeep, E., 1991. *Tourism Planning: an integrated and sustainable development approach*, Van Nostrand Reinhold, New York.
6. McIntyre, G., 1993. *Sustainable Tourism Development: Guide for local planners*, World Tourism Organisation, Madrid.

7. O'Reily, A.M., 1986. "Tourism carrying capacity: concepts and issues", *Tourism Management*, 7(4): 254-258.
8. WTO, 1993. *Sustainable Tourism Development: Guide for local planners*, World Tourism Organisation, Madrid.
9. Wall, G., 1983. "Cycles and Capacity: A Contradiction in Terms!" *Annals of Tourism Research*, 268-269.
10. Lucas, R.C., 1964. "Wilderness perception and use: the example of boundary water canoe area", *Natural Resources Journal* 3(3): 394-411.
11. Wagar, J., 1964. "The carrying capacity of wildlands for recreation", Forest Service Monograph, *Society of American Foresters*, 7: 23.
12. Mathieson, A. and Wall, G., 1982. *Tourism: economic, physical and social impact*, Longman, London, p. 38.
13. Fennell, David A., 1999. *Ecotourism: An introduction*, Routledge: London.
14. Bhattacharya, A.K. and Banerjee, Suchitra, 2003. "Relevance of Carrying Capacity and Eco-Development Linkages for Sustainable Ecotourism", *Indian Forester*, 129 (3).

17

Biodiversity Conservation and Ecotourism

Lessons from Sunderban Tiger Reserve

A.K. Bhattacharya, Saurabh Jha and *Ankur Dave*

Background

Tourism is "one of the most important economic, social, cultural and political phenomena of the twentieth century" (Ceballos–Lascurian, 1996)[1]. Today tourism is described as the world's biggest industry on the basis of number of jobs created, number of clients served and its contribution to global Gross Domestic Product (GDP). Statistics produced by the World Travel and Tourism Council (WTTC, 2002)[2] indicate that tourism generates 11 per cent of global GDP, employs 200 million people, and transports nearly 700 million international travellers per year—a figure that is expected to double by 2020. The size of the industry and its rate of growth present both opportunities and threats for biodiversity conservation.

Biodiversity Considerations and Ecotourism

The Convention on Biological Diversity (CBD) defines biodiversity as "the variability among living organisms from all sources including, *inter alia* terrestrial, marine and other aquatic ecosystems and the ecological complexes of which they are part; this includes diversity within species, between species and of ecosystems" (CBD, 1992)[3]. In layman's language it means all the living things on earth and the ecological processes associated with them (Vermeulen and Koziell, 2002)[4].

Biodiversity conservation is essential to human development because of the goods and services it provides. Its components can be used directly as food, medicine, building materials and indirectly in the form of environmental regulation, soil conservation, and pollution control. Though these services are not widely recognized in economic terms but one of the studies estimated the combined economic values of 17 ecosystem services at US$ 16.54 trillion per year (Costanza *et al.* 1997)[5]. According to an estimate, 40 per cent of the global economy is based on biological products and processes (WEHAB Working Group, 2002)[6].

The International Ecotourism Society (TIES) defines ecotourism as: "purposeful travel to natural areas to understand the culture and natural history of the environment, taking care not to alter the integrity of the ecosystem, producing economic opportunities that make the conservation of natural resources beneficial to local people" (TIES, 2002)[7]. The major underlying assumption is that visitors' can provide the necessary economic incentives for conservation and development thus agreeing to the claim that ecotourism "is a mode of ecodevelopment which represents a practical and effective means of attaining social and economic improvement for all countries" (Ceballos–Lascurian, 1991)[8].

Ecotourism in Conservation

The Tourism can contribute to sustainable development by having less impact on environment in comparison to other industries, by playing a positive role in awareness generation for conservation, by providing economic incentives to protect habitat and since it depends on environment as part of its product, it is motivated to protect them (WTTC and IHRA, 1999)[9]. Though it holds good at an industry level—but less at an individual operator level, whose business is affected by seasonality (Buckley, 1995)[10].

In recent years there has been concerns over global environmental degradation and ways to combat them. Several approaches like laws, advertising, economic incentives etc., can be used to check environmental degradation and tourism has been perceived as one of the potential tools in global conservation (Buckley, 1995)[10]. The idea is to persuade people to see such places whom upon returning home act as advocates for the area they have visited. The intentions are good but the concept of ecotourism still hasn't agreed upon measures of a successful ecotourism project, "who are the beneficiaries and how the benefits be distributed", which has ultimately translated into unrealized conservation goals and reduced benefits for many of the sectors which ecotourism seeks to support (Wells and Brandon, 1992)[11]. It is clear that in any project it is not always possible to satisfy all stakeholders simultaneously and hence conservation benefits should be always considered as the bottom line.

The key benefits for conservation from ecotourism can be clustered into five areas (Brandon, 1996)[12]:

- a source of financing for biodiversity conservation.
- providing economic justification for protection.
- providing economic benefits to local community.
- awareness generation to promote conservation.
- involving private sector for biodiversity conservation.

The Conservation of Sunderban Tiger Reserve

This chapter discusses conservation and ecotourism development strategies in Sunderban Tiger Reserve (STR), India, which was inscribed on the UNESCO world heritage list in year 1985 and recognised as a Biosphere Reserve in the year 1989. STR is particularly known for being the home of the famous Royal Bengal Tiger. Moreover, it is the largest delta in the world and home to large variety of species of flora and fauna (STR Management Plan, 2000). Over the past decades, the marine environment and forest resources came under increasing pressure from local community and commercial enterprises which has led to overexploitation and marine habitat destruction. While marine resource extraction has been regulated, adequate alternatives have thus far not been provided to residents living on the periphery of the reserve (STR Management Plan, 2000). Moreover, benefits from tourism have not yet materialised and are unlikely to provide feasible livelihood options in the future, as the industry is set to remain externally controlled and operated. Such restrictions on the use of the resources that local community depended upon for their livelihood has led to resource use conflicts with reserve management which is challenging the conservation objectives of the STR. The primary objective of the STR management is biodiversity conservation and sustainable use of the natural resources. Now sustainable use can be 'consumptive' forms of resource extraction like fishing and honey collection, to ecotourism, which is often seen as 'non-consumptive', as it is observed to be non-extractive (Brandon, 1998)[13].

Ecotourism, as a non consumptive way of resource extraction can be considered a viable option. With resource extraction restricted or regulated, the involvement of local community in ecotourism development can be crucial to meet the livelihood needs of the community and also involving people in the conservation efforts.

Ecotourism Development

Ecotourism has become a buzzword after 1990s in biodiversity conservation and protected area management context. It is distinguished from conventional tourism by its socio-economic implications and potential to self-finance the destinations to contribute to conservation efforts (Brandon, 1996)[12].

In the context of STR, the management plan stresses on conservation of biodiversity and awareness generation with no written guidelines for providing economic benefits to local community and involving private sector for biodiversity conservation. The overall conception of 'ecotourism' being practiced in STR is more of traditional nature tourism what has historically been under control and ownership of outsiders and most of the revenue generated leaks outside the local economy. The type of jobs generated is labour related jobs on seasonal basis in the lodges outside the reserve. There hasn't been increase in the locally managed business as the local people lack the requisite skills, and the private operators are unwilling to invest time and money to train them. It has also not given impetus to traditional arts and culture of the area. Thus as poor and disadvantaged suffer under resource use restrictions, the outsiders attain to gain benefits by paying minimally for it.

One of the biggest benefits of ecotourism is that it has potential to raise money from visitors' and act as a major source of financing for conservation. Though a minimum fee is charged to visitors, but the entire money collected goes to the State exchequer. The fee is kept minimum to keep STR open to all citizens, even the poor. But almost every study (Brandon, 1996) [12] done of protected areas recommends capturing revenue to maintain protected areas and to offset the costs of visitor use. In short, "the money generated by ecotourism does not necessarily go towards maintaining biological diversity or management of Parks themselves" (Wells and Brandon, 1992)[14].

One of the biggest potentials of ecotourism is to generate awareness about the destination areas among the visitors and hence build support for the conservation of the area. In STR, this aspect has not been effectively as a high percentage of visitors are dissatisfied over guides' performance and standard of information dissemination. Though there is an interpretation center in STR at Saznakhali Wildlife Sanctuary but Forest department data shows that only 24 per cent of the tourists coming to Sunderbans availing its facility. Thus one of the primary objectives of ecotourism development in STR (to create awareness among people about Conservation of Biodiversity) stands defeated.

Future tourism development inside the reserve is restricted and tourism infrastructure will continue to be developed outside the reserve. While West Bengal Tourism is running the tourism infrastructure within the reserve, the facilities outside the Park are run and owned by the private sector and the national and local elite. This further suggests that the revenue generated will leak from the local economy. Also the high rate of growth of visitors in the area leading to more facilities being created which are unchecked by any kind of legal restrictions might encourage the lodge owners to adopt environmentally sensitive designs and operations. The problem arises from the fact that the construction is carried on the Panchayat land and there is no legal framework to regulate the construction of these facilities within the acceptable limits. Moreover, there has been complains from STR management, local community and villagers about the laxity shown by launch owners during their tours. The major complaint has been regarding launches and boats being used in the region operating above permissible sound limits, whose effects though not monitored seem to affect the ecology of the area. Also efforts of tour operators in waste management is not adequate enough and as the reserve lacks full strength staff it is not easy to monitor every launch and boat moving in the area.

Conclusions and Recommendations

There is great interest in ecotourism as a source for conserving biodiversity, achieving socio-economic development and maintaining cultural harmony of an area. One of its major advantages is that it has the potential to be ecologically and culturally sensitive but to achieve these objectives, a proper planning needs to be laid, followed by ecotourism initiatives and involvement of all stakeholders and management controls to monitor the activities. Some of the key factors that are to be kept in mind for successful implementation of ecotourism projects are:

Mechanism for Controlling the Arrival of Visitors within Acceptable Limits

Key to conservation is that ecological and cultural impacts are kept to minimum, which helps, in sustainable use of the resource from which the benefits are being derived. This can only be achieved by using the concept of carrying capacity whereby an appropriate level of visitation in an area is determined to keep in check the impact of tourism on ecology. This can be achieved by careful analysis of the local conditions, by defining a clear management objective, using authority and incentives to impose limitations on visitations and using tactics like zoning by which the impact is limited to an area.

Leakage from the Local Economy

At most ecotourism sites, the businesses are generally controlled by outsiders or local elites which has not led to sufficient benefits for the local community to support conservation. The local community generally ends up with small jobs, which is not enough to motivate local community to conserve biodiversity. The resource extraction is already being prohibited which has affected the livelihoods of the people and when the community doesn't get any benefit from tourism it ends up in resource use conflict with the management.

Visitation and Conservation Education

Generally tourists are fickle and want to see wildlife, especially tiger has high tourist appeal, but if their sighting becomes difficult then visitors' satisfaction is low and eagerness to come back is reduced. But good guiding and interpretation involves the ability to make other resources attractive and educate visitors' about its ecological and cultural importance. Such interpretation Programmes should try to make visitors' sensitive towards ecology and culture of an area and its aim should be make visitors' 'brand ambassadors' of the area thereby strengthening the cause for its conservation.

To improve upon the visitation and conservation education, substantial investments need to be made to strengthen the management capacity of the authorities to design and implement sustainable ecotourism.

Thus to overcome this problem following points should be considered

- A high 'staff to visitor ratio' be maintained in the ecotourism zone which can be achieved by hiring trained staffs on contract basis.

- The activities taking place in the interpretation area should be properly publicized and the activities be conducted to involve visitors during evening hours.
- The guides in the area should be properly trained to make them proficient in languages and interpretation skills.

Facilities and Services

The facilities and services depends on "zoning, combined with an analysis of the type of tourist the Park wants to attract, the proximity of alternate facilities, acceptable levels of impact, and the revenue the Park wants to generate" (Brandon, 1996)[12]. A careful examination needs to be done on kind of tourists the destination area wants to attract depending on the ecological and cultural sensitivity of the area and the kind of impact which the area can sustain.

The problem can be tackled by:

- Application of "Ecologically Sensitive Area" ruling to the peripheral areas of the destination area. This ruling restricts the activities of infrastructure development within the area (as set by the authorities) and hence in future further facilities development can be regulated using this ruling.
- The management of tourism facilities will be on the basis of suggestions from experts in the fields of wildlife management, tourism and land laws. The responsibility of control and monitoring of construction will rest on the authority that controls the tourism activity.
- Environmental Impact Assessment (EIA) report should be made mandatory if the number of rooms and the floor area of construction exceed a limit. These limits can be set after discussion with the experts in the field of EIA.
- A monitoring mechanism should be developed for the area, which monitors the impact of infrastructure development and visitation to the area on the ecology and culture of the area.

Community Based Ecotourism (CBE) Activities Should be Promoted

The villages around the protected areas should be encouraged to take up tourism related activities at community level and thus earn an alternative means of livelihood. Local people should be encouraged to form groups such as cultural group or co-operative committees, so that they can set up shops to sell locally made products and handicrafts to tourists.

Certain recommendations have been made before embarking on CBE:

- A detailed analysis of what the community can supply to the tourism enterprise is needed before investments are made.
- Funding entities must understand the organizational structure of the community, review various models for community tourism ownership before making any investments, ensure that an appropriate model for ownership and accounting of funds is implemented by the community itself, and provide on-site training in accounting skills.
- A Visitors' database should be maintained to know the demands and preferences of visitors for community based ecotourism.

It can play a vital role in not just the environmental conservation, but may hold the answer for livelihood issues in resource rich, yet economically poor rural set up near by the ecotourism destinations.

Shift from Marketing of Protected Areas only for tiger Sighting

The emphasis must shift from tiger sighting and extend to wildlife sighting, wilderness experience, scenic beauty and cultural heritage. This is important to reduce visitors' expectations and inform visitors about the rich biodiversity in the reserve.

The marketing strategies should be taken into consideration a holistic view of forests and its nearby areas and must not brand the tourism as tiger tourism. The next step of changing this model towards an ecotourism activity is diversifying in to adventure activities, camping, developing alternate tourist areas outside the reserve, etc.

All the above mentioned aspects are inherent to the concept of Ecotourism and should be taken care of, so that in the future the tourist activities do not adversely affect the environment of the region and also positively contribute to the local community.

NOTES

1. Ceballos-Lascurain, H., 1996. *Tourism, Ecotourism and Protected Areas*, Gland, Switzerland: IUCN.

2. WTTC, 2002. *Corporate Social Leadership in Travel & Tourism*. London UK, World Travel and Tourism Council.
3. CBD/Convention on Biological Diversity, 1992. http://www.biodiv.org/doc/legal/ cbden. pdf
4. Vermeulen, S. and Koziell, I., 2002. *Integrating Global and Local Biodiversity Values: A Review of Biodiversity Assessment*, London: International Institute for Environment and Development
5. Costanza, R., d'Arge, R., de Groot, R., Farber, S., Grasso, M., Hannon, B., Limburg, K., Naeem, S., O'Neill, R.V., Paruelo, J., Raskin, R.G., Sutton, P., and van den Belt, M., 1997, *"The Value of the World's Ecosystem Services and Natural Capital."* Nature 387:253–260.
6. WEHAB Working Group, 2002. *"A Framework for Action on Biodiversity and Ecosystem Management."* NewYork: United Nations. www.johannesburgsummit.org/html/documents/summit_docs/wehab_papers/wehab_biodiversity.pdf.
7. The International Ecotourism Society, 2002. *Ecotourism statistical fact sheet*. (General Tourism Statistics).
8. Ceballos-Lascurain, H., 1991. Tourism, Ecotourism, and Protected Areas. *Parks* 2(3): 31-35.
9. World Travel and Tourism Council (WTTC) and International Hotel and Restaurant Association (IHRA), 1999. *The Global Importance of Tourism*, Background Paper no. 1, prepared for the Commission on Sustainable Development, seventh session, 19–30 April 1999.
10. Buckley R., 1995. *Where Tourism and Ecology meet*, Ecotourism conference, Bangkok, 6–8 February 1995.
11. Wells, M., and Brandon, K., (with Lee Hannah), 1992, *People and Parks: Linking Protected Area Management with Local Communities*, Washington, DC: The World Bank.
12. Brandon, K., 1996, *Ecotourism and Conservation: A Review of Key Issues*, Environment Department Paper no. 33. Washington, DC, World Bank.
13. Brandon, K., 1998. Perils to Parks: The social context of threats. In K. Brandon, K.H., Redford and S.E. Sanderson (eds), *Parks in Peril: People, Politics, and Protected Areas*, Washington, D.C.: Island Press, pp. 415-439.
14. Wells, M., and Brandon, K. (with Lee Hannah), 1992, *People and Parks: Linking Protected Area Management with Local Communities*, Washington, DC: The World Bank.

18

Biodiversity Conservation in Madhya Pradesh

Policies and Strategies

Suchitra Banerjee and *A.K. Bhattacharya*

Background

Biodiversity is providing the basis for life on earth which includes the variability of animals, plants and microbes. It includes *inter alia*, lower plants (including bacteria, viruses and mycoplasma like organisms), higher plants (herbs, shrubs and trees), animal breeds including fish, birds and invertebrates. Based on natural resource categories, it is further classified into Forests, Croplands, Rangelands and Aquatic environment. All these species are used directly or indirectly for food and agriculture, feed for domestic animals and also for the provision of essential raw materials and services for life support such as fibre, fuel, fertilizer and pharmaceuticals. Hence, biodiversity makes ecosystem stable, functional and environmentally sustainable.

The very biodiversity that nurtured human cultural diversity since times immemorial now stands threatened. In fact, the present era is often described as an era of species extinction. The rate of species loss is as fast today as it was at the time of extinction of dinosaurs some 65 million years ago. The species loss is a threat to mankind next only to thermo-nuclear war. Of the estimated 10 million species, only 1.4 million species have been catalogued. At the current rate of tropical deforestation (17 million hectares per year), 4-8 per cent of the rain forest would be facing extinction by 2015 AD and 17-35 per cent by 2040 AD. Nearly 15 per cent of Earth's species will be threatened over the next 25 years.

Biodiversity in Madhya Pradesh

The state of Madhya Pradesh has an extremely large forest area—1,55,414 sq. kms., out of the total geographical area of 4,43,446 sq. kms., i.e. 34.8 per cent. The habitat diversity of the state is immense with the elevation varying from 61 msl to 1438 msl. The temperature in summer exceeds 45°C except in hilly region. The temperature during winters varies from 5° C to 25° C except in the plains of Chhattisgarh, where the winter is not so pronounced. The rainfall ranges from 1500 mm in the eastern and southern regions to less than 800 mm in the western zone. On account of its size and varied habitats, its environment supports a unique and wide variety of plant life. Almost every plant family is represented in India's rich flora. Out of these many of the species have medicinal value. Biodiversity of Madhya Pradesh is extremely rich with a wide variety of species. The conservation and sustainable use of this Biodiversity is fundamental to ecologically sustainable development.

As per official statistics the forests of Madhya Pradesh are broadly classified into Reserved Forests (80,996 sq kms), Protected Forests (69,083 sq kms.) and Unclassified Forests (5,335 sq kms).

By composition, the forests of Madhya Pradesh are classified into following categories:

Teak Forests *(Tectona grandis)*,	27,783 sq. kms	17.88 %
Sal Forests (*Shorea robusta*)	25,704 sq. kms	16.54 %
Miscellaneous Forests	1,01,927 sq.kms	65.58 %
Total	1,55,414 sq.kms	100.00 %

Source: Madhya Pradesh Forest Department.

The Botanical Survey of India has identified more than 47,686 species of plants occurring in India. The details of the group-wise plants species are as under:

- Flowering Plants 17,000
- Algae 2,500
- Fungi 23,000
- Lichens 1,600
- Bryophytes 2,564
- Pteridophytes 1,022
- **Total 47,686**

Source: Indian Drug Manufacturers Association Bulletin-Volume XXVII (16).

Madhya Pradesh boasts of a wide variety of wild animals inhabiting the vast expanse of its forest areas. It has 11 national parks and 35 sanctuaries covering nearly 11 per cent of the forest area of the state. Madhya Pradesh has nearly 25 per cent of the tiger population of the country. Herbivors like the *chital, sambar,* black buck, *chinkara, nilgai,* wild boars etc., roam about freely in the forests of the state. But this wildlife is gradually getting threatened in the wild because of poaching and dwindling of habitat.

The animal genetic resources make a large contribution to food and agriculture production but this resource is now being threatened. Important diversity is being lost, thereby reducing the options for achieving sustainable agriculture and food security. Animals are intricate part of agro-eco-systems, providing food, fibre, manure and fuel. They provide 70 per cent of draught power for the rural communities and can provide a critical safety net for farmers and communities when crops fail. For many farmers and herders, animals also represent an asset that provides security.

The state has a total of 467 lakh animals out of which cattle and buffaloes constitute 79 per cent, goat 17 per cent and remaining 4 per cent animals include pig, sheep and other animals. Important recognized breeds of cattle considered as native breeds of Madhya Pradesh are 'Malwi' and 'Nimari'. 'Gaolao' breed is found all over 'Vidarbh' region of Maharashtra and adjoining areas of Madhya Pradesh (Chhindwara district) and 'Kankattha' breed of cattle in Panna district. Similarly, 'Bhadawari' breed of Buffalo is found in abundance at Bhind and Gwalior districts. 'Jaloni' breed of sheep is found in Tikamgarh and Shivpuri which forms the border with Jhansi and Jalaun districts of Uttar Pradesh. 'Jamnapari' breed of goat is found in various villages of Bhind district situated near chambal river. 'Malwi camel' are also found in Mandsaur district of Madhya Pradesh. In the poultry sector 'Kadaknath' is the native breed of Madhya Pradesh found at Jhabua and Dhar districts whereas 'Aseel' is found in abundance at Bastar district of Madhya Pradesh.

The State of Madhya Pradesh is a predominantly a tribal state. The tribal population as per the 1991 census was 1,19,87,030, i.e. 22.97 per cent out of a total population of 6,61,80,470. The major tribal groups inhabiting the state are the Bhils, Bhilalas, Baigas, Gonds, Korkus, Saharias and Oraons etc. Out of the 71,000 villages in the state nearly 30,000 villages are within or on the fringes of forest areas. It is evident that most of these tribals groups inhabit the remote forested tracts of the region. Biodiversity constitutes a resource on which the local communities depend for their livelihood security but due to the ever-growing Indian population, this resource is threatened by overuse and consequent depletion. It is, therefore, obligatory on part of the state to initiate action for the conservation, restoration and judicious use of biodiversity.

There is also the need to recapture local knowledge about farming systems, environment, biodiversity and its sustainable use. A synthesis of traditional wisdom and ecological prudence of tropical agricultural modern advances in technology will help higher use, efficiency of inputs and sustainable use of biodiversity.

Establishment of the Biodiversity Board in Madhya Pradesh

Considering the importance of Biodiversity and the need for its conservation, the State Government took the advice of M.S. Swaminathan, who had chaired the National Commission on Biodiversity. On the basis of discussions, it was felt that the Government of Madhya Pradesh should set up a State level *"Madhya Pradesh Bio-diversity Board"* in order to promote the conservation and enhancement of biodiversity, its sustainable use and to provide for equitable sharing of benefits. The board is chaired my Chief Minister of MP and is constituted of 34 membership with representation from different ministries, department, planning commissions, non-government organizations and individuals. A Biodiversity Board was established on 13th January 2000 by the state government under which three standing committees were constituted to provide policy oversight to the programmes relating to biodiversity encompassing its conservation, sustainable use and equitable sharing of benefits. The standing committee on sustainable use of biodiversity has been assigned the following tasks.

- Promote the evaluation, characterization and cataloguing of genetic resources in a manner that protects the interests of communities and the State Government.
- Foster the use of biodiversity for improving agricultural productivity and stability.
- Promote a green health movement based on herbal drugs.
- Promote research on micro-flora and micro-fauna.
- Identify micro-organisms useful in bio-remediation and bio-monitoring.
- Promote agro-forestry to meet the fodder and fuel wood needs of rural and urban communities.

Integrated Strategy for Sustainable Use of Biodiversity

The integrated strategy while recognising the need to conserve all biological resources (including genetic resources) also underlines the need to distinguish between "critical priorities" and "overall priorities". While critical priorities centre on the need to target "provenly endemic" biological resources of Madhya Pradesh, the "overall priorities" would cover all biological resources, both endemic and non-endemic, which qualify for conservation on account of their scarcity value. The salient features of this strategy are as follows:

Promote the Evaluation, Characterization and Cataloguing of Genetic Resources in a Manner that Protects the Interests of Communities and the State Government

Evaluation, characterisation and cataloguing of the genetic resources

- Rapid Biodiversity Assessment (RBA) and biological indicators, keystone species can carry out these surveys. Biodiversity indexing, documentation and its sustainable use is a long term continuous process. RBA will help in indexing and formulating action plan for the evaluation, characterisation and cataloguing of genetic resources in a manner that protects the interests of local communities. Geographical Information System (GIS) will serve as a prerequisite for RBA work in the following manner:
- Remote sensing maps and their interpretation for selecting enriched, moderately degraded and severely degraded sites and/or using International Union for Conservation of Nature (IUCN) criteria to assess the conservation status of sites to collate information that will be used for each agro-climatic zone of Madhya Pradesh.
- Identification of benchmark sites of one square kilometre area under each category of sites (viz. enriched, moderately and severely degraded) in each agro-climatic zone will be done. Pilot surveys in each category will be carried out by taxonomist (flora and fauna), in already identified benchmark sites in each agro-climatic zone. Information generated from aforesaid work will help in enlisting biological indicators indicating extent of biodiversity degradation. Assessment of the quantities that can be sustainably harvested from the sites will be done in order to arrive at the potential of the sites for various species.

- During bio-diversity assessment care will be taken to document the interest of the local communities. No shift from cultivation of landraces will be proposed (rather incentives will be proposed for funding such efforts). No addition of exotic strains or genotype will be encouraged. Efforts will be made to conserve and document ethno-knowledge of local communities and habitats from which they harvest non-timber forest produce. Efforts will also be made to identify areas in which the threatened and endangered species existed in the past for future restoration of species and their rehabilitation.
- Several Central and State Institutes/Organisation/NGOs have generated valuable information based on in-depth investigation on biodiversity and/or extent of degradation of natural resources. These include Survey of India, Zoological/Botanical/Forest/ Geological Survey of India, Mining Corporation, Educational Institutes, State Forest Research Institute, Tropical Forest Research Institute, Jabalpur, and Central Research Institute of Ayurveda, Gwalior. But this information is extremely scattered and need to be consolidated under a single organisation in a systematic manner. This forum will be provided by the formation of the State level Grid. Information on biodiversity indexing documented by aforesaid organisations will save energy and time for identification of biological indicators and keystone species. It will also provide a base for enhancing the sustainable use of biodiversity.

Location specific participatory approach involving communities, NGO's and other concerned institutes and organisations is proposed for documentation of biodiversity and factors affecting sustainable use. The parataxonomist in the area will help to a great extent in the identification and documentation of biodiversity of the area. The above information gathered from different agro-climatic zones will be compiled in a state level data-base network. These data bases will be available at bio-informatic centres established in each agro-climatic zone.

Animal Biodiversity

Keeping in view the importance of conservation of genetic resources in the animal sector (including avian, insects and aquatic fauna), the following action points are proposed:

- Establishment of state Informatics centre on indigenous animal genetic resources and creation of data bank with capacity to appropriately interlink at national and international level.
- Network programme on characterization, evaluation and conservation of animal genetic resources through field survey.
- Characterization of indigenous animal resources using genetic markers and their discrimination with allied and exotic types.
- To foster sustainable use system in the field/farmers herd through *in situ* conservation of animal genetic resources.
- *Ex situ* conservation of farm animal genetic resources through establishing animal germplasm repository/gene banks at different locations.

Technical Programme

For evaluation, characterization and cataloguing of animal/genetic resources, Madhya Pradesh will be divided into three broad zones, each zone will have a head quarter at three Veterinary Colleges, viz. Mhow, Jabalpur and Anjora (Durg). Under each zone following recognized breeds will be covered.

Particulars	*Zone-I*	*Zone-II*	*Zone-III*
Colleges	College of Vety. Science and A. H. Mhow, J.N.K.V.V., Jabalpur	College of Vety. Science and A. H. Jabalpur, J.N.K.V.V., Jabalpur	College of Vety. Science and A.H. Anjora, Durg, IGKVV, Raipur
Zones (Districts)	Malwa Plateau (Ujjain, Shajapur, Rajgarh) Narmada Valley (Khandwa, Khargone, Barwani)	Satpura Plateau (Chhindwara, Rewa) Gird zone (Gwalior, Bhind, Shivpuri)	Chhatisgarh Plains (Raipur, Bilaspur) North Region and Chhatisgarh (Shahdol, Bastar)
Breed of Animal and Poultry	"Malwi" and "Nimari" breed of cattle "Malwi Camel" "Kadaknath" breed of poultry	"Gaolav" breed of cattle "Kenkatha" breed of cattle "Badawari" breed of buffalo "Jamnapari" breed of goat	"Aseel" breed of poultry

Efforts will be made for cataloguing other local breeds of animal and poultry found in other zones of Madhya Pradesh.

(A) Field survey in establishing the status of breed and creation of breed descriptors.

- Precise information will be generated through survey Programme on population, demographic distribution trend, production system used, availability of input, socio-economic conditions of farming communities and performance of the animals.
- Appropriate tools including questionnaires and breed descriptors proformae will be developed as per guidelines by the National Bureau of Animal Genetic Resources (NBAGR), Karnal, to understand the status and development of breed descriptors in their natural habitat. This will provide a comprehensive picture of all animal breeds, their present use and future prospects.
- Survey of breeds/animal types will include mandatory recording of the following information:
    - demographic and geographical distribution.
    - the native environment.
    - enumeration of population of breeds in terms of age and gender.
    - management practices and utilization.
    - qualitative and quantitative characterization of breeds in relation to morphological traits, production potential and reproductive status studies.
    - qualitative and quantitative description of unique animals and elite producers.

(B) Collection of blood samples for genetic evaluation preferably from the unrelated animals of that breed. Attempts will be made to generate breed specific markers like

- Cytogenetic markers
- Biochemical polymorphic markers and
- Molecular and DNA marker.

— Location specific participatory approach involving state veterinary department, NGO's other concerned organisations is proposed for survey and documentation of biodiversity starting from village, block, district, zone and state level.

— Cytogenetic and molecular evaluation through blood samples is proposed at zonal head-quarters, i.e. three

veterinary colleges of the state, for which specific required facilities will have to be created at all the three zonal centres.
- Collaboration with National Bureau of Animal Genetic Resources (NBAGR), Karnal, Wildlife Institute of India, Dehradun, Bombay Natural History Society, Mumbai, Indian Institute of Forest Management, Bhopal and NBPGR, New Delhi is proposed.

Foster the Use of Biodiversity for Improving Agricultural Productivity and Stability

In order to foster the use of biodiversity for improving agricultural productivity and stability, it is essential to conserve and sustainably use the biodiversity and arrest the constant depletion of germplasm. In order to achieve this objective the following activities are proposed to be initiated—

- *Establishment of Core Collection Centres for Forest Areas in each Agro Climatic Zone*: These core collection centres will be identified and protected in each agroclimatic zones. These will be created in each forest type, rangeland and wetland. This will help to conserve the representative germplasm of the area having potential of practical utility in the future. It will also serve as an *in situ* gene bank and a wild species repository.
- *Participatory Approaches for Biodiversity Utilization*: The most important way of utilising bio-diversity is to identify and evaluate options in each agro-climatic zone in consultation with local communities for conservation of bio-resources as well as participatory piloting of programmes to convert dormant genetic wealth (under-exploited/unexploited) into dynamic products for sustainable livelihood. This will help improve agricultural productivity in harmony with the local environment without having any adverse effect on the stability of the delicate production pyramid of the area.
- *Linkages between ex situ and in situ Conservation Sites by Community Gene Banks (CGB) and Community Seed Banks (CSB)*: It will be necessary to establish a direct link between the *ex situ* and *in situ* conservation sites by CGB and CSB. This will help communities to have a stake in the conservation and development of various species and they can be benefited by propagating their local varieties/breeds.

- *Community Field Gene Bank (CFGB) at field level (in situ on farm) for land races*: These will provide seed multiplication and seed regeneration in Community Gene Management systems (CGMS).

The linkages will be as follows:

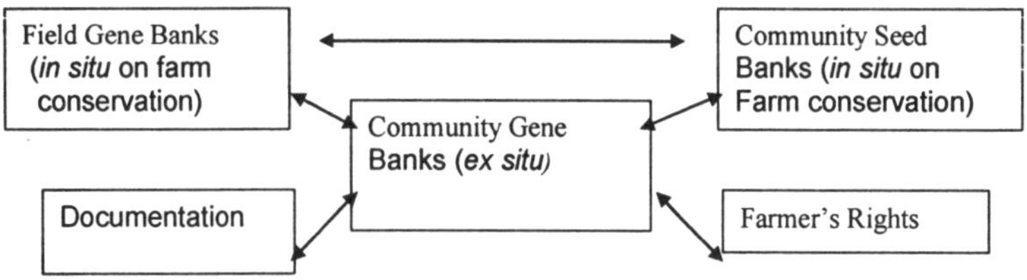

- *Establishment of State level Active Germplasm Site and/or Repositories* : These will be identified for major species in different agroclimatic zones. These will be responsible for collection, conservation, characterization, documentation and enhancement of germplasm by pre-breeding. These sites will have effective liaison with the national repositories and National Gene Bank at the NBPGR, New Delhi, NPFGR, Lucknow and NBAGR, Karnal.
- *Action Plan for Conservation and Sustainable Use of Animal Bio-Diversity*:

— Each zone of Madhya Pradesh will identify and enlist all animal and poultry breeds and set priorities for breed characterization and conservation.

— The conservation will be done by *in situ* and *ex situ* technique.

(A) *In situ* conservation for sustainable management of biodiversity will be done by:

(i) maintenance of grassland and efficient management of feed and fodder resources like silvi-pasture, multi-tier vegetation, controlled grazing of forest for balancing grassland ecosystem and animal population etc.

(ii) creation of public awareness for conservation and provide incentives to farmers/institutions maintaining purebreds.

(iii) *in situ* conservation at farm level by modern techniques like maintenance of effective population size and open nucleus breeding system, nuclear herd and improving the management system.

(B) *Ex situ* conservation by in-vitro methods to regenerate endangered breed, to recognize and conserve local breeds evolved through natural selection, DNA studies and gene mapping. This will be done through:

(i) Cryo-preservation of gametes, embryos and DNA.

(ii) Some new approaches like establishment of embryonic stem cell lines and somatic cell conservation.

A Detail plan of work along with budgetary requirements including both the mandates may be worked out for various species of animals.

Promote a Green Health Movement Based on Herbal Drugs

Around 70 per cent of India's medicinal plants are found in tropical forests. Less than 30 per cent of the medicinal plants are found in temperate forests and higher altitudes. Micro studies show that a large percentage of medicinal plants occur in the dry and moist deciduous. The forests of Madhya Pradesh fall under the category of tropical dry and moist deciduous type and are extremely rich in medicinal flora. The group-wise plants species occurring in India with known medicinal value are as under:

S. No	*Group*	*Total No. of Species*	*No. of Species with known Medicinal value*
1.	Flowering Plants	17,000	8,000
2.	Algae	2,500	650
3.	Fungi	23,000	750
4.	Lichens	1,600	650
5.	Bryophytes	2,564	150
6.	Pteridophytes	1,022	200
	Total	*47,686*	*10,400*

Source: IDMA Bulletin-Volume XXVII (16).

The scenario of Medicinal Plants in India was enumerated as under:

- No. of Plants used in Ayurveda — 1000
- No. of Plants (Wild Species) used by tribals as their daily requirement — 9500

Total — **10,500**

Source: IDMA Bulletin-Volume XXVII (16), Pgs 469-470,1996

The plants used by tribals have been classified as under:

- No. of plants having direct medicinal use 7500
- No. of plants having edible utility 3900
- No. of plants giving new leads/claims 950
- No. of plants providing fiber and cottage industries raw material 525
- No. of plants used by tribals as pesticides 300
- No. of plants used by as Bio-pesticides 175

Source: ORG and Industry estimates FRLHT Research Dept. —August 1997.

About 2000 drugs that have been used in curing human elements in India, about 1500 are of plant origin, out of which 1000 medicinal plants are used in Indian System of Medicine (ISM), though about 60 species are in larger demand. Madhya Pradesh is enriched with about 625 medicinal flora representing 123 families. About 181 medicinal herbs (used in ISM) are found abundantly in Madhya Pradesh.

The proposed programme of work is as follows:

- Establishment of medicinal plants conservation reserves *in situ* and *ex situ* with ethno-medicinal garden and nursery network in each agro-climatic zone.
- Regulation on wild collection of non-timber forest produce based on ecological status survey and slow but regular switch over to herbal drugs and non-timber forest produce based agro-forestry systems in revenue and waste land areas around villages.
- A committee of the state level grid will be formed with SFRI, TFRI and centres of JNKVV, IGKVV working on medicinal and aromatic plants with the pharmaceutical and ethno-medicinal knowledge rich tribal societies to cater herbal wealth for health care at grassroots level. Production technologies and post harvest processing technologies generated at ACRIP-M and AP, GOI—Adhoc Projects at JNKVV as well as at SFRI, TFRI and NGO's will be made available to the communities organised under the village level committees like the JFM committees and herb collecting and dispensing centres will be established for post harvest value addition to the raw material for own use and trade. These centres can be manned by local healers who will be identified and registered by the DISM and not only help the

local communities sell value added herbal products but also dispense local medicines to the local people for minor ailments.

- CBD has provided a new opportunity for upholding the sovereign rights of nations on their bio-resources. It stipulates conditions for access to the resources and suggestions for fair and equitable sharing of benefits resulting from the use of such resources and the associated knowledge systems. By contrast the WTO upholds the principle of the protection of IPR through its agreement on Trade Related Intellectual Property Right (TRIPS). By virtue of its maxim of rewarding new and innovative inventions, the TRIPS tends to ignore the role played by farmers and tribal communities in conserving biodiversity and in breeding today's landraces. Farmers' rights over genetic resources should be recognised and enshrined. There are pharmaceuticals and traders utilizing raw material/knowledge on drug herbal wealth of the state. Economic instruments could be explored so that funds can be generated from them for conservation of herbal wealth. Madhya Pradesh has an innovative *"benefit sharing model"* that is being implemented in the Tendu Patta Trade. The same model can be expanded to meet the requirements of article 8 (1) of CBD.
- A medicinal plants conservation and indigenous knowledge fund is proposed to be created to receive contributions from Pharmaceuticals and traders. This fund should be established under the Biodiversity Board. This fund can be used onward to revitalise local health traditions in rural areas and also to promote conservation projects.
- *Conservation of health care herbal species*: Conservation of existing health care herbal species in their natural habitat will be given utmost importance keeping in view the IUCN standards and other rapid assessment systems. Advantage will be taken of work already done at the TFRI, Jabalpur on indexing of medicinal species in the forest area and on threatened species.
- *Health care at grassroots level* : Attempts were made by SFRI and TFRI to popularize 16 medicinal plants under the scheme *'Dadima ka Batua'* which can take care of common health problems at grassroots level. In addition, abandoned forest nurseries will be revived in each agro-climatic zone as health care herbs nurseries and repositories. They will provide planting material to tribal and village communities for their court yards, kitchen garden, community garden.

- *Post-Harvest value addition* : Project on post-harvest value addition to medicinal raw materials, non timber-minor forest produce is under progress at TFRI. Work done on these aspects by the Madhya Pradesh Minor Forest Produce Federation (MPMFPF) should also be taken into account before formulating a viable strategy for the state. The recommendation package will be popularised amongst the growers etc. by different extension methods through extension agencies. The MPMFPF can act as a nodal agency for popularizing these methods through its network of primary forest cooperatives and the JFM committees.
- Promote cultivation through organic resources as per accepted standards of APEDA.

Promote Research on Micro Flora and Micro Fauna

Soil is a habitat for a vast complex and interactive community of micro-organisms whose activities largely determine chemical and physical properties of soil. It is important to determine optimum diversity of soil microbial populations of both natural and agricultural systems for their sustainable management. Microbial biodiversity indices can function as biological indicator to show the community stability and describing the ecological diversity of a community.

Bio science department of R.D.V.V., Jabalpur has already indexed flora of Jabalpur, Mandla and Pachmarhi. The Department has also documented ethno-knowledge which rests with tribes of Madhya Pradesh on herbal drugs.

The Central Regional Station of Zoological Survey of India, Jabalpur and other organisations and institutes in the state have worked on bio-diversity of fauna series of Madhya Pradesh. Attempts will be made to form a joint team of Scientists/Taxonomists (micro and macro flora) and Scientists from Zoological Survey of India for documentation of floral species with which macro-faunal have association and inter-dependency in different agro-climatic zones to asses the interrelationship of species and principles of their co-existence.

Identify Micro-Organisms Useful in Bio-Remediation and Bio-Monitoring

There is a need to utilize natural resources in a sustainable manner without affecting environment. Biodiversity indexing of macro/micro flora can be a better way to predict environment degradation with special reference to flora (micro/macro) in a particular habitat.

Documentation of work done on bio indicators at different institutes followed by indexing of bio indicators in bench mark sites will help to assess the factors regulating sustainability of natural/physical resources.

Bacteria are among nature's chief recyclers. Because of its ability to break down a variety of compounds into their basic elements, bacteria are extensively used in environmental biotechnology. One of the applications where bacteria are gaining greater use is in the oil industry. Two of the major families in which these microbes are found are *Pseudomonas* and *Bacillus*. It has been demonstrated that a special process that utilizes a strain of *Pseudomonas* can cost-effectively remove contaminants from oil industry wastewater. This process has been further tested by the Saskatchewan Research Council (SRC), using the *Pseudomonas* strain and another type of oil-eating bacteria currently found in bio-remediation lagoons. Both types of bacteria were able to remove the hydrocarbons from the wastewater in far less time than it takes to treat the water in a lagoon. Tests have shown that using the bio-film reactor of bio-reactor, is a very efficient way to enable bacteria to do their work.

A naturally occurring bacteria called *Nitrospira* converts the harmful form of nitrogen in water (nitrite) to a less harmful form (nitrate). This could help to clean wastewater from abattoris, sewage plants and aquaculture farms before its release to rivers and streams. This discovery was made at the Co-operative Research Centre for Waste Management and Pollution Control, Australia and revolutionizes scientific thinking about the biological breakdown of nitrogen in water. It is to be noted that until now it was believed that *Nitrobacter* was the bacterium of choice for nitrate conversion.

A new method of de-contaminating waste-water consists of applying a fungus, *Rhizopus arrhizus* as the organism is a superior metal accumulator for metal ions when compared with certain bacteria and yeast (adsorption land biomass). Heavy metal uptake of chitin can be attributed to the chemical changes occurring at the outer layer of the organism during biosorption.

Identification of aforesaid genus and other microbes having activities as bio-remediation, bio-monitoring and bio-termination can help in a long way in reducing the pollutants effects on bio-diversity.

Considering the fact that patent regimes on micro-organisms are in the offing, greater efforts are needed to identify the organisms and their beneficial uses as stated above with a view to establish priority. There is also a need to establish a state level repository.

Promote Agroforestry to Meet the Fodder and Fuelwood Needs of Rural and Urban Communities

Forest of Madhya Pradesh come under the category of "Tropical dry and moist deciduous types" with the productivity of 1.05 cu.m./hectare which is almost double the country's average (0.7) as compared to the world's average of 2.1 cu.m./hectare. There is tremendous potential to increase the forest productivity per hectare by conserving the existing flora *in situ* and shifting the fuel, fodder and NTMFP needs to different agroforestry systems in waste land and revenue area around the villages as well as tribal villages within forests.

Strategies for Enhancing Transfer of Proven Technologies and Feed Back on Different Agroforestry Systems

Multi Purpose Tree Species (MPTS) for different end uses are well defined for different eco-zones. Agroforestry systems depending upon the kind of components involved are also formulated. There is a need to translate, modify and transfer the aforesaid technologies based on location specific demands in different sectors of each agro-climatic zone of Madhya Pradesh. Chronological steps are suggested as follows:

- Pilot scale demonstration on different agroforestry systems and watershed in collaboration with different state departments and NGOs.
- Post harvest management of agroforestry products along with post-harvest value addition for improved economic advantage.
- Identification, selection, screening and multiplication of germplasm of over exploited/under exploited and non-exploited aromatic, medicinal, dye and industrial herbs, shrubs and tree species existing in each agro-climatic zones.
- Establishment of seed orchards of promising species and standardization of techniques for their multiplication.
- Performance evaluation in terms of productivity, quality and credit worthiness of these species in unified agroforestry systems.
- Identification and selection of suitable plant species for afforestation in mining areas have already been done. There is

need to evaluate and modify the list in view of the existing floristic composition of locations of mines, so as to re-establish flora already existed prior to the mining areas of Balaghat, Panna, Baster, Katni etc.

Some vital issues that need to be addressed to empower the personnel and communities handling the sustainable use of Biodiversity are:

- Human resource development issues be identified in selected areas of importance for capacity building. Training may be given both at the national and International level.
- The state may establish regulatory mechanisms to implement the proposals contained in this strategy paper
- The state may address legal issues involved in the sustainable use of biodiversity.
- The state may establish mechanisms of biodiversity conservation and utilisation through state level education system.
- The state government may undertake a programme for educating panchayat level, block level and district level functionaries on the various provisions of the National Biodiversity Legislation in the interest of promoting enlightened 'decision making'.

Proposed Grid for Sustainable Use of Bio-Diversity

It is proposed that the Agro Climatic Zones of Madhya Pradesh shall be the basis for implementation of the strategy for the sustainable use of biodiversity. The proposed *Biodiversity Grid* will be a four-tier structure with a State Level Technical Committee, an Agro Climatic Zone Level Implementation Committee, the District Level Committees and the Block Level Committees. The various committees will be constituted as under

State Level Technical Committee

The head quarters of the State level technical committee is proposed at Bhopal. It will have the following members. The State Government should nominate an eminent person in the field of biodiversity to chair the Grid. The member secretary would be the Chief Wild Life Warden, Madhya Pradesh. The members are the Principal Chief Conservator of Forests, Madhya Pradesh, Directors, Agriculture/Horticulture/Veterinary/

Fisheries/Tribal Welfare/ISM and the Department of Women and Child Welfare, Managing Director, MPMFP Federation, the Vice-Chancellors of the A.P.S. V.V. Rewa, M.G.G.V.V., Chitrakoot, J.N.K. V.V., Jabalpur, R.D.V.V. Jabalpur, I.G.K. V.V., Raipur and D.A. V.V., Indore, Director, N.R.C., Indore Directors of the Regional Stations of the Z.S.I., B.S.I., S.I., Jabalpur, Directors of the S.F.R.I., T.F.R.I., Jabalpur, I.I.F.M., RRL and National Law College, Bhopal and Principal, Indore School of Social Work, Director General, MAPCOST, EPCO, Bhopal, Director, IGRMS, DISM, Bhopal, Member Secretary of the Biodiversity Board in Madhya Pradesh, Two representatives from Pharmaceuticals/Traders actively engaged in trading medicinal and aromatic plants nominated by the government, Two representatives of the local Tribal communities/JFM committees, one of them to be a woman, Two representatives of eminent NGOs engaged in environmental development issues and working in the conservation of Biodiversity, One Social Worker, one Ethno-botanist and a practitioner of Traditional medicines.

The Forest Department shall be the nodal agency for the grid. It will endeavour to establish an independent wing to look after the work of the Grid under the Chief Wild Life Warden.

A sub-committee of the State level Grid will be formed with SFRI, TFRI and centres of JNKVV, IGKVV working on medicinal and aromatic plants along with the representatives of leading pharmacautical houses/ companies. This sub-committee will also have the local healers chosen from amongst the ethno-medicinal knowledge rich tribal/local societies. This committee will suggest ways and means to harness the herbal wealth for health care at grass root level.

It is also proposed to establish a IPR cell at the State level to ensure equitable sharing of benefits. The IPR cell will bc formed under the aegis of the grid and can be located at the MAPCOST, Bhopal who have the necessary facilities. The IPR cell will organise Workshops and seminars at the local level to educate the masses and the members of the various committees. It will also have a right to get information from all levels of the grid directly and initiate action.

Zonal Committees

The Committee will be constituted in each agro-climatic zone with following set up:

The committee will be chair by conservator of forest, and member secretary would be Divisional Forest Officer, Territorial, members of the committee are Dean Agriculture College/Associate Director Research, One representative each of the Agriculture/Horticulture/Veterinary/ Fisheries/Tribal/Women and Child Welfare and the Department of the Indian System of Medicine nominated by the government, One representative from Pharmaceuticals/Traders actively engaged in trading medicinal and aromatic plants nominated by the government, Two representatives of the local Tribal communities/JFM committees, one of them to be a woman, One representative of NGOs engaged in environmental development issues and working in the conservation and use of biodiversity, One Social Worker, one Ethnobotanist and a practitioner of Traditional medicines.

District Level Committees

1. *Chairman*: Divisional Forest Officer, (other than the member secretary of the zonal committee).
2. *Member Secretary*: Sub-divisional Forest Officer.
3. *Deputy Directors*: Agriculture, Horticulture, Veterinary, Fisheries, Tribal and representative of the department of the Indian system of medicine.
4. One representatives from Pharmaceuticals/Traders actively engaged in trading of medicinal and aromatic plants nominated by the government.
5. Two representatives of the local Tribal communities/JFM committees, one of them to be a woman.
6. One representative of NGOs engaged in environmental development issues and working in the conservation and use of Biodiversity.
7. One Social Workers, one Ethnobotanist, one Parataxonomist and a practitioner of Traditional medicines.

Block Level Committee

1. *Chairman*: Range Forest Officer (territorial).
2. *Member Secretary*: deputy ranger.
3. Agriculture Extension Officer, Horticulture Officer, Veterinary Officer, Fisheries Officer, Tribal Officer, Revenue Officer, representative of the department of the Indian system of medicine and the department of women and child welfare.

4. One representative from Pharmaceuticals/Traders actively engaged in trading of medicinal and aromatic plants nominated by the government.
5. Two representatives of the local Tribal communities/JFM committees, one of them to be a women.
6. One representative of NGOs engaged in environmental development issues and working in the area of conservation and use of Biodiversity.
7. One Social Worker, one Ethnobotanist, one Parataxonomist and a practitioner of Traditional medicines.

The committees will be structured as under and will co-ordinate the functions of the Grid. The member secretary and/or his nominee will be the ex-officio member of the next higher committee of the grid. He/she will also co-ordinate between the two committees. The state level committee will strive to link all agricultural and other universities and research institutions and make efforts to initiate the coordination of research and conservation efforts for the overall conservation, utilisation and improvement of Biodiversity. The state level committee will also act as a link for smooth transfer of technology and its implementation at the grassroots level through the various committees and also receive proposals from the local level.

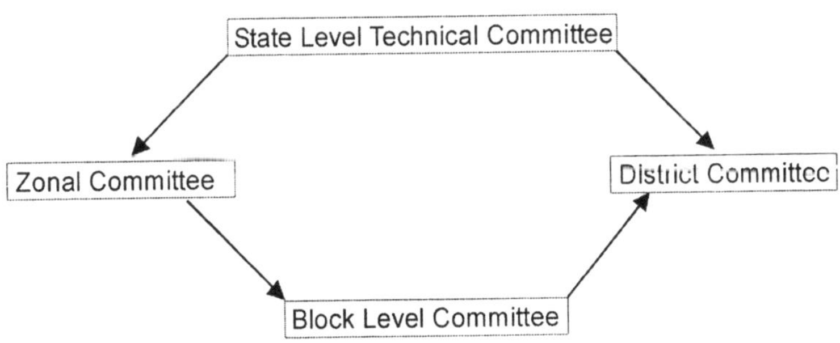

19

Wildlife Conservation Versus Human Populace

The Legal Challenges in Protected Area Management in Orissa

A.K. Bhattacharya and *Samir Stephan Kujur*

Background

India has a long tradition of wildlife and forest conservation. The ancient Hindu scriptures or *Vedas* directed people to protect their environment and wildlife. In about 242 B.C. the emperor Ashoka's fifth pillar edict gave protection to fish, animals and forests. And before that, in the treatise on Statecraft called the *Arthasastra* (attributed to Kautilya in 300 B.C.), there is clear reference to the establishment of *Abhayaranayas,* or forest sanctuaries (Gee, 1962[1]; Rao, 1988[2]; Singh *et al.*, 1990[3]). The early conservation concept was primarily based on recreation and entertainment of the ruling class, however, the later one aims at sustainable development of natural resources for the well being of human society.

The Orissa is located of 17°41′ to 22°34′ (N) latitudes and 81°29′ to 87°29′ (E) longitudes. Once upon a time Orissa was famous for its rich and diverse flora and fauna. Commercialisation, expansion of agriculture land, biotic pressure and above all, multifarious use of forests has led to severe degradation and decline of forest and wildlife population considerably. At the time of independence it had over 40 per cent of its geographical area under forest. At present the actual forest cover is about 30.3 per cent (about 47,205 sq. km.) of the States geographical area (FSI, 1991)[4]. There are two National Parks and 18 Sanctuaries with the total area of 9174.25 sq. km including water body of 1408 sq km. This is about 5.9 per cent and 19.47 per cent of the State's geographical area and the estimated forest cover respectively (OFD, 1997)[5].

According to conservationists the wildlife, like the human beings, has also a right to live. The Protected Areas should be devoid of human habitation and wildlife have to be kept in a natural environment without or minimum human interference (Karanth *et al.*, 1998)[6]. The conservation of wildlife and their habitat is an important goal for Wildlife Department, and a mechanism has to be evolved in which the relative strengths of all sections, especially of local communities, government agencies and conservation groups, can be put together (Kothari and Pathak, 1998)[7]. People in general perceive that the authorities are mainly concerned with the well being of wildlife and their habitat only and they feel that the authorities are not interested in the villagers' point of view and concerns (John *et al.*, 1986)[8]. Therefore, the question of co-existence and continuation of rights within the limits and effective management of the Protected Area is the ongoing debate between the Conservationists, Non-Governmental Organisations, Governments and people in general.

There is no doubt that the State has taken some concrete steps for protection, conservation and enhancement of fauna to ensure a natural habitat for them, however, the attempts are yet to make the satisfactory interpretations. The reasons are discussed in the findings of the study.

Historical Management Regime

During medieval period the *Zamindars* and feudatory kings had their game shooting reserves for hunting wild animals. The tribal communities had the practice of *Akhandparidhi* (indiscriminate mass hunting of wild animals within a time span). To regulate shooting wild animals the rules were framed. As per rule during April to September the shooting of wildlife was prohibited. In 1907, the Lt. Governor of Bengal had formed the rule to prohibit shooting of Bison, hornless male deer, deer with horn in velvet and the females of all deer species in the districts of Angul and Puri. This rule was also applicable in the Protected Areas. During 1930s, Badrama, Raigoda, Chandka and Balukhand Sanctuaries were created to protect wildlife. At critical points anti poaching check gates were set up. The reward system was started to encourage the staff for protection works. The Orissa Reserve Forest Shooting Rule, 1938 was enacted after the formation of Orissa as a separate State and wildlife protection activities were managed under Indian Forest Act, 1927.

After independence, only Chilka (Nalaban) was the rich estuarine/ Marine fauna in the State and the largest brackish wetland in the country. This was first Sanctuary to be declared under Orissa Forest (Shooting) Rule, 1972 because of an ideal habitat for more than 94 species of migratory birds and endangered Irrawady *Delphinus delphis*. A new era started in the history of wildlife protection and conservation under a joint venture of Orissa Forest Act, 1972, Wildlife (Protection) Act, 1972, Orissa Forest (Shooting) Rule, 1972 and Orissa Wildlife (Protection) Rule, 1974. The Wildlife (Protection) Act, 1972 was promulgated on 19th September 1972. As amended up to 1991, it has seven chapters. Chapter I deals with definitions and explanations of the terms used in the Act. The appointments of authorities are covered in Chapter II. The Chapter III spells the prohibition of hunting animals mentioned in schedules I, II, III and IV. It also deals with grant of hunting permits under special circumstances. Chapter III-A talks about the protection of specified plants (i.e. prohibition of picking, uprooting, cultivation, etc.) and permission for special purposes and dealing of specified plants without licence. Chapter IV articulates the process of declaration of Sanctuary, National Parks and Closed Areas, settlement of rights inside the Sanctuary, restrictions on entry in the Sanctuary and other controls. Chapter IV-A paves the way for formulation of Central Zoo Authority and includes the guidelines and standards to be followed for management of Zoos in the country. Chapter V deals with trade or commerce in wild animals, animal articles and trophies and includes guideline for trade in certain wildlife articles. The prohibition of trade or commerce in trophies, animal articles, etc. derived from certain animals is covered in Chapter V-A. The crime control and investigation are defined in Chapter VI. The Chapter VII is miscellaneous, and deals with power of Central and State Government to make rules, declaration of certain wild animals to be vermin, etc.

The Wildlife (Protection) Rule, 1974 came into action after adoption of the said Act by the State Government in 1974.

In 1976, an independent Wildlife wing was framed under the leadership of Chief Wildlife Warden. Under the Chairmanship of the Forest Minister a State Wildlife Advisory Board was constituted. Initially two Wildlife Conservators (based at Bhubaneswar and Chandbali) and respective Territorial Divisional Forest Officers were operating as Wildlife Wardens. Presently, under the Wildlife Wing there are five Wildlife Divisions and one Manager Forest Division respectively. The Simlipal National Park, Sanctuary, Tiger Reserve and Nandan Kanan Zoological Park and Sanctuary have one Conservator of Forest each. The National Park and Tiger Reserve in Simlipal are managed by DFOs (Divisional Forest Officers) and ACFs (Assistant Conservator of Forests) of Baripada and Karanjia Territorial Forest Divisions.

The Gharial Research and Conservation Project (Tikarpada), Saltwater Crocodile Research and Conservation Project (Bhitarkanika), Nandan Kanan Crocodile Captive Breeding Project (Nandan Kanan) and Mugger Crodile Research and Conservation Project (Ramtirtha, Jashipur) are the research stations, which were established to study the crocodile with the support of FAO (Food and Agricultural Organisation) and UNDP (United Nation's Development Programme). Apart from that the sea turtle research and conservation project, Black buck research projects, wildlife in Simlipal and Elephant Management projects are going on with the support of MoFE (Ministry of Forest and Environment), Govternment of India. The Wildlife wing has collaborated with Wildlife Institute of India, M.S. Swaminathan Research Foundation, Nature and Wildlife Conservation Society of Orissa, Indian Statistical Institute, Calcutta, Orissa University of Agriculture and Technology, Utkal University, Regional Research Laboratory and Zoological Survey of India for doing scientific studies.

During 1973 and 1979, six Sanctuaries were declared with an area comprising 3214.27 sq. km, whereas eight Sanctuaries and one National Park (proposed) with an area of 2582.59 sq. km were notified between 1980 and 1985. Two sanctuaries and one proposed National Parks were brought under Protected Area Network between 1987 and 1988 with an area of 1279.03 sq. km. The Gahirmatha Marine Sanctuary was notified on 27th September 1997 having 27sq. km of landmass including Reserve Forest (RF), Mangrove, mud flats and accreted sandbars. This comprises an area of 1435 sq. km including 1408 sq. km waterbody. Thus in the 90s only two Protected Areas have been notified including the Marine Sanctuary. Besides Protected Areas, there is one game reserve for Black buck, one closed area, one Zoological Park and 14 deer parks and zoos. Not only that, excluding the Sanctuary Area the Chilka Lake has been notified as a closed area from 16th December 1997 for a period of five years.

Methodology

The study was taken up at Regional Centre for Development Cooperation (RCDC), Bhubaneswar during May-July 2002 for partial fulfillment of Post Graduate Diploma in Forestry Management Course at Indian Institute of Forest Management, Bhopal.

A sample of six Protected Areas (Lakheri-Valley Wildlife Sanctuary, Kotagarh Wildlife Sanctuary, Gahirmatha Wildlife Marine Sanctuary,

Bhitarkanika Wildlife Sanctuary and National Park and Balukhand-Konark Wildlife Sanctuary) in Orissa were selected for the study. The selection was based on their difficult accessibility in comparison to others, representative samples of almost all type of wildlife and their habitat and no significant research work has been carried out till date in these Protected Areas. Orissa. The accessibility of Gahirmatha Wildlife Marine Sanctuary, Lakheri-Valley Wildlife Sanctuary and Kotagarh Wildlife Sanctuary were quite difficult due to poor infrastructure and inadequate as well as availability of minimum transportation facilities. Gahirmatha Wildlife Marine Sanctuary and Bhitarkanika Wildlife Sanctuary and National Park represented the Marine and Mangrove wildlife habitat respectively, whereas Lakheri-Valley Wildlife Sanctuary and Balukhand-Konark Wildlife Sanctuary were selected for the representation of Elephant and Black buck population respectively. Kotagarh Wildlife Sanctuary represented the mix population of carnivore and herbivore like Tiger, Leopard, Sambhar, etc.

The survey was conducted in two steps. First, an interactive discussion was held with local level conservationists, Non-Governmental Organisations and Government officials and then a survey was administered open as well as close-ended questionnaires to know their perception about existing legal challenges in Protected Areas management in Orissa. Besides interviewing and the administration of questionnaires, the Focus Group Discussions were also conducted with people living in and around the Protected Areas. It is well known that a structured survey, when completed with the alternative approaches such as Focus Group Discussions, can provide many useful insights into the motivations, beliefs and values, which influence behaviour (Campbell, *et al.* 1999)[9]. This provided valuable information in developing better understanding of the respondents' perception about legal challenges in Protected Areas management in Orissa.

Results and Observations

(A) Procedure under Section 19 to 25 of Wildlife (Protection) Act, 1972

Till now, the procedure under Sections 19 to 25 of Wildlife (Protection) Act, 1972 has been completed by the Government of Orissa only for the Bhitarkanika National Park, Baisipali, Badrama, Khalasini, Debrigarh, Kuldia and Hadgarh (Keonjhar) Wildlife Sanctuaries. Almost one decade ago, the Sanctuaries in Orissa had been declared but the procedure under Section 19-25 for the final notification under Section 26-A of Wildlife (Protection) Act, 1972 is yet to complete. According to a writ petition by Centre for Environmental Law and WWF-I *versus* Union of India and others, the

Hon'ble Supreme Court had asked all the State Governments to complete the necessary procedure for the final notification on or before 21-8-1998. Since in most of the Sanctuaries the required procedure has not been completed, the Government of Orissa had sought for extension of one more year from the date of hearing to complete all procedures leading to final notification. Technically, six finally notified sanctuaries don't need any procedure under Section 19-25 of the said Act. Four of them (i.e. Badrama, Baisipali, Kulasini and Debrigarh) are completely inside reserve forests. Gahirmatha Marine Sanctuary is the Government land because it is partly in the territorial water in Nalabana, an island inside Chilka. All such Sanctuaries are deemed Sanctuaries under Section 66 (4) of the said Act. Therefore, in response to the writ petition was the completion of the procedure under Section 19-25 in Lakhari-Valley, Kotagarh, Sunabeda, Kuldiha, Balukhand-Konark and Keonjhar portion of Hadgarh Wildlife Sanctuary. Tables 19.1, 19.2, 19.3 and 19.4 show the present status of the Kotagarh Wildlife Sanctuary, Lakheri-Valley Wildlife Sanctuary, Balukhand-Konark Wildlife Sanctuary and Bhitarkanika Wildlife Sanctuary respectively.

Table 19.1: Brief of Kotagarh Wildlife Sanctuary

Particulars	*Kotagarh Wildlife Sanctuary*
Area (in Sq. km)	399.05
Notification No. and Date	30253 dt. 03-12-1981
Location	Between longitude 83°12′ to 84°00′ (E) and latitude 19°0′ to 19°56′ (N)
Climate	Northern tropical moist deciduous zone
Boundary Demarcation (Core and Buffer zone)	Not yet completed.
Forest type and Nos. of Reserve Forest, Proposed Reserve Forest, Demarcated Protected Forest, Un-demarcated Protected Forest, etc.	*5 Reserve Forests*: (i) Madagoda—36.02 sq. km (ii) Haripur—42.66 sq.km (iii) Lassery—72.96 sq. km (iv) Bonduru—3.67 sq. km (v) Supamaha—7.55 sq. km. *3 Proposed Reserve Forests*: (i) Killangi—6.073 sq. km (ii) Subarnagiri—30.04 sq. km. (iii) Guma (S)–14.04 sq. km

(*Contd.*)

(Contd.)

	2. *Demarcated Protected Forests*: (i) Sri Rampur 'B' block –45.8544 sq. km (ii) Guma (N)—10.644 sq. km
Procedure of Sections 19—25 of WPA, 1972 (as amended up to 1991)	Collector Kandhamal has forwarded the revised proposal for final notification of Kotagarh Wildlife Sanctuary to Chief Conservator of Forests (Wildlife) and Chief Wildlife Warden, Bhubaneswar Letter No. 1092, dated 20-08-1998
Revenue villages and their areas	Belgarh Police Station—15 villages with 16.444 sq. km, Kotagarh Police Station—44 villages with 88.72 sq. km, Baringawadi Police Station—6 villages with 24.83 sq. km
Final declaration No. and date	Not yet completed the procedure.

Source: DFO Office, Balliguda.

Table 19.2: Brief of Lakheri-Valley Wildlife Sanctuary

Particulars	*Lakheri-Valley Wildlife Sanctuary*
Area (in sq. km)	185.87
Notification No. and Date	2333 dt. 8-2-1985
Location	Between longitude 84°15′ to 84°25′ (E) and latitude 19°15′ to 19°25′ (N)
Climate	Sub-tropical non-monsoon type
Boundary Demarcation (Core and Buffer zone)	Core zone: 64.14 sq. km. Buffer zone: 121.72 sq. km.
Forest type and Nos. of Reserve Forest, Proposed Reserve Forest, Demarcated Protected Forest, Un-demarcated Protected Forest, etc.	1. Reserve Forest: Chandragiri Reserve Forest—111.775 sq. km 2. Proposed Reserve Forests: (i) Alara Ramaguda—59.453 sq. km (ii) Dhobabhobani—14.64 sq. km Total—185.88 sq. km
Procedure of Section 19-25 of WPA, 1972 (as amended up to 1991)	Not completed.
Final declaration No. and date	Not yet completed the procedure.

Source: DFO Office, Paralakhemundi.

Table 19.3: Brief of Balukhand-Konark Wildlife Sanctuary

Particulars	*Balukhand-Konark Wildlife Sanctuary*
Area (in sq. km)	71.72
Notification No. and Date	9013 dt. 23-4-1984 and 15216 dt. 1-9-1987
Location	Between longitude 85°52′ to 86°14′ (E) and latitude 19°48′ to 19°54′ (N)
Climate	Winter temperature is approx. 10°C and Summer temperature above 40°C
Boundary Demarcation (Core and Buffer zone)	No work has been carried out after notified as a Sanctuary
Forest type and Nos. of Reserve Forest, Proposed Reserve Forest, Demarcated Protected Forest, Un-demarcated Protected Forest, etc.	1 Reserve Forest: Balukhanda—15.667 sq. km. 7 Proposed Reserve Forests: (i) Balighai—11.713 sq. km. (ii) Liakhia—5.197 sq. km. (iii) Konark West—13.196 sq. km. (iv) Konark East—15.23 sq. km. (v) Golara—6.80 sq. km. (vi) Nadiamath—3.32 sq. km (vii) Sarlakecut—0.604 sq. km Total—71.72 sq. km.
Procedure of Section 19-25 of WPA, 1972 (as amended up to 1991)	Not completed.
Final declaration No. and date	Not yet completed the procedure.

Source: RO Office, Balukhand and Konark.

Table 19.4: Brief of Bhitarkanika Wildlife Sanctuary

Particulars	*Bhitarkanika Wildlife Sanctuary*
Area (in sq. km)	672.00
Notification No. and Date	6958 dt. 22-4-1975
Location	Between longitude 86°30′ to 87°60′ (E) and latitude 20°30′ to 20°50′ (N)
Climate	Winter temperature is approx. 10°C and Summer temperature above 40°C
Forest type and Nos. of Reserve Forest, Proposed Reserve Forest, Demarcated Protected Forest, Un-Demarcated Protected Forest, etc.	10 Proposed Reserve Forests: 25 Forest Blocks Bhitarkanika (N)—70.99 sq. km Bhitarkanika (S)—88.56 sq. km
Procedure of Section 19-25 of WPA, 1972 (as amended up to 1991)	Not completed.
Final declaration No. and date	Not yet completed the procedure.

Source:DFO Office, Rajnagar.

Final Publication and Disposal of ROR (Record of Rights) in Protected Areas

The Hon'ble Chief Minister of Orissa had given a public statement on 1st April 2000 (on the occasion of Utkal Dibasa) that all pre-1980 Forest villages could be regularized within six months. This statement was regularized by Revenue Department of Orissa under No. GE (GL)-S-17/2000-21060/ R dt. 4-5-2000 to take a decision to regularize such hamlets who have not any claims in the revenue record of rights (also known as encroachments), which have certain eligibility criteria before 25-10-1980 (Govt. of Orissa, 2000)[10]. The pre-1980 procedure has not been completed which was supposed to be regularized before 1-8-2000. The concerned department has made the joint verification report for regularization of encroachment '*Jakesi*' inside the

Table 19.5: Pre-1980 Encroachments Inside Kotagarh Wildlife Sanctuary

District		*Kandhamal*
Tahsil		Balliguda
Name of the encroached habitat		Srambi
Name of the forest with legal status		Srirampur 'B-block' Proposed Reserve Forest
No. of encroachers		44
Category	Scheduled Caste (SC)	1
	Scheduled Tribe (ST)	42
	Others	1
Population	Male	117
	Female	117
	Total	234
Land encroached prior to 25-10-1980 (in ha)	Homestead	0.704
	Agricultural	68.122
	Total	68.826
Year and Evidence of encroachment		OR No. 8 of 1979-80 of Kotagarh Range
Category of encroachers		Both landless and home-stead less
Area recommended to be regularized (in ha).	Homestead	0.704
	Agricultural	61.600
	Communal	6.522
	Total	68.826

Source: DFO Office, Balliguda.

Sanctuary, which was already declared as revenue village by the Revenue Department in May 1995. The Committee has regularized another encroached habitat 'Srambi' inside the Sanctuary with the total No. of 44 encroachers, however, there is no final publication and disposal of ROR (Record of Rights) made by them for these 44 encroachers (ref. Table 19.5 and Table 19.6).

Table 19.6: Some Pre-1980 encroachers of the encroached habitat, 'Srambi'

Name of the encroacher		*Dalapa Pat Majhi*	*Gomesh Singh*	*Dera Pat Majhi*	*Tamaku Pat Majhi*	*Daud Pat Majhi*	*Arsin Pat Majhi*
Father's Name		Pingo Pat Majhi	Michael Singh	Pasora Pat Majhi	Daboca Pat Majhi	–	Sakadhara Pat Majhi
Category		ST	SC	ST	ST	ST	ST
Area (in acres)	Agriculture	1.03	3.74	5.17	2.58	1.56	7.05
	Vegetable	0.6	2.84	2.575	5.11	2.95	2.15
	Homestead	0.8	0.3	0.11	0.06	0.16	0.21
Population	Male	1	1	1	1	1	1
	Female	1	1	1	1	1	1
	Child	1	10	3	1	2	6
Settlement (in years from 25-10-1980)		>15	>15	>15	>15	>13	>13
Land detail		It is only the prior approval made by the Joint Verification Committee					

Source: Focus Group Discussion with encroachers of Srambi.

Inter-Departmental Conflicts

The lack of inter-departmental coordination in the State is one of the major confusion and fear within the forest dwellers residing within the limits of the Sanctuaries. If we take a case of Kotagarh Wildlife Sanctuary, in 1995, the Revenue Department had expressed its desire on the basis of '17-point Programme' to give status of revenue village to the 23 hamlets who had records of revenue offences much before commencement of the Forest Conservation Act, 1980. But the area was within the limits of the Sanctuary. So as per the joint verification of Forest Department and Revenue Department, only those encroachment habitats were given the status of a revenue village, against whom there were evidences of revenue as well as forest offences before the said dead line. According to the joint verification on the 16th May 1995, the Revenue Department shows the total area of the four villages (i.e.

Jakesi, Bandeka, Bandapipili and Srambi) including homestead and agricultural 1050 acres, whereas it is only 339 acres according to the Forest Department.

In Lakhari-valley Wildlife Sanctuary, the 'Olanda' villagers are unsure of their tomorrow due to the strategies developed by the Forest Department and Revenue Department but are flooded with permanent structures by the Block authorities (DFO Paralakhemundi, 2000)[11]. Similarly in Kotagarh Wildlife Sanctuary, the Ora panchayat is getting a huge amount of support for infrastructure development (e.g. Anganbari, roads, schools etc.). The poverty alleviation programmes like Jawahar Rojgar Yojana (JRY), Integrated Rural Development Programme (IRDP) and Indira Awas Yojana (IAY) are also continuing to provide development extension facilities to allow them a stronger root.

Human-Animal Interface

As per said Act, there is apparently no bar on the continuation of revenue villages in the 'buffer' zone. But it is very difficult task for the Protected Area Managers, especially in Sanctuaries like Kotagarh and Balukhand-Konark Wildlife Sanctuary, where the 'Core' and 'Buffer' zones are yet to be demarcated.

Due to high human population and livestock in and around the PAs, there is frequent interface with the wild animals and their habitats. Generally, it is seen that the attack on human beings have always occurred whenever they entered into territories of wild animals. However, during the summers wild animals enter into villages to get drinking water and fodder and create problems for the local communities. Some degree of antagonism between Sanctuary authorities and people exists due to crop damage by wild animals, felling of trees and offences.

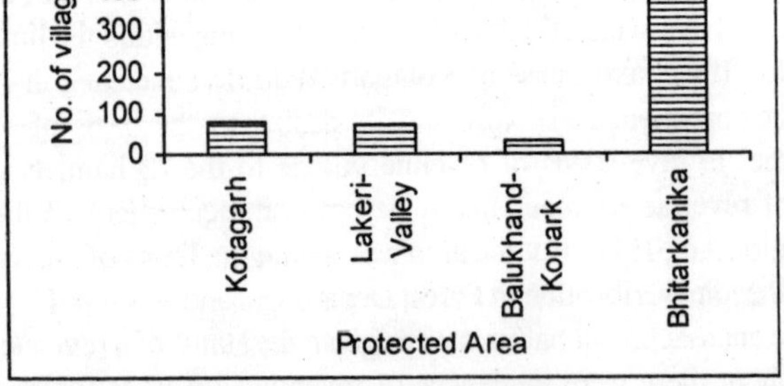

Graph 19.1: Villages (Revenue and Encroachments) inside and Adjoining the WLS

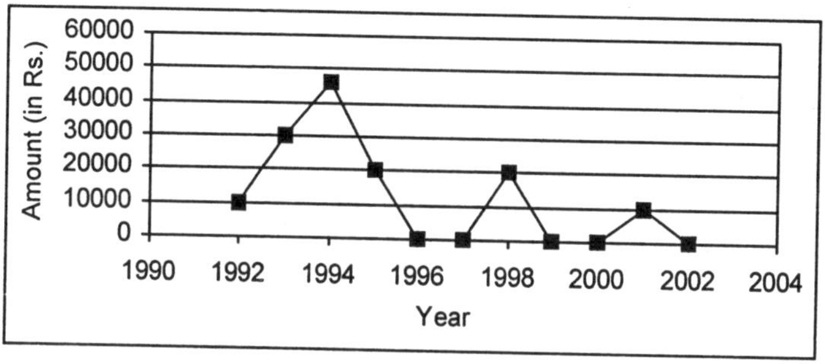

Graph 19.2: Compensation paid in Kotagarh WLS

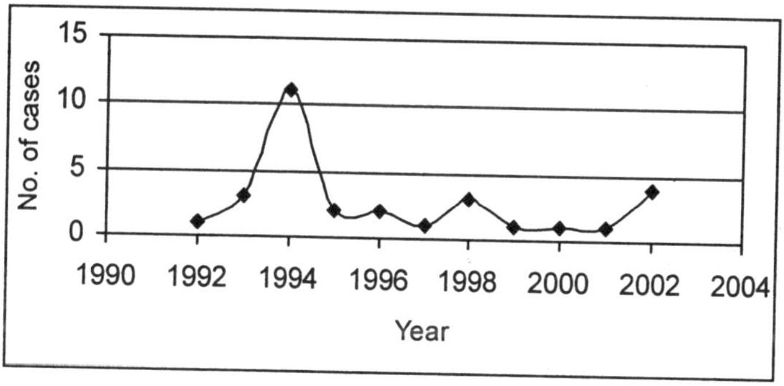

Graph 19.3: Human-Animal Conflict from 1992 to January 2002 in Kotagarh WLS

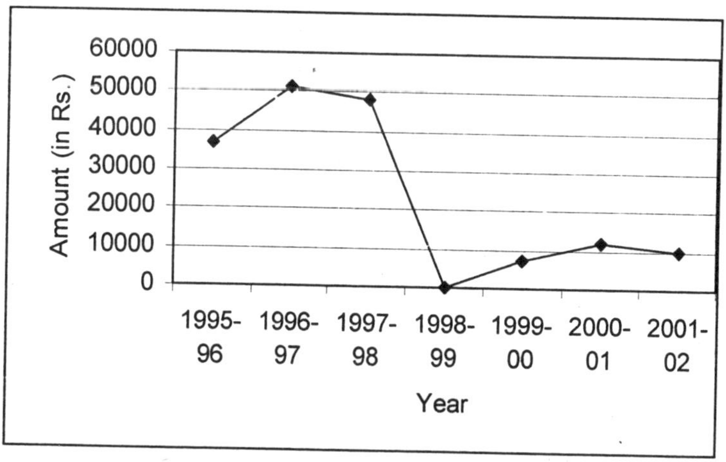

Graph 19.4: Compensation paid in Bhitarkanika WLS

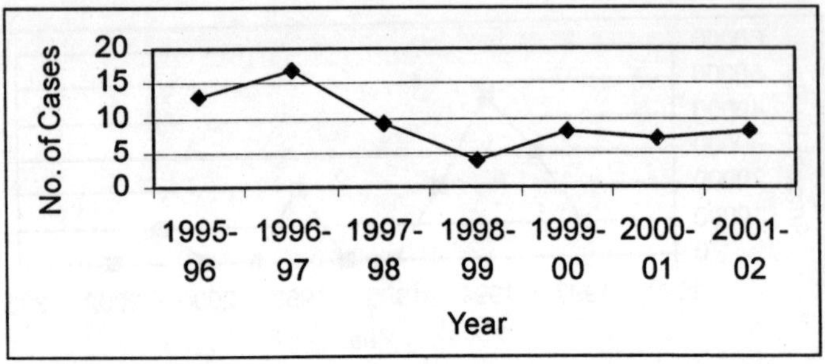

Graph 19.5: Human-Animal Conflict from 1995-96 to 2001-02 in Bhitarkanika WLS

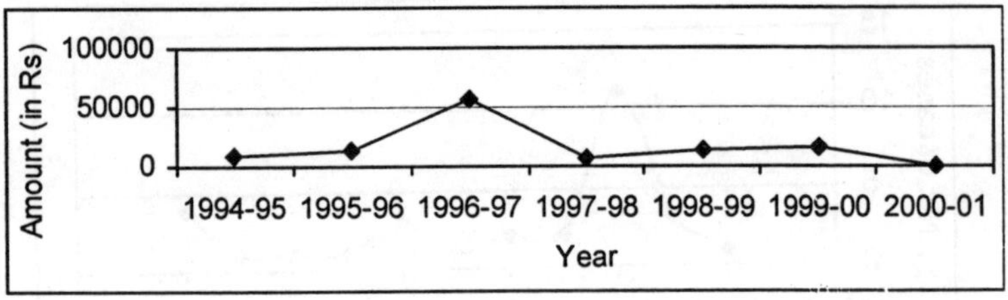

Graph 19.6: Compensation paid in Lakheri-Valley WLS

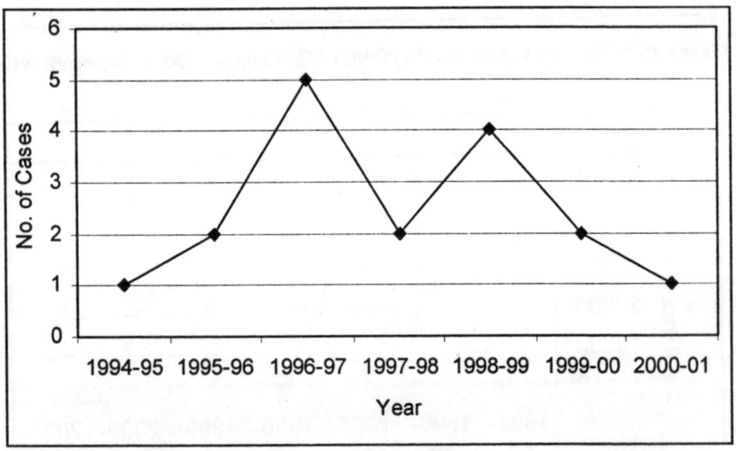

Graph 19.7: Human-Animal Conflict from 1994-95 to 2000-01 in Lakheri-Valley WLS

The livestock of villages are dependent on the forest in absence of worthwhile pasturelands. While grazing they enter into the territories of carnivores and are attacked by them. Thus a number of livestock are injured or killed. If the compensation is not paid immediately who lose their livestock, turn hostile to the wildlife and the Sanctuary authorities. This can be seen in graphs of the respective years of the Human-Animal conflict *versus* Compensation paid (Graphs 19.2-19.7).

Conclusion

Theoretically, the Wildlife (Protection) Act, 1972 is well documented to protect the fundamental rights of lives and homes of the wildlife, however, there are many ground difficulties and challenges to implement it effectively in its true essence. The awareness among the people about their roles towards conservation of wildlife and need of their protection are the major challenges for managers of the Protected Areas.

Under Sub-section (1), Section 18 of Wildlife (Protection) Act, 1972 the State Government can notify any area as a Sanctuary, this may include any part of territorial waters or reserved forests, which are suitable for wildlife and their habitats. Thus there should be wider scope for public consultation before notifying the area as a Sanctuary. The forest dwellers, which have been living there for age old and dependent on existing natural resources for their sustenance, feel estranged from their rights over the surrounding natural resources.

In most of the cases after the first notification as Wildlife Sanctuary, the Forest Department has started imposing various restrictions and prohibition on the people like Non-Timber Forest Produce / Minor Forest Produce collection. If we examine the clause (a), Sub-section (1), Section 26A of WPA, 1972 minutely, the Government cannot impose restrictions on the people unless all the claims are settled down and the final notification is issued in black and white.

The aim of Government's Protected Area Network Plan for wildlife is to save around four per cent of India's geographical area. However, in India it is very difficult to find this chunk of land, which devoid of human existence. Therefore, it would be a large-scale human displacement to exclude the forest dwellers from Protected Areas. The rehabilitation and resettlement process for exclusion of the forest dwellers may contribute to marginalize these vulnerable people.

Conservation of wildlife and their habitat is an important goal for Wildlife Department of Orissa. The basic need of Protected Area management in Orissa will be for planning and the management of Protected Areas based on a healthy interaction between man and nature, especially from traditional practices, which have helped to conserve and sustainable use of natural resources. They will have to consider ways in which the relative strengths of all sections, especially of politicians, bureaucrats, resource managers, developmental agencies, resource users and local communities can be put together.

The management of Protected Areas covers a wide range of activities, which is impossible for the management authorities to carry out all conservation related functions by themselves. It must delegate some duties and coordinate closely with other agencies. Thus the interests of two or more stakeholders are in line and these agencies can benefit from an alliance to promote their joint efforts.

NOTES

1. Gee, E.P., 1962. "National Parks in India", in V.H. Cahalane's (ed.), *National Parks—A World Need*, Special Publication, M-14, American Committee for International Wildlife Protection; New York.
2. Rao, K., 1988. *Management Planning for Protected Areas in Developing Countries*, Wildlife Institute of India; Dehra Dun.
3. Singh, S. and Rodgers, W.A. 1990. "India: National Parks and Natural Reserves", in C.W. Allen's, (ed*). International Handbook of National Parks and Natural Reserves*, Greenwood Press; New York.
4. Forest Survey of India (FSI), 1991. *The State of Forest Report*, FSI; Dehra Dun.
5. Orissa Forest Department (OFD), 1997. *Wildlife Orissa at a Glance*, Nature and Wildlife Conservation Society of Orissa; Bhubaneswar.
6. Karanth, K.U., Sunquest, M. & Chinnappa, K.M., 1998. "Long-term Monitoring of Tigers: Lessions from Nagarahole" in J Seidensticker, S. Christie and P. Jackson (eds), *Riding the Tiger: Tiger Conservation in Human Dominated Landscapes,* Cambridge University Press; Cambridge.
7. Kothari, A. & Pathak, N., 1998. "Sharing Benefits of Wildlife Conservation with Local Communities: Legal Implications", *Economic and Political Weekly,* October 3: 3603-10.
8. John Kathy, 1986. *Managing Protected Areas in the Tropics*, Natraj Publishers; Dehra Dun.
9. Campbell, Oona, Cleland, John, Martin, Collumbien & Southwick, Karen, 1999. *Social Science Methods for Research on Reproductive Health*, WHO; Switzerland.
10. Government of Orissa, 2000. *Pre-80s Settlement letters and Guidelines*, DFO, Rayagada.
11. DFO, Paralakhemundi, 2000. *Pre-80s Settlement letters and list of Villages*, DFO, Paralakhemundi.

20

A Study on Socio-Economic Aspects of Wildlife Offences in Few Districts of Madhya Pradesh (India)

Suchitra Banerjee and *A.K. Bhattacharya*

Introduction

The socio-economic dimensions of the wildlife offences have hardly been paid any attention in the wildlife management. Subsistence hunting is considered less serious because it is normally for the common animals (Eltringham, 1988)[1]. In tribal dominated forest areas, specially in Madhya Pradesh, wildlife hunting on certain occasions like festivals and marriages have traditionally been considered as their rights (Banerjee, 2002)[2]. The wildlife offences committed by the communities because of their traditions of the wildlife dependent food habit and the hunting practices are difficult to control unless the related livelihood issues and the socio-economic dimensions are addressed.

The following five phases are clearly discernible, so far as the past management of wildlife is concerned (Kiss, 1993[3]; Banerjee, 2002[2]):

(a) The period of abundance and plenty (up to 1910).
(b) Wild animals and Man coexisting in a happy balance (between 1910 to 1945).
(c) Exploitation or over-exploitation of resources (After Second World War to 1970).
(d) Protection and Conservation (1970 to 1990).

(e) Community linked wildlife management (eco-development approach)–to-date.

The literature survey (Banerjee, 2002[2]; Mathur *et al.*, 1999[4]) indicates that no exclusive study has so far been made to analyse the socio-economic dimensions of the wildlife offences. Through the present study an attempt has been made to examine the socio-economic aspects of the wildlife offences.

The study deals with the socio-economic factors responsible for poaching and trade, the strengths and weaknesses of the existing rules, regulations and acts pertaining to wildlife poaching and trade, recommendations for improvement in wildlife management with respect to poaching and trade, possible remedies for wildlife offences and wildlife management conflicts.

Study Area

The study was carried out in four districts of Madhya Pradesh, viz. Betul, Bhopal, Hoshangabad and Vidisha, which are rich in biodiversity, forests dependent tribal population and also have potential threat of forest offences. The details of the geographical area, forest area (FSI, 1999)[5] and the wildlife census and the wildlife offences of these districts are given in Table 20.1:

Table 20.1: Wildlife Census and Wildlife Offences

S. No.	District	Geographical Area (Sq. km)	Forest Area (Sq. km)	Wildlife species* number	Registered Wildlife offences Number (Period)
1.	Betul	10,043	3,600	**	**
2.	Bhopal	2,772	285	56	15 (1978–2001)
3.	Hoshangabad	10,037	3,311	104	52 (1995–2001)
4.	Vidisha	7,371	776	57	37 (1995–2001)

* Mammals, birds and reptiles

** Field data were collected from three districts, viz. Bhopal, Hoshangabad and Vidisha, whereas focus group discussions were conducted in Betul and Hoshangabad.

Source: Working Plans of respective Forest Divisions.

Methodology

The methodology included participatory methods, tools and techniques including collection of secondary and primary data through structured questionnaire survey, focus group discussions and field surveys. Secondary data were procured from the Forest Department at Division and Range levels followed by discussions with the forest officials. Questionnaire survey was done for the collection of information from the forest officials and villagers through prescribed scheduled forms.

Focus group discussions with Forest officials and village level communities were organised through workshops. After briefing the objectives of the study, discussions were conducted one to one as well as groups. Attempt was made to arrive at some unanimous opinion about the information furnished. The focus groups consisted of all stakeholders including forest officials of all levels, representatives of village level forest committees and representatives of PRIs.

Random field surveys were made in the study area. PRA and RRA techniques were adopted to obtain the information from the target groups formally and informally, specially to involve the villagers to know their viewpoints.

Results and Observations

The observations and the results are confined to faunal species and based on the information collected from various circulars, documents and records of the Forest Department regarding the wildlife offences / protection and first hand information collected through the individual and focus groups discussions with the communities/villagers and the staff of the Forest Department (Banerjee, 2002)[2].

Modus Operandi (*Methods Adopted in Wildlife Offences*)

The experience in Madhya Pradesh shows that the methods are area specific (Menon & Kumar, 1998[5] and 1999[7]). The reported methods are listed below (Murthy, 1999)[8]:

Table 20.2: Modus Operandi Methods

S. No.	*Districts*	*Poaching Methods*
1.	Seoni and Chindwara	Run down the animal by dogs then kill. Electrocution.
2.	Balaghat and Mandla	Telephone wire nooses. Poisoning with Potassium cyanide.
3.	Raipur	Gill traps. Poisoning.
4.	Sagar and Damoh	Small explosives. Gunning down. Animal traps. Snares and nets. Chasing the animal by the dogs.
5.	Shivpuri	Poisoning waters for fishes. Gunning down.
6.	Bastar	Hunting with bows.
7.	Shahdol and Umaria	Small explosives. Gunning down. Animal traps. Snares and nets.
8.	Jabalpur, Dindori and Katni	Small explosives. Gunning down. Animal traps. Snares and nets. Chasing the animal by the dogs. Poisoning with Potassium cyanide.

Observations of the Focus Group Discussions

Kamti Eco-Development Committee (EDC)

Along the northern bank of the Tawa is Kamti, the core of Satpuda, which is very rich in bio-diversity. The discussions with the forest staff and the villagers in groups and individually revealed following information:

- Apart from routine wildlife there are "Pangolin" (seen in Satpuda range in year 2000 first time after for 20 years) and Blackbuck population, which is a sure sign of habitat improvement (White, 2000)[9].
- No preliminary offence report about killing or poaching of any animal by the villagers in the National Park has been recorded (FOCR, 2001)[10].
- The EDC is very active and effective with lots of voluntary services extended by the villagers. From the funds allotted by the Forest Department, it extends loans to the villagers at low rates of interest (2 or 4%) for the acquisition of materials and resources such as irrigation equipments, tube-wells, biogas plants, sewing machines etc. The scheme was to provide an alternative source of income to

villagers and promoting self-sufficiency to reduce their dependence on the forest and wildlife offences.

The success of the Kamti EDC is a noticeable exception. Before the formation of the EDC, villagers resented restrictions and Forest Development upheld the law, resulting into conflicts between villagers and Forest Department. After the constitution of the EDC, the staff went to the villagers, sat with them and asked how best they could help and what kind of resources were most needed. These are the evidence of the vastly improved working relationship between villager and Forest Department. The villagers protect forests and wildlife from fire and poaching (White, 2000)[9].

- Some particular plant species have been raised by EDC for wild boars and foxes so that they need not destroy fields and became eco-friendly with human beings.
- With a positive view it could be possible to manage the wildlife through community based wildlife management. The lessons are as follows:
  - Local commuters should be involved in wildlife management.
  - Villagers should be made aware of the strictness of the laws.
  - The success of community based wildlife programmes will ultimately depend on the viability of wildlife utilization as an economic enterprise, specifically on a demonstration that it is more profitable and beneficial to landholders (whether individual or communities).

Interaction with Village Level Forest Committees of Betul District

In order to have first hand information, a workshop was organised at Shahpur, Betul, which was participated by members of eight village level forest committees including Vikrampur, Moorah, Katawadi, Lonia, Bhowra Dhana, Bichhua, Dhar and Sarni. All these villages have tribal population mainly belonging to the Gond tribe.

The results of the discussions are as follows:

- Villagers of Katawadi village reported that there was no poaching case in this village. People occasionally eat wild animals like Hare, Wild boar, Deer and Sambhar as food. One of the major factors of killing animals for food is that there is no agriculture activity in the area. Villagers relish country liquor and boar meat.

They trap Hares, Boars and Jungle Fowl using indigenous trap called 'Phanda'. During tribal festivals only the community is observed, killing Teetar, Hare etc.

- Wild boars reportedly destroy the agricultural crops repeatedly. The boars are driven away from the fields using sound bombs and air gases etc.
- There are poaching activities in certain areas of North Sarni. The people involved are reportedly the employees of MPEB. However big mammals like Sambhar, Cheetal, Sonkutta are available in the area, but they kill some small mammals like hares.
- Villagers of Pathai recall a case that took place in 1995. 35 peoples were arrested on trapping a 'Sambhar'. Since then there is no case of poaching being registered.
- Members of Moorah Committee as well as forest staff reported that 95 per cent of poaching activities had been controlled. There are no P.O.R. in the jungles of Moorah Kota.
- In Bhownra Dhana-Sarni, wild animals have run away because of construction of powerhouse.
- The villagers of Vikrampur village have adopted an indigenous method for controlling Boars. They make rope barricades with nails and Palash *(Butea)* leaves and coat them by lime.
- A small group of Muslim population, most of them being MPEB employees, are fond of shooting water birds and other small animals frequently using air guns.
- The interrogation revealed following facts:
  - — In some areas there are no wild animals left hence there is no poaching.
  - — Wild animals that are poached for eating include Sambhar, Wild Boar, Cheetal, Jungle Fowl and Teetar, Bater.
  - — Animals that harm their poultry and crop are Son Kutta (wild dog), Hyena, Fox and Wild Boars.
  - — It seems that the centuries old culture and traditions of the tribals are not commercialized. They are unaware of the demands of the wildlife commodities in the trade.
  - — The villagers do not know methods of smuggling and special smuggling techniques.
  - — Almost every villager of Gond tribe is a flesh eater.
  - — Methods of poaching like Pit poaching, Electrocution, Poisoning, Harpooning etc. are not known to them.
  - — Instead, they practice certain methods to protect their crops and field without causing any harm to the wild animals.

Statistics of Wildlife Offence Cases in the Study Area

The details of districtwise and yearwise number of wildlife offence cases are as follows:

Year	*Districtwise number of cases*			
	Vidisha	Bhopal	Hoshangabad	Total
1978-1994	29			29
1995	02		04	06
1996	03		04	07
1997	02	06	05	13
1998		06	06	12
1999	01	03	10	14
2000	—	—	11	11
2001	—	—	12	12
	37	15	52	104

The details of monthwise registration of cases are as follows:

Month	*Mon*	*Feb*	*Mar*	*Apr*	*May*	*Jun*	*Jul*	*Aug*	*Sep*	*Oct*	*Nov*	*Dec*	*Total*
Cases	13	08	07	08	05	17	10	06	04	05	09	10	104

Wildlifewise numbers of cases are as follows:

S. No.	*Wildlife*	*District wise number of Cases*			
		Hoshangabad	Vidisha	Bhopal	Total
1.	Barking Deer (Bhedki)	1	1	—	2
2.	Bear	4	—	—	4
3.	Black buck	3	—	—	3
4.	Cheetal	18	12	3	33
5.	Chinkara	1	3	—	4
6.	Lion cubs	2	—	—	2
7.	Neel gai	—	5	3	8
8.	Panther	2	3	—	5
9.	Peacock	2	9	4	15
10.	Pheasants	—	—	6–Teetar 2-Bater	8
11.	Porcupine	—	—	1	1
12.	Python	1	—	—	1
13.	Rabbit	2	1	1	4
14.	Sambhar	17	—	1	18
15.	Tiger	—	2	—	2
16.	Tortoise	1	—	—	1
17.	Whistling Teel	—	—	5	5
18.	Wild boar	3	1	2	6

The Districtwise, wildlifewise and parts-wise cases are as follows:

District	*Wildlife*	*Parts*	*No of Cases*
Hoshangabad	Tortoise	Entire	1
	Sambhar	Genitals	1
		Ovaries	1
		Meat	6
		Bones	5
		Skin	10
		Gorus	1
	Panther	Skin	1
	Chinkara	Antler	1
	Cheetal	Skin	4
		Antler	3
		Meat	2
	Bear	Paws	1
	Blackbuck	Skin	1
		Antler	1
	Sanda	Entire	1
	Peacock	Meat	1
	Bhedki	Skin	1
		Meat	1
	Boar	Meat	1
	Rabbit	Skin	1
		Meat	1
Bhopal	Rabbit	Meat	1
	Neelgai	Meat	1
	Teetar / Bater	Entire	2
	Wild boar	Meat	2
	Porcupine	Entire	1
	Sambhar	Antler	1
	Peacock	Legs	1
	Whistling Teel	Entire	1
Vidisha	Cheetal	Skin	4
		Skull	1
		Meat	4
	Tiger	Skin	1
	Peacock	Feathers	2
	Neelgai	Meat	4
	Boar	Meat	1
	Rabbit	Meat	1

The reasons-wise numbers of cases are as follows:

Reasons	*No. of Cases*
Livelihood	62
Food	29
Rituals and others	12

Discussion

104 cases were recorded and analysed from three Districts, namely Bhopal, Hoshangabad and Vidisha.

This majority (71.4%) of cases pertain to the period of six years, i.e. from 1995 to 2001; only few cases belong to the period between 1978 and 1994.

The number of cases of wildlife offences is maximum in Hoshangabad (50%) followed by Vidisha (35.6%) and Bhopal (14.4%). This pattern manifests the richness in the biodiversity in the districts. Hoshangabad has got the maximum biodiversity and number of wildlife species is maximum (48%) in this district. But the ratio of the offences to the total number of wildlife species in the respective district is maximum in Vidisha District (Fig. 20.1).

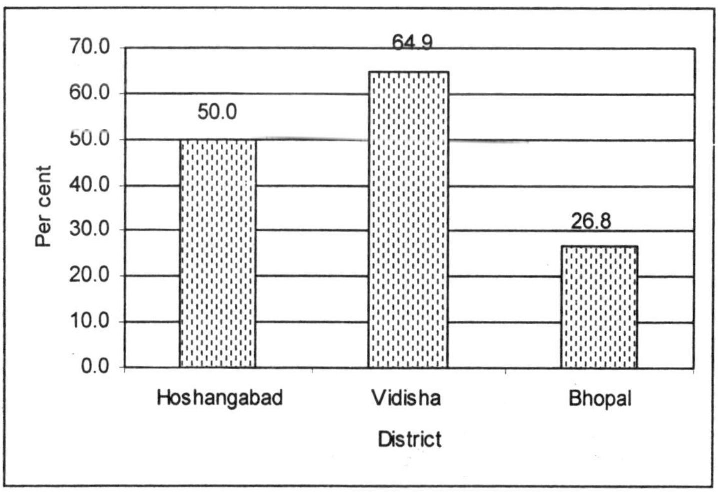

Fig. 20.1: Ratio of Wildlife Offences to Total Wildlife Species

This trend is substantiated by the inventories of the wildlife reported in the Working Plans of the respective Districts. As per the Working Plans the number of the major faunal species reported from these districts are 56 (25.8%), 104 (47.9%) and 57 (26.3%) respectively from Bhopal, Hoshangabad and Vidisha.

The reports of wildlife offences indicate that maximum cases have been reported in June (16.3%) and then in January (12.5%) and December (9.6%).

Wildlife for which cases have been registered mainly included eighteen species. The study indicates (Fig. 20.2) that most of the offences for wildlife have been committed for Cheetal (31.7%), followed by Sambhar (17.3%) and Peacock (14.4%) and parts for which the offences have been committed are meat (35%) followed by skin (31%) and antlers (8%).

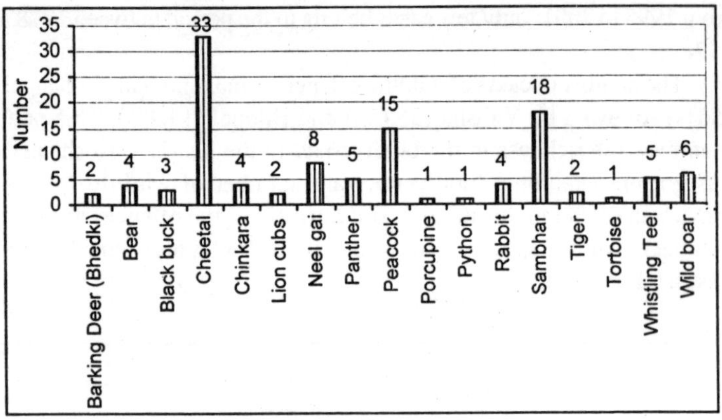

Fig. 20.2: Number of Offences as per the Wildlife Species

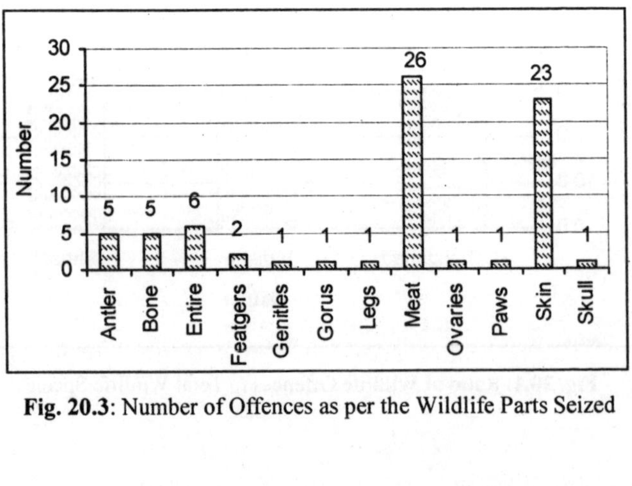

Fig. 20.3: Number of Offences as per the Wildlife Parts Seized

From the analysis of cases (Fig. 20.3), it may be concluded that most of the offences have been committed for personal consumption. These cases of poaching can be said to have been committed for livelihood or economic reasons. However, killings have been done for other social reasons like festival and marriage.

The reasons for which the offences have been committed include livelihood, which may be by selling the wildlife material (60%) or by direct consumption of the animal as food (28%) followed by rituals and other reasons (12%).

Livelihood poaching may have some direct or indirect influence of the commercial wildlife crimes, as it might form a forward linkage for the poaching and encourage the villagers to commit the wildlife offence.

Out of total registered offence cases, 27 cases are pending with the Department for inquiry, whereas 67 cases are pending with the court for decisions. Only eight cases have been finalized, out of which in three cases offenders were released on benefit of doubt, whereas in five cases fines were imposed for Rs 3500.

The individual and the focus group discussions with the members of the village forest committees indicate that—

- The hunting and the poaching of the local animals have been a deep-rooted traditional practice. Although restricted by the acts, rules and regulations killing of the animals cannot be stopped completely.
- Wildlife plays an important role in the livelihoods (including the food value) of the local people.
- There is significant lack of awareness about the importance of the wildlife in the ecosystem and the damage being caused by the hunting or poaching.
- Even the commercial traders may use them sometimes, but they are not aware about the organized illicit trade of the wildlife articles.
- Although they have not expressed openly about the wildlife killing and trade practices, but there are indications that hunting and poaching are a continuous process, sometimes need based and sometimes for change of taste.
- The experience of Kamti EDC of Hoshangabad is a positive indication that local people can be the best custodians for the wildlife and forest protection.

Conclusions

- During the study period twenty varieties are reported to have poached, out of which Cheetal and Sambhar are the most hunted species. The species are hunted mainly for their meat and skin.
- Out of 104 cases, maximum numbers of cases recorded are 17 in the month of June and next in the month of January.
- Maximum cases recorded are from the District Hoshangabad followed by Vidisha.
- Rise in the population of the Black Bucks in Kamti is a sign of habitat improvement.
- Maximum number of cases seem to be pending in the court. Only in five cases fines have been imposed.
- Villagers do not seem to be aware of the stringentness of the laws. They need to be made aware of the laws and the legislation.
- Most of the cases have been committed for the meat for personal consumption, for livelihood and economic reasons. Killings have been done for other social reasons like festivals, marriages as local traditions and religions, as apart of their culture.
- Some communities depend heavily on wildlife as a source of meat and other products. They hunt only ritualistically. In many communities hunting is a part of the local culture. Whenever hunting is a part of culture community cohesion, hunters often enjoy high social status.
- Different wildlife species play different and important role in local tradition and religions. People's attitudes towards wildlife are not based solely on economics even though the wildlife in the local economics is an important factor that is often overlooked or undervalued.
- Management and conservation of wildlife can improve the livelihood of the rural communities without contributing to environmental degradation (Kiss, 1993)[3]. Kamti is a good example of wildlife resource management with local participation. In order to minimise man–wildlife conflict the community people have raised some food species for large ungulates, so that they may not damage their crops. The success of community based wildlife programmes will certainly lead to reduction in wildlife crimes.

Recommendations

- A new approach is to be adopted towards the problem of destruction through poaching with indigenous methods like snares, traps and poison of wildlife. The demand for meat protein for communities who depend heavily on wildlife as a source of meat and other products has to be satisfied, and in these areas adequate stock will have to be first built up by restocking with animals brought elsewhere or bred in captivity. Based on demand it may be examined to harvest annual surplus of the meat in the shape of a fixed number of buffalo, antelope, Nilgai, Sambahar and to supply it to the local population provided they agreed to stop their illegal killing, which in many cases destroys more than what can be eaten.
- A sense to initiate a community-based wildlife management should be developed. Training and education are essential elements in the community based wildlife management. Wildlife officials must be sensitized to the sociological aspects and be trained to communicate to work with local communities. Local people must be educated about the present and potential value of the wildlife and the natural habitat and be trained in the technical, financial organisational and management skills needed to secure these benefits for themselves and their communities.
- The traditional and the need based practices of the hunting and the poaching need to be examined under the existing acts, laws, rules and regulations and provisions for acceptable alternatives need to be seriously worked out and properly implemented. Commercial wildlife used can be replaced by livestock husbandry and examination of trends of ecological and economic data can reveal that wildlife systems have significant advantages in ecosystems in both earning capacity and sustainability. In Zimbabwe and African countries these measures have already been implemented (Kiss, 1993)[3]. The idea of livestock development is worth considering in Madhya Pradesh where large tribal groups exist in or near forests and consider the wildlife as their traditional rights.
- Capacity building through training and extension should be of the priority areas. An effective education programme must promote traditional attitude and initiate actions to cause changes in attitudes where needed. This is a difficult and challenging

task, which is too often overlooked. It is essential and specialised function requiring trained and dedicated staff.

NOTES

1. Eltringham, S.K., 1988. *Wildlife resources and economic development*, International Book Distrubutors.
2. Banerjee, S., 2002. 'Studies on socio-economic and legal aspects of wildlife Offences'. Minor Research Report, University Grants Commission, March 2002.
3. Kiss, Agnes, 1993. *Living with Wildlife—Wildlife Resource Management with Local Participation in Africa*, The World Bank, Washington DC.
4. Mathur, P.K., Mukherjee, S.K., Sankar, K., Badola, Ruchi, Hussain, S., Rana, M.S., Sardar, A.K., Agrawal, S., Verma, Y.S., 1999. *Bibliography–Wildlife and Protected Area Management in Madhya Pradesh*, Wildlife Institute of India, Dehradun.
5. FSI, 1999. *State of Forest Report 1999*, Forest Survey of India, Dehradun.
6. Menon, Vivek and Ashok Kumar, 1998. *Wildlife Crime—An Enforcement Guide*, First Edition, Natraj Publishers, Dehradun.
7. Menon, Vivek and Ashok Kumar, 1999. *Wildlife Crime—An Enforcement Guide*, Second updated edition, Wildlife Protection Society of India, Near Delhi, Natraj Publishers, Dehradun.
8. Murthy, R.S., 1999. *Hand Book of Wildlife Crime*, Prevention, Detection, Investigation and Case Processing (A field guide to Frontline Staff–private circulation).
9. White, G.D., 2000. *Surprises in Satpura. Sanctuary Asia*, Vol. XX.
10. FOCR, 2001. *Forest Offence Case Register*, Matkuli Range Office, Satpuda National Park.

21

The Illegal Fisheries Case in the Totladoh Dam in Pench Tiger Reserve, Seoni, Madhya Pradesh

A Case Study

A.K.Bhattacharya, Shobna Boyle and *Samita Vasudevan*

Background

The Pench Tiger Reserve (PTR) is located in the Southern Slopes of the Satpura ranges in the Seoni and Chhindwara districts of Madhya Pradesh. The Reserve was created in 1992 with an area of 757.89 sq. km, making it the 19th Tiger reserve of the country. PTR comprises the Pench National Park (area 292.85 sq. km.), the Mowgli Pench Sanctuary (area 118.31 sq. km.) and a buffer division (area 346.73 sq. km.). To the south of the tiger reserve lies the Pench Tiger Reserve of Maharashtra. The river Pench, which splits the Park area into two running from the north of the Park to the south, forms the lifeline of the Park.

The Pench hydroelectric dam on the river Pench, straddling the Maharashtra-Madhya Pradesh boundary was constructed between 1973 and 1988. The Totladoh dam and reservoir, built as part of the hydel power project is situated inside the National Park and comprises a total area of 72 sq. km. Out of this 54 sq. km. falls under the Madhya Pradesh reserve while 18 sq. km is under the Maharashtra tiger reserve.

The Case

Construction of the Totladoh dam was completed in 1990, but, the Madhya

Pradesh Fisheries Development Corporation (MPFDC) was involved in the culture and trade of fish in the reservoir since 1981 itself. The location of the dam inside the National Park made this activity illegal as per the Wildlife Protection and Forest Conservation Acts. Because of the abundance of fish in the area and the huge amount involved in the illegal trade, a fish-mafia was active in the area. The mafia network had developed due to the connivance of a number of traders, politicians and top officials in the Forest Department. Instead of the usual system wherein the fishermen paid a royalty of Rs.12 directly to the MPFDC[1], a network of traders set up a system where they paid the royalty and gave the fishermen Rs.10 along with a supply of boats and nets. The traders would then supply the fish to the local markets and to the markets of Nagpur, Bhopal, Seoni and Chhindwara at a market rate of around Rs.50 per kg there by earning themselves huge profits.[2] At the peak of illegal fishing in 1994 the annual recorded catch was 700 tonnes of fish which represents US $240000 in royalties and a market value of US $1million. The Forest Department staff however believed that this was just about half the total catch and that the profit from the unrecorded catches was split between the politicians, administrators and the local police.

The Impact of Illegal Fishing

The practice of illegal fishing inside the Park area was the cause of immense disturbance to the wildlife inside the Park and their habitat. Official records show that between 1992-96 eight tigers and sixteen leopards had been poached in the reservoir. Large numbers of poaching cases were also reported on the prey species like Chital and Sambhar during the same period. Apart from the aquatic life, the variety of amphibians like crocodiles and turtles found in the reservoir also faced the threat of being wiped out from the area due to the massive indiscriminate fishing activity.

The disturbance to the Park and its inhabitants was caused by various factors. Access to the reservoir was sought through a number of illegal routes to the Park. The fishermen would then spread their nets in the reservoir in the evenings and late at night to collect the catch early in the mornings. Having collected the catch in the mornings the fishermen would then use cycles and motorbikes to transport the same to the distribution centres using various routes through the Park. The fishermen used to set up temporary tents on the islands in the reservoir where they cooked their meals and littered the area with garbage and polythene bags. Their activity along the peripheral shallow areas also prevented vegetation along the coastline. Fishermen also illegally extracted timber from the Park area. There were even cases of areas of the forest being set on fire by these very fishermen and poachers in retaliation against action taken by the department. The water birds as well as the birds which used dead and dying trees and small islands in the reservoir as their roosting and resting sites were disturbed as the fishermen uprooted such dead and dying trees to clear their path for movement of their boats. (Personal communication, DD, PTR, SDO, Chinndwara district, PTR). The activities associated with illegal fishing thus had a negative impact on the feeding and migratory habits of the Park wildlife causing serious degradation of their habitat. The cause of conservation began facing a serious threat.

Legal Aspect

Indian legislation protects its Tiger reserves and National Parks under its Wildlife Protection Act and Forest Conservation Act. Despite this the traders, politicians and Forest Department officials had colluded to allow commercial fishing inside PTR.

In 1995 the then collector of Chhindwara district in his letter of 7th June 1995 addressed to the Secretary Forest Department, Government of Madhya Pradesh, in connection with the issuing of a final notification for the establishment of Pench National Park had asserted that displaced persons from the four villages, namely Palaspani, Umarighat, Chhindewani and Chhedia had traditional fishing rights in the Pench river. The letter reiterated that after displacement these people had not been rehabilitated systematically. No agricultural land had been made available to them, and they did not have any other means of livelihood except catching fish, which was their traditional occupation. He, therefore, recommended recognition of traditional rights of 332 families of these four villages. In view of these reports, the State Government stated on affidavit that it was satisfied that the traditional rights of the fishermen had not been settled. Instructions were accordingly given to the chief wildlife warden for issuing permission for fishing to 305 local fishermen, the number being derived from the electoral list of the above four villages.

In the same year the Animal and Environment Legal Defence Fund filed a writ petition in the Supreme Court challenging this decision.

Obstacles faced by the Forest Department in its Combat against Illegal Fishing

As the case was fought in the court, the conflict between the forest department and the fish mafia snowballed. There were incidents of transfer cases of officials who tried and opposed the political maneuvers of the illegal fishing network. There were also incidences of violence against the department staff on patrolling duty. The department faced numerous problems in checking the illegal fishing activity. The reservoir area is large and has common inter-state boundary with Maharashtra. The fishermen hailed from the local villages and possessed considerable knowledge of the different routes for accessing the reservoir in the Park. The long coastline of the Pench river made it easy for them to access the reservoir and difficult for the department with its small manpower to monitor and control the illegal acitivities. The limited resources and manpower of the department put a further constraint on their ability to discharge their duties. The involvement of big traders and unsocial elements in the fishing and trade activities also posed major problems.

Supreme Court Judgement

In its judgement of March 1997[3] the Supreme Court banned all fishing activities in the Totladoh reservoir. The decision, however, allowed the 305 fishing permits to be issued to the affected families as pleaded by the defendant, under tight restrictions, contingent upon settlement of the rights of the families through compensation.

Management Intervention

Despite the Supreme Court decision the illegal fishing in the reservoir continued unabated. The fish mafia continued to use violence and the 305 permit holders as a screen for their illegal activities. Many permit holders were forced to sell their permits to middlemen. It was manually becoming increasingly difficult for the staff to check the illegal activities.

In the year 2001, with the new field director and Assistant Conservator of Forests Chhindwara division assuming office, efforts to solve the problem of illegal fishing were undertaken with renewed vigour. The collector Chhindwara district with the help of a committee set up for identifying the actual right holders finally identified 145 families as having traditional nrights to fishing. Accordingly a compensation of Rs. 72 lakhs was decided upon to be divided among these 145 families, the division of which was made on the basis of the fish catch they had over the last four months. By 2001-02 the respective amounts were paid to the various families and their rights thus settled.

Thereafter the management began intensive efforts with careful planning and formulation of an action plan to check the illegal fishing. Detailed information of the illegal fishing and trade was obtained. The cooperation of the department staff at all levels was obtained through regular meetings and discussions. As per the action plan patrolling was intensified. A contour road to make access to the reservoir coast easier, was also constructed. As a step towards rehabilitating the affected families and gaining their cooperation a few of the fishermen were now employed for patrolling or as watchmen. These people helped the department track the route the fishermen used for accessing the reservoir and for leaving after having collected their catch. The vehicles like used in fishing were now caught and confiscated. Boats and nets were seized. In the year 2001-02, a total of 6 jeeps, 7 motorcycles, 60 cycles, 7 boats, 103 nets and a record catch of 38.16 quintal fish was seized.

A World Bank project was already underway in the 99 villages surrounding the tiger reserve. The development activities under this project were carried out through the village Eco-Development Committees (EDCs). The activities focused on empowering the socio-economic status of the local communities and reducing the biotic pressure on the environment. Accordingly the activities undertaken included distribution of pressure cookers, establishment of biogas plants, provision of LPGs at nominal rates, formation of grain banks etc. The landless were provided credit for taking up income generating activities like rope making and establishing tea stalls, small shops etc. These development activities helped the department in securing the goodwill and support of the villagers in combatting the illegal fishing and trade.

Conclusion

Continuance of the planned efforts and intensive patrolling beyond March 2002 helped the department gain significant control over the illegal fishing and trade. The department through its efforts has been able to achieve a complete cessation of the illegal fishing activities. The rehabilitation efforts for the fishermen have seen the employment of some of them with the FD for patrolling along the river side, or as watchmen in the guest houses. Those owning land have been encouraged to take up agricultural activities and the landless have been given money under the respective EDC to set up their own small business. Still others have been provided training in alternate employment activities like rope making. There has also been a suggestion for giving the genuine fishermen rights to fish in the ponds being constructed in the villages. Future proposals for the Park Management also include providing employment to these fishermen as boatmen and guides in the watersports activities planned for the Park.

In view of the commendable efforts in checking the illegal fishing activities the S.D.O Chhindwara division, was awarded the Amritadevi award which carries with it a medal and cash prize.

The results of the efforts directed towards controlling the illegal fishing activity is evident in an improvement in the bio-diversity status of the area. Villagers now admit to being able to see herds of Chital, Sambhar and Gaur collecting at the reservoir coast in the evenings and at dawn. Census records show an increase in the wildlife count. Furthermore there have been no reports of fire, illegal felling and poaching in the past one year. At the same time management plans to continue with its patrolling and rehabilitation efforts with the same dedication as shown in the last two years.

NOTES

1. *Source*: Primary data.
2. *EIA Report*: Fisth Mafia in PTR.
3. Supreme Court Judgement, 1997, Writ Petition (Civil) No. 785 of 1995.

22

Ecotourism as a Management Tool in Pench Tiger Reserve, Seoni, Madhya Pradesh

A Case Study

A.K.Bhattacharya, Shobna Boyle and *Samita Vasudevan*

The Park

Pench Tiger Reserve named after the Pench river, which flows from north to south, through the Reserve is located in the southern reaches of the Satpura hill range in the Seoni and Chhindwara districts of Madhya Pradesh. The terrain is undulating, with most of the area covered by small hills, steeply sloping on the sides. An extensive forest belt extends in three directions, north, east and south, covering forest tracts of Seoni, Balaghat and Nagpur districts.

The Reserve was created in 1992, becoming the 19th Reserve in the Project Tiger Network. The core zone of the Reserve, Pench National Park was created in 1983. This was carved out of the Pench Sanctuary created in 1977 with an area of 449.39 sq. km. The area of the Park is 292.85 sq. km., which divided almost equally in two districts—145.36 sq. km. in Seoni is all Reserve Forest whereas the area in Chhindwara includes 138.24 sq. km. Reserve forests, 6.26 sq. km. Protected Forests and 2.78 sq. km. revenue land. The buffer zone, 465 sq. km., comprises 333.05 sq. km. of R.F., 102.36 sq. km. of P.F. and 29.59 sq. km. of revenue land. The total area of the reserve is 757.86 sq. km. A dam has been constructed on the Pench river on the southeastern boundary of the Reserve.

Management Challenges

The Park was opened to tourists in 1983, but the number of tourists coming for recreation and other purposes inside the park was initially very meagre. This situation changed gradually and it was not until the late 90s that tourism started picking up. However, the influx was still miniscule compared to the tourism in other major national parks and tiger reserves. Insufficient availability of tourism infrastructure and other tourist facilities, non-implementation of interpretation programme and the absence of adequate publicity were some of the reasons for the dismal growth of tourism in the Park.

The park also faced challenges like grazing by cattle from surrounding villages. Grazing posed problems of adverse effect on regeneration, and the threat of disease spreading from domesticated cattle to the wild ungulates. The menace of illicit felling mainly for fuelwood also loomed large. The villagers on the other hand faced the problem of their crops being destroyed by the ungulates from the Park.

Another serious challenge was that of illegal fishing in the Totladoh reservoir on the Pench river. Construction of the dam was completed in 1990, but the fishing activity had begun since 1981 itself. The location of the reservoir inside the National park prompted a Supreme Court decision banning further fishing in the area on the basis of writ petition. In spite of this the illegal trade continued with the connivance of the traders, top officials and politicians. The adverse impact of fishing inside the park was reflected in an increase in the incidence of poaching and littering in the area with polythene bags left by the fishermen. Efforts to check the illegal fishing caused conflict between the department staff and the fishermen which sometimes escalated into violence. The Park also faced the threat of fires caused by angry villagers or the fishermen. A holistic approach with coordination of efforts of the concerned departments was necessary to create the optimum set of inputs required to ensure the success of ecotourism.

Management Interventions

Efforts Towards Park Promotion

In June 2001, with the new field director of the park assuming office, efforts to promote the Park as a prime tourist destination were undertaken with renewed vigour. A brief research into the history of the area and perusal of books and pamphlets available on the region revealed that the Seoni region is the original setting of the famous work of Rudyard Kipling '*Jungle Book*'. Rudyard Kipling borrowed heavily from Robert Armitage Strendale's books—'*Seonee*'. '*Mammalia of India and Ceylon*' and '*Denizens of the Jungle*' for descriptions of the topography, the wildlife, and its ways. Mowgli was inspired by Sir William Henry Sleeman's pamphlet, '*An account of wolves nurturing Children in their Dens*' which describes a wolf-boy captured in Seoni district near the village of Sant Bori in 1831. Many of Jungle Book's locations are actual locations in Seoni District like Waingunga river, Kanhiwara village and the Seonee hills. The mass popularity of the character of '*Mowgli*' was consequently sought to be used to the advantage of the Park. PTR has since been promoted as '*Mowgli Land*' not only through hoardings and advertisements in travelogues but also in various articles in leading magazines like '*India Today*' and T.V. channels like '*Aaj Tak*' that have featured PTR. The result is evident in the increase in tourists from 5,288 in 2000-01 to 10,488 in 2001-02. The rise in tourist numbers can be attributed to permission of access to diesel vehicles, publicity by the media and the improved infrastructure.

Involvement of the Private Sector

The increase in tourists visiting the Park prompted improvement in facilities and development of new facilities. Roads, guest house, dormitories were maintained in good condition to cater to the demands of the tourists. With the increased publicity, the Park attracted the attention of private hoteliers and resort onwners who soon set up their establishments near the Park entrance. There are at present three such tourist complexes operating in the area, one of these owned by the M.P. Topurism Board, and a few more proposed to be set up. These private hoteliers brought with them their contracts with travel agents and tour operators in major tourist units like Delhi and Mumbai. So far the tourists visiting the Park had been mainly from the nearby areas especially Nagpur and Jabalpur. The involvement of the private sector, could lead to an increase in the promotion of tourists from the rest of India and abroad and a subsequent increase in the Park income. Another pertinent aspect of the involvement of the private sector was the generation of income for the local people. The presence of the tourist complexes also has encouraged income generating activities like poultry farming and cultivation of vegetables on substantial scale in the nearby villages by ensuring a ready market for the produce of the same.

Interventions for Park Improvement

To maintain and improve upon the tourist situation special attention has been focused on the needs of the Park and the tourists. For better Park management various management interventions have been undertaken. Considerable planning and resources has been devoted to water management inside the park. This is in the interests of the animals as well as for better opportunities for animal sightings for the tourists. In addition to waterholes the park earlier had a system of bore wells, whereby water would be pumped out manually and filed in the surrounding trench for animals to drink from. The Park also had concrete saucers, in which water would be filled from tankers. One of the main objectives of ecotourism is conservation of biodiversity in its natural state. Efforts are, therefore, concentrated on providing a natural habitat for the animals to as great an extent as is possible. It was perceived that the system of handpumps and concrete saucers was incongruous with the surrounding widerness and aesthetics of the forests.

Moreover with rising temperatures the water stored in the concrete saucers used to become too hot for consumption by the animals. The management, therefore, decided to gradually phase out the system and replace it by constructing earthen ponds (*talab*) and shallow wells (*jhiriyas*). A number of these facilities have been provided after careful selection of the site taking into account aspects like slope of the area, access for animals, proximity of the area to some natural source of water, etc. The construction was undertaken by employing the services of labourers from the nearby vilalges. The result is evident in lesser reports of migration of animals from their areas and increase sightings of animals. The Park is now constructing an interpretation centre in collaboration with the Centre for Environment (CEE) to spread awareness and sensitise the people towards conservation. To cater to the demands of the higher income segment the Park has also constructed new accommodations designed to blend with the surrounding wilderness.

Combating the Challenges

The problem of illegal fishing was solved to a significant extent when the management undertook intensive efforts at all levels to combat the threat posed by the fish mafia. Comprehensive data on the system of illegal fishing and trade was obtained. The cooperation of the department staff and personnel at all levels was sought through meetings and discussions. The cooperation of the village ecodevelopment committees was also solicited. After complying with the Supreme Court Order of compensating the families affected by the ban, patroling along the reservoir was intensified and the motor boats, cycles, motorcycles and other vehicles used in the trade were confiscated. Efforts towards rehabilitation of the affected families saw 25 of the fishermen being employed in patrolling activities, and as watchmen. The impact of the resultant cease in illegal fishing was reflected in an increase in the flora and fauna count, which had seen considerable disturbance due to the fishing activities. The vacated areas are now covered with grasses improving the aesthetics of the area and attacting large herds of herbivores.

Community Participation and Development Activities

An essential component of ecotourism is the participation and socio-economic development of the local communities. Efforts towards Park development and conservation cannot be successfully undertaken without soliciting their participation, since they are one of the main stakeholders. EDCs in Pench had been formed way back in 1997-98 but not much activity had been undertaken by them. The department still lacked support from the local people. Gradually, however, the situation began changing as development activities began to be implemented. The management undertook activities like distribution of pressure cookers to all households under EDC, distribution of LPG at nominal rates (25% of the actual rate), setting up of biogas plants, introduction of training centres for women which trained them in sewing and making papads agarbatti, etc. The EDCs further led to the formation of Self-Help Group (SHGs) which has undertaken activities like formation of grain bank, Diary Cooperative etc. There is also a scheme whereby the landless are given money and training to take up activities like rope making, setting up of tea stalls etc. These activities provided the people with income generating sources thus reducing the biotic pressure on the forests. The introduction of bio-gas facilities and LPG saw the fuelwood demand fall drastically (65-70%)[1]. The EDCs also undertook construction of ponds and game proof walls to prevent the Park animals from crossing over and destroying their fields. The construction was undertaken by employing labourers from the villages on a rotation basis. The youth from the villages adjoining. The Park have been engaged as guides to benefit from their knowledge of the area while generating employment. The official canteen and the guest houses inside the Park also provide employment to the local people. The EDCs in some villages have undertaken pasture development for grazing and ponds for providing water for their livestock. This has considerably reduced the grazing pressure inside the Park and dissuaded the cattle from entering the Park in search of water. The success of these activities can in part be attributed to the cooperation and coordination achieved between the various government departments which was in itself a result of the convergence of their schemes.

Encouraged by these development activities people now are forthcoming to provide cooperation to the department in its conservation efforts.

To give due recognition to the efforts made towards conservation and park maintenance, rewards have been given to the concerned department employees and villagers. This has boosted their morale and secured their commitment to conservation.

Apart from the activities undertaken through the EDCs, the Park management has also involved itself in health care activities. The villagers are given basic information on nutrition and hygiene. Medicinal plants with instructions on its dosage and effects, are provided to the villagers. Most instance the herb Chiraita is distributed among the villagers during outbreak of malaria. Diseases like rheumatism and juandice are also sought to be similarly contained. This has resulted in a significant control of the disease in the area. The provision of such health care has also helped reduce the expenditure of the households on medicines.

Future Proposals

Attempting to further explore the untapped potential of the Park the future proposals for the Park envisage a number of activities. The increased tourist influx to the region has made crowd management a crucial aspect of the overall Park management. Diversification of activities is perceived to be one of the effective tools for this purpose. Introduction of activities like boating, rafting, trekking through Nature trails developed in the buffer zone are part of the future proposals. The proposed natural trail would give an opportunity to the tourists to get a feel of the jungle and educate and sensitize them towards the concerns of the Park. Development of eco-camps would provide a platform for showcasing and marketing the local arts and handicraft. Tourists would also be witness to the local culture and folk songs and dances of a few selected villages. This centre would provide an opportunity for the local people to share and earn from their indigenous knowledge, while allowing the tourists to gain an insight into the same.

It is believed that the introduction of these activities would increase tourist satisfaction thereby leading to extension of their stay, generate local employment and subsequently increase the income of the Park and the private sector. Ecotourism, as it has developed in PTR in the recent years, has been able to conserve and enhance the bio-diversity of the area while also generating local participation in conservation efforts. The economic incentives created from tourism have helped decrease the biotic pressure on the Park. The future proposals seek to maintain and improve upon the current situation.

NOTE

1. *Source*: Primary data.

23

Environmental Auditing in Ecotourism

A Study on Visitors' Management in Van Vihar National Park, Bhopal, MP (India)

A.K. Bhattacharya, Vandana Saksena and *Suchitra Banerjee*

Introduction

Tourism can bring both benefits and problems to an area. If well planned, developed and managed, tourism generates local jobs and income and provides opportunities to local entrepreneurs to establish tourism enterprises, all of which lead to improved living standards of residents. Tax revenue generated by tourism can be used to improve community facilities and services.

Tourism requires improved infrastructure, which can be used by residents. Tourism can stimulate the expansion of other economic activities such as agriculture, fisheries, manufacturing and craft production.

Tourism can provide the justification and help, pay for conservation of local natural areas, archeological and historic sites, arts, crafts, and cultural tradition and overall improvement of environmental quality because these are attractions for the tourists. However, if not well planned, developed and managed properly, tourism can result in congestion, pollution and other environmental problems (Bramwell, 1990[1]; Buhalis & Fletcher, 1995[2]).

Uncontrolled use by the tourists of nature areas and historic sites may lead to their deterioration. Over commercialization of cultural traditions can result in degradation of the areas cultural heritage (Bhattacharya & Banerjee, 2003)[3].

Tourism and environment are closely interrelated. The natural and built environmental provides many of the attractions for the tourists and tourism development can have both negative and positive impacts on the environment. Sustainable tourism development depends on protecting the environmental resources for tourism (Boo, 1991[4]; Butler, 1991[5]).

The partners for sustainable tourism development are the tourism industry, owners and managers of tourism commercial enterprises, the environmental supporters, advocates for environmental conservation and the community residents, community groups and leaders and the local authority.

The interaction among these partners are necessary to achieve improved quality of life for the community while still achieving conservation of environmental resources for tourism and reasonable profits for the tourism industry.

Keeping these points in view, a new kind of tourism has come into being and is known as Ecotourism. Thus, Ecotourism is a responsible travel to natural areas that conserves the environment and sustains the well being of local people.

Ecotourism

The term Ecotourism was coined by Hector Ceballos Lascurain in 1983. The term was used to describe the nature-based travel with emphasis on education, management and development of sustainable tourism product and activity.

World Tourism Organisation has defined Ecotourism as "tourism that involves travelling to relatively undisturbed natural areas with specified object of studying, admiring and enjoying the scenery and its wild plants and animals, as well as any existing cultural aspects (both of past and present) found in these areas".

The Ecotourism Society (TES) defined Ecotourism as "responsible travel to natural areas that conserves the environment and sustains the well-being of the local people". Its main aims are ecological and socio-cultural integrity, responsibility and sustainability (Cater, 1994)[6]. Ecotourism has developed from the traditional nature tourism and sustainable development concepts.

Principles and Characteristics

- It should be small scaled, locally controlled, land sensitive and ecologically oriented.
- It should be based on environmental ethics with a concept of small is beauiful, doing more with less and treated with fostering resource integrity.
- It should be sustainable environmentally, socially, culturally and economically.
- It should enrich experience and should be educational.
- It should have the participation of the local communities.
- It is bio-centric rather than homocentric in philosophy. In fact Ecotourism should accept nature largely on its terms and not trying to significantly transform nature for personal convenience.

Ecotourism is emerging as an important component of the Indian tourist industry in general and the Madhya Pradesh tourism in particular. The significant growth in nature tourism and the numerous tourist operators bear adequate testimony to this. Though the term "Ecotourism" is popular, many rather loosely use it. Hence, it is imperative to distinguish this from general mass tourism.

Ecotourism has been considered here as sustainable, equitable, community based endeavour for improving the living standards of indigenous host communities. Marketed as a form of nature based tourism. It also serves as a sustainable development tool. Ecotourism or nature tourism is distinguished from resort tourism or mass tourism by requiring lesser infrastructure development and a low impact on the environment (Bhattacharya *et al.,* 2003)[7].

Environmental Auditing

Concern Regarding Environmental risk has led to the development of the field of Environmental Auditing.

Environmental auditing is a way of checking whether a multitude of new environmental laws and regulations are being complied by the organisation, company or protected areas. More recently, it is used as a tool for assessing environmental management systems, policy, and equipment.

The International Chamber of Commerce defines an environmental audit as "a systematic, objective method of identifying and verifying the laws, regulating procedures and guidelines for protection and conservation of environment, health, occupational hygiene, safety, and emergency preparedness standards being followed."

The International Organisation for standardization (ISO) defines environmental audit as "a systematic, documented verification process of objectively obtaining and evaluating audit evidence to determine whether specified environmental activities, events, conditions, management systems, or information about these matters conform with audit criteria, and communicating the results of this process to the concerned party."

Environmental Audit does not stop at compliance with legislation. Nor is it a green washing public relations. Rather, it is a total strategic approach to the organization activities.

Environment auditing is a management tool compromising a systematic documented, periodic and objective evaluation of how well the management systems are performing with the aim of waste prevention and reduction, assessing the compliance with regulatory requirements, facilitating control of environmental practices by a Company's Management, and placing environment information in the public domain.

Environmental Auditing is which is described as "means to an end by which environmental considerations are integrated into tourism business decisions, and, is therefore highly suited to ecotourism (Brown, 1994[8]; Buckley, 1990a[9]).

Relevance of Environmental Audit

Environmental audits are excellent indicators of performance, allowing companies to assess performance and its impact on the environment and identify ways to increase efficiency in environmental performance (Goodall, 1995[10]; Wells, 1995[11]). Audits also verify compliance with relevant environmental legislation and corporate policy and minimise human exposure to risk (Goodall, 1995)[10]. They can be used as benchmarks of management systems, such as the Environmental Self Assessment Programme (ESAP) developed by the Global Environmental Management Initiative (GEMI), which provides a structured format for an organisation to compare its management with the principles of sustainable development established by the International Chamber of Commerce (ICC) (Cahill, 1995[12]; Fitzgerald, 1995[13]; Wells, 1995[11]).

Furthermore, the existence of an environmental auditing programme not only boosts consumer and investor confidence, but can also help reduce insurance premiums and provides evidence of due diligence. For the Environmental Impact Assessments (EIA) required in projects of environmental significance monitoring should be conducted in order to assess the impact of developments on the environment. Monitoring of this type may be conducted within the framework of an environmental audit which would not only ensure that monitoring is being conducted, but also recognise if any limits were being exceeded (Brown, 1994[8]; Buhalis, 1995[2]).

Integration of Environmental Auditing into Ecotourism

The World Travel and Tourism Council regard on-going environmental audits of tourism activities as an integral management tool. The concept of sustainable tourism implies sustainable practices at two levels. First, there is the need to secure the sustainability of tourism's primary resources at the destination level, and secondly, in a wider global context, the need to ensure the practice of environmentally sustainable tourism (Goodall, 1995)[10].

Tourism is generally perceived as a private sector activity, so environmentally sustainable practices are viewed as the responsibility of individual firms or departments. Moreover, there are increasing pressures upon companies to act in an ethical and responsible manner towards the environment.

Environmental auditing has also been recognised for providing the means to assess and identify areas where compliance and efficiency could be increased and ensuring sustainable practices. Sustainable tourism and in particular the tourism industry must take action in order to embody the tenets of sustainable development. Environmental auditing is a pragmatic tool for tourist operations to use with the goal of attaining sustainable development (Stabler 1995)[14].

Despite the benefits of applying environmental auditing to tourism having being recognised, there has been no recognition of the potential benefits that environmental auditing may bring to ecotourism. The only exception to this is Best Practice Ecotourism, produced by the Australian Commonwealth Department of Tourism in 1995.

Integration of environmental auditing into ecotourism is a significant and necessary step that needs to be taken by operators in reaction to both internal and external forces. Internal forces include corporate policy, staff influence, community concern for the environment, marketing advantages and the prospect of financial savings resulting from environmental auditing programmes. External reasons why environmental auditing should be incorporated into ecotourism include codes of conduct set by industry associations, accreditation schemes and government rules, regulations and legislations (Cater 1996[6]).

The use of environmental auditing in the ecotourism industry was also being recognised as important as it allows ecotourist operations to reduce environmental impacts, compare environmental performance whilst at the same time providing clients with a better understanding of an operation's commitment to the environment. This may in turn result in marketing advantage. Other benefits of applying it to ecotourism included the fact that environmental auditing is cost efficient and provides suggestions for how to improve environmental management (Buckley, 1990a[9], 1990b[15], 2001[16]).

The incorporation of environmental auditing into the ecotourism industry, particularly through environmental management standards such as ISO-14000 is important as it allows the cumulative and long-term aspects of environmental impacts that may be incurred by ecotourism to be addressed. The ISO-14000 series is one of several industry responses to the increasing public interest in sustainable development that came in the wake of 1992 RIO Earth Summit. While ISO-14000 is a family of standards. ISO-14001 contains an EMS standard against which a business, regardless of its size, product, service or sector, is certified. Certification to ISO-14001 means that a company's environmental management system (EMS) confirms the specifications of the standard, as verified by an environmental audit process.

It may be noted that ISO does not do auditing: it simply facilitates the development of EMS standards. It includes requirement for public participation, corporate disclosure of its environmental statements, and compliance with government laws and regulations (Krut & Gleckman, 1998)[17].

Environmental auditing may play an important role in determining whether operations are *bona fide* ecotourism. According to the Ecotourism Strategy the three key elements of ecotourism are that it is nature-based, involves education and is operated on a sustainable basis. Environmental auditing plays an important role in determining whether the third key element of ecotourism, sustainable practice, is being applied and therefore, whether or not the operation may be labelled as ecotourism (Blamey, 1996)[18].

Based on the above definitions and taking into account the impact of environment, Van Vihar National Park has lot of potential for Ecotourism development which has yet to be realized in its right perspective. This unique and smallest national Park with its richness in fauna and flora and near pollution free environment has provided an excellent opportunity to travel loving people a wonderful destination for Ecotourism to enjoy experience, understand and appreciate nature and the need to conserve wild-life (Bhattacharya & Saksena, 2003)[19].

Objectives of the Study

For the present study environmental auditing has been used as a potential tool in ecotourism with special reference to assess the visitors' management in Van Vihar National Park, Bhopal (Bhattacharya and Saksena, 2003)[19].

This study for Visitors management in Van Vihar is relatively a new concept of Ecotourism in the light of the assessment whether various indicators stand the test of codes of conduct of Ecotourism, environmental standards pertaining to air, water, noise and solid waste pollution (Jim, 1989)[20].

The specific objectives were to:

- examine the existing codes of conduct, rules and regulations, acts, circulars guidelines for ecotourism in the Park.
- evaluate the strengths and weaknesses of these regulations/ legislations.
- assess the awareness of visitors about these regulations/ legislations.
- assess the expectations of visitors for the Park.

Methodology

- Collection of primary data in a structured questionnaire format.
- Collection of secondary data from records available for the purpose of verification.
- Survey– focus group discussion and individual with the visitors and other stakeholders.
- Monitoring of ambient air quality and noise pollution using HVS and Noise meter.
- Quantification and qualification of solid waste.
- Analysis of data to examine the impact of Ecotourism in the Park.

The data regarding the visitors' response have been collected through a questionnaire survey format. The information were collected through

— Questionnaires directly supplied to visitors.
— Direct observations of the visitors' activities (spot survey).
— Questionnaire survey in individual and in-group.
— Questionnaires supplied to the forest staff.

Study Area

Van Vihar, the smallest National Park of the country is situated in Bhopal, the capital of Madhya Pradesh which lies in the eastern edge of Malwa Plateau and situated in the central part of state. The name "Bhopal" has been derived from "Bhojpal" or "Bhoja's dam" the great dam which now holds up the famous Bhopal city lakes, and is said to have been built by a minister of Raja Bhoj, the Parmara ruler of Dhar. Initially the lake was quite big but as time elapsed only a small portion of it has remained to be seen as the upper lake popularly called *"Bada Talab"*, which is now used as one of the main sources of drinking water to the urban population. On the other side, there is the lower lake popularly called *"Chhota Talab"* which is used for bathing, washing and other miscellaneous purposes. Bhopal enjoyed a status of capital till the state was merged in Indian Union in November, 1956 to form the new State of Madhya Pradesh.

Established in the year 1983, Van Vihar National Park is not only the smallest National Park but is also a unique one in the world as it is spread over a small area of only 445 hectares of degraded and denuded hillock enclosed with fencing adjoining the populous city of Bhopal. Most of the Park area is hilly and undulated varying from gentle to steep slopes. Around 65-75 per cent of the Park area is covered by hills and plateau; rest of the terrain is plain.

Though provided with legal protection of a National Park, the area has been developed on the line of a modern Zoological Park, where wild animals have been kept in near natural condition of space and habitat setting to serve as a repository of genes. The landscape of the area has led to the designing of enclosures in a fashion that serves the objective of fostering a positive man-animal relationship.

Van Vihar National Park, a fertile place of greenery, is a living example that reflects this capacity of Nature. The nature has risen to live for another cycle. It has tremendous capacity to bring itself back into use. The adjoining upper lake adds immense beauty to its landscape. The lake enlivens particularly when thousands of migratory birds land in the lake at the onset of winter providing tremendous joy of the glories of nature to the visiting tourists and visitors.

Results and Discussion

In all 175 questionnaires were supplied to the visitors entering to the Park. They were requested to co-operate in giving proper replies to the questions.

The demographic information about the visitors such as their age, sex, education, profession, occupation, place of origin and their income profile have been presented.

Age and Sex Ratio

75 per cent of the visitors to the Park are in the age group of 18-30, followed by 62 per cent in the age group of 35-50. 32 per cent visitors in the age group of 50 and above come to the Park. The minimum arrivals is of the visitors below 18 years. Seventy-two per cent of males and nearly 30 per cent females visit the National Park.

Occupation

The result of visitors' survey indicates that 20 per cent of them are government employees belonging to teaching, LIC, Banking sectors. Twenty-four per cent are from the private sector companies, while 16 per cent are businessmen who are involved in export and in travel business. Remaining 40 per cent are students, housewives and retired persons who are frequent visitors to the Park as told by them personally.

Qualifications

Of the total visitors coming to the Park eleven are educated below 12th standard, most of them are students, and forty-three per cent are graduates and postgraduates. Thirty-five per cent visitors were found to be technically and professionally qualified including engineers, diploma holders, doctors, lawyers, chartered accountants etc.

Income

While 71 per cent visitors responded to the queries of income, 29 per cent were reluctant to respond. Of the 71 per cent, 56 per cent come from income group of more than Rs.1 lakh annually, while 11 per cent come from the income group of Rs. 50,000-1,00,000 per year. A very small section of visitors (four per cent) belong to the income group of less than Rs. 50,000.

Mode of Transport

The survey indicates that 46 per cent people visit Van Vihar in two wheelers, i.e. scooters, motorbikes and mopeds, thirty per cent visit in four-wheelers, while about 12 per cent visitors come in three wheelers (auto-rickshaws). Of the remaining 12 per cent, 8 per cent enjoy the visit inside the Park in Van Vihar owned battery operated bus and only 4 per cent see the Park walking without using any vehicle. The battery-operated bus provides an environmental friendly atmosphere to the visitors.

Visitors' Perception: The visitor's perception was obtained on the basis of their interest towards ecotourism.

Purpose of Visit: Response to the purpose of visit was based on ranking system. The result indicates that 88 per cent people come to Van Vihar to view wild-life. Eighty per cent gave priority to love for nature. Rest go to Van Vihar for scenic beauty followed by outing and close by picnic spot. The local residents of Bhopal specifically bring their guests, relatives and friends to the Park for showing wild animals, birds, reptiles and the scenic beauty of hills and the lake. This occupies the last rank of the purpose of visit under tourist perception.

Number of Visits Per Year: The sample data indicate that 42 per cent of the people visit Van Vihar once in a year, followed by 22 per cent twice a year, 6 per cent thrice a year, and 4 per cent four to six times per year. Ten per cent local people visit this place more than 10 times a year. Only 2 per cent visitors could not respond properly.

What Fascinates the Visitors Most?: On the question of liking raised to the visitors, they gave priority to wild life, followed by natural beauty, lake view, surroundings of the area etc. The need of a well equipped interpretation centre was felt by most of the visitors in view of getting proper information about Van Vihar and the wild-life.

Awareness about Ecotourism: Only 48 per cent visitors responded in positive about awareness towards Ecotourism. According to them "Ecotourism is travelling to nature, tour to ecology, visiting natural places having beautiful landscape, helps awareness about ecotourism in developing human concern to protect nature and ecosystem, love towards nature and nearness to its surroundings which includes the attraction of wildlife". For some it may be a good source of earning foreign exchange by attracting tourists from abroad. Significantly 44 per cent of the visitors had no idea of ecotourism, while remaining 8 per cent did not show any interest in responding to this question.

Awareness about the Codes of Conduct of Ecotourism: From the survey it was clear that very few people had knowledge of codes of conduct of ecotourism. According to them codes of conduct comprise the rules and regulations of Wildlife Protection Act, Environment Protection Act pertaining to air, water, noise and solid waste, non-disposal of plastics and polythene bags, littering, noise free environment, and promotion of plantation. Majority (60%) of the people do not have any idea about such codes. Only 10 per cent visitors are aware about the codes of conduct of ecotourism. Thirty per cent did not respond to the question.

Guidelines and instruction by the Park staff: Forty-eight per cent visitors confirmed of having received the guidelines/instructions regarding tourism in the Park, while 26 per cent people strongly denied. Remaining 26 per cent did not respond.

Food Brought by Visitors in the Park: In reply to the question "what do you bring to eat during your visit?" Forty-six per cent replied they bring fast food. Eight per cent people bring cooked food while 43 per cent visitors use containers to bring eatables.

Use of Wrappers /Carry Bags/Polythenes: 46 per cent visitors who bring cooked and fast food said that they put the left over food and used paper, polythene etc. into the dustbins. 6 per cent throw these waste material by the side of the lake, 12 per cent carry them back. The remaining 34 per cent throw them here and there or anywhere.

Facilities in Van Vihar: Sixty-six per cent visitors feel that the facilities inside the Van Vihar are inadequate, 26 per cent are satisfied with the adequacy of facilities and rest 8 per cent did not respond.

Guide: Ninety per cent visitors say that they do not get any guide who could guide or instruct them properly about the Park, ecotourism and wild animals. Only 5 per cent visitors told that some Park staff did guide them a little bit about the wild animals and natural surroundings. Five per cent did not respond.

Willingness to Pay for the Guides: The survey indicated that the visitors felt great need of guides in the Park. 75 per cent visitors were willing to pay any fee that may be charged by the Park management for making the guides available during their visit. Only 22 per cent visitors showed reluctance for payment to guides. Three per cent did not respond.

Awareness Generation: According to 78 per cent visitors, Van Vihar is a proper site of awareness generation as well as getting education about wild-life conservation. Of the remaining people about 18 per cent do not think so. They come only for recreation purpose or just for outing. Rest of the people did not respond.

Information from the Interpretation Centre: Of the total people visiting the Park, 18 per cent visitors get adequate information from the interpretation centre known as "Vihar Vithika"; 54 per cent get partially adequate information, while 24 per cent are totally unsatisfied from it. They complained that there was none to either welcome or guide them or provide useful and interesting information about the Park.

Involvement of Public in Park Management: Most of the visitors (about 80%) showed their interest in Park management if they were given chance to do so. Some people did not show any willingness to involve themselves in Park management. Only 10 per cent did not respond.

Merits and Demerits of the Park

Merits	*Demerits*
Hilly and undulated natural beauty	Ticket cost is high
Good wildlife	Inadequate drinking water kiosk
Near to Lakeview	Absence of guide
Close by picnic spot	Absence of a canteen
Water beauty—an additional attraction	Poor maintenance of wild-life
Calm and quite atmosphere	Air and Noise pollution due to entry of vehicle inside the Park
Neat and clean roads	Very shabby and dirty toilets
	No proper resting and sitting places

Facilities

The visitors suggested the following facilities in ranking order in the Park:

S.No.	*Facilities in Order of Ranking*
1.	Trained guide
2.	Proper drinking water facility at short distances
3.	Need of eco-friendly canteen in the Park

(*Contd.*)

(*Contd.*)

S.No.	Facilities in Order of Ranking
4.	Camera should be allowed inside the Park free of cost
5.	Battery operated bus fare should be reduced
6.	A cyber café and telephone booths should be introduced
7.	Pictures of wild-life should be put on sale

Suggestions

Suggestions and opinions of the visitors towards conservation values and improvements in the Park are mentioned below in following table:

S.No.	Suggestions by the Visitors for Better Management of the Park
1.	A money collection box should be kept in the Park for collection of the wildlife fund so as to enable the visitors to donate or contribute generously towards it.
2.	Nature camps for awareness generation should be organized on a regular basis.
3.	In order to earn foreign exchange proper publicity be made to attract foreign tourists towards ecotourism attraction in the Park.
4.	The permission fee to carry cameras should not be charged.
5.	Signboards and hoardings should be put up at prominent places inside the Park carrying useful instructions and messages displayed properly in Hindi and English for the visitors.
6.	An officer of a senior rank from the Forest Department should keep regular vigil over the activities and quality of food supplied to the wild animals.
7.	Fare of the battery operated bus should be reduced by 50 per cent.
8.	The present visiting time to the Park should be increased by an hour in the afternoon.
9.	Educated trained guides should be appointed in the Park.
10.	Forest staff should take daily action for progress and maintenance of Park and up keeping of wild-life.
11.	A flower garden should be developed by the side of the lake.
12.	Diesel vehicles should be strictly prohibited to enter the Park.
13.	The number of dustbins should be increased and distance between two bins should be conveniently close.
14.	The toilets of the Park should be cleaned daily and properly.
15.	There should be resting huts for visiting public.
16.	The Park should remain open for visitors through out the day.
17.	There is need of a canteen in the Park.
18.	There should be audio-video film shows on wildlife nature attractions including the lakeside beauty.

Perception of the Park Staff

Ecotourism Codes of Conduct

Only 25 per cent staff have some idea about the codes of conduct pertaining to Ecotourism but their understanding only confines up to "do's and don't do's" about the wild animals and vegetation.

Departmental Circulars and Guidelines

Regarding the departmental circulars and guidelines and Park management, 62 per cent have no idea about any circulars and guidelines concerning ecotourism management. Only 38 per cent have some idea about them. When asked about attending the workshops and seminars on ecotourism most of the staff members replied in negative. A very small section of senior staff members attended some or the other sort of training and seminars related to ecotourism.

Expectations from the Visitors

About 62 per cent staff expect the visitors to follow rules and regulations as well as instructions about Van Vihar besides keeping the Park clean and green without spreading dirt here and there. Responsible staff (37.5%) wants visitors to co-operate with them with decency and decorum without creating scenes by entering into avoidable discussions. About 12 per cent expect that people visiting the Park should not make loud noise and avoid littering while moving. They also expect that the visitors should not approach nearer to the animals or tease them. Instead, they should help conserving the wild life besides promoting tourism.

Potential Threats to the Environment of Van Vihar

According to the Park staff the potential threats of environment are polythene bags, police firing range near the Park, horn blowing and noise pollution, diesel vehicles, smoking inside the Park, *Gutka* pouch, intoxication, growing of *Parthenium* and Lantana, animal abuse by the visitors, littering and rash driving.

Suggestions to Promote Ecotourism in Van Vihar

The forest staff suggest that ecotourism can be promoted in Van Vihar national Park by plantation, resting place, safe drinking water, slide show, up-to-date website of Van Vihar National Park, special tourist buses, co-ordination with state tourism corporation for the promotion of ecotourism, creation of ecotypes, trained guides, advertisement through TV, newspapers and hoardings.

Pollution—Ambient Air and Noise Monitoring

A. Location—White tiger enclosure
Date—7-10-02: Time (3 pm to 7:15 pm)
Peak time (Open to visitors)

	5.25 pm	*5:30 pm*	*5:35 pm*	*5:40 pm*	*5:45 pm*
Noise level (dB)	78.3	92.5	88.5	77.3	94.9
SPM (µg/ m^3)		74.2			
SO^2 (µg/ m^3)		7.0			
NO_x (µg/ m^3)		13.7			

B. Location—Yellow tiger enclosure
Date 7-10-02: Time (3 pm to 7:15 pm)
Peak time (Open to visitors)

	6:07 pm	*6:10 pm*	*6:13 pm*	*6:15 pm*	*6:17 pm*
Noise level (dB)	72.6	87.8	90.8	89.9	87.5
SPM (µg/ m^3)		80.2			
SO_2 (µg/ m^3)		7.0			
NO_x (µg/ m^3)		13.7			

C. Location—Yellow tiger enclosure
Date 13-10-02: Time (3 pm to 07:15 pm)
Peak time (Open to visitors)

	4:35 pm	*4:37 pm*	*4:40 pm*	*4:42 pm*
Noise level (dB)	61.4	54.6	74.2	57.1
SPM (µg/ m^3)		101.2		
SO_2 (µg/ m^3)		7.0		
NO_x (µg/ m^3)		15.5		

D. Location—White tiger enclosure
Date 13-10-02: Time (3: pm to 7:15 pm)
Peak time (Open to visitors)

	5:47 pm	*5:50 pm*	*5:52 pm*	*5:55 pm*	*5:57 pm*
Noise level (dB)	62.7	62.2	62.8	60.6	62.4
SPM (µg/ m^3)		101.2			
SO_2 (µg/ m^3)		7.0			
NO_x (µg/ m^3)		15.5			

Interpretation of Pollution Analysis Results

On the basis of the readings obtained in pollution monitoring as shown in the tables above the following interpretations can be drawn:

1. Noise level exceeds the prescribed limits as per above tables. Higher level of noise pollution may affect the behaviour of the animals.
2. SPM level marginally exceeded the prescribed limits as shown in table of location B on 13-10-02, due to the influence of the visitors on holiday. Number of vehicles was also comparatively high.
3. Excessive dust in the environment may affect the photosynthesis of the macro-vegetation.
4. Dust in the environment and exhaust from the vehicle may affect the respiratory system of the visitors as well as the inhabitants.
5. Concentration of gases, viz. SO_2 and NO_x was observed within the prescribed limits.

Conclusion

Although a small National Park in size, Van Vihar has quite a good potential for Ecotourism development no matter it be on a small-scale level. It has been observed that the pollution level inside the Park and immediate adjoining areas is well within the prescribed limits. However, for the purpose of waste disposal there is need of environment friendly energy and waste management technologies for which a small incinerator in accordance with CZA guidelines should be installed. An Ecotourism Management Plan involving the public in Park management including indigenous people, women, youth, businesses, scientific communities and NGOs at all levels of decision-making should be prepared. Management strategies within the framework of National and State Wildlife Action Plans and Tourism Policies of the Central and State Governments should also be formulated. Besides, it will be worthwhile to set up a snake Park and a flower garden near the lakeside. An effective interpretation centre should be developed and supported by knowledgeable persons to provide appropriate information to visitors in English as well as in Hindi on "what to see and where to see", "dos and don't dos" and ecotourism codes of conduct.

NOTES

1. Bramwell, B., 1990. Green Tourism in countryside. *Tourism Management*, 11 (4): 358-360.
2. Buhalis, D. & Fletcher, J., 1995. Environmental impacts on tourist destinations. In: *Sustainable Tourism Development*, Eds. Coccosis, H. & Nijkamp, P., Avebury, Sydney. pp 3-24.
3. Bhattacharya, A.K. & Banerjee, S., 2003. Relevance of CC and Eco-development linkages for SET. *Indian Forester,* 129 (3), 332-340.
4. Boo, E., 1991. Making Ecotourism Sustainable: Recommendations for Planning, Development and Management. In: *Nature Tourism: Managing for the Environment*, Ed. Whelan, T., Island Press, Washington, D C. pp 187-199.
5. Butler, R.W., 1991. Tourism, environment and sustainable development. *Environmental Conservation*, 18 (3): 201-209.
6. Cater, E., 1994. Introduction. In E. Cater and G. Lowman (Eds.) *Ecotourism: A sustainable option?* (pp. 3-17). Chichester: John Wiley.
7. Bhattacharya, A.K., Banerjee, S. & Saksena, V., 2003. "Local initiatives to localize Ecotourism—An Exploratory study in Kerwa-Van Vihar National Park Catchments. *Tourism Recreation Research*, Vol 28(1) 2003.
8. Brown, G., 1994. *Environmental Audit Guidebook.* Centre for Professional Development, Kew, Victoria.
9. Buckley, R.C., 1990a. *Environmental Audit: What Directors and Managers need to know?* Environmental Auditing. International Business Communications, Sydney.
10. Goodall, B., 1995. Environmental Auditing: a tool for assessing the environmental performance of tourism firms. *The Geographical Journal*, 161 (1): 29-37.
11. Wells, R.P., 1995. Auditing for Compliance is Only the Beginning: Lessons from Leading Companies. In: *Auditing for Environmental Quality Leadership,* Ed. Willig, J.T., John Wiley and Sons, Brisbane. pp. 11-21.

12. Cahill, L., 1995. Evaluating Management Systems as Part of Environmental Audits. In: *Auditing for Environmental Quality Leadership,* Ed., Willig, J.T., John Wiley and Sons, Brisbane. pp. 23-30.
13. Fitzgerald, C., 1995. GEMI's Environmental Self-Assessment Programme (ESAP). In: *Auditing for Environmental Quality Leadership,* John Wiley and Sons, Brisbane. Ed. Willig, J. pp. 41-50.
14. Stabler, M., 1994. An economic critique of tourism environmental auditing. *Tourism Research and Education in Australia: Proceedings from the Tourism Research and Education Conferences,* Eds Faulkner, B., Fagence, M., Davidson, M. & Craig Smith, S. Bureau of Tourism Research, Canberra.
15. Buckley, R.C., 1990b. Environmental Audits. In: *Pollution Law*, Calcutt Watson and Associates and Blake Dawson, Waldron, Sydney.
16. Buckley, Ralf, 2001. "Ecotourism Accreditation in Australia", in Tourism Ecolabelling: Certification and Promotion of Sustainable Management, edited by X. Font and R.C. Buckley (Oxon, UK: CABI Publishing 2001)
17. Krut & Gleckman, 1998. *Pacific Institute, Managing a Better Environment: Opportunities and Obstacles for ISO-14001 in Public Policy and Commerce* (Oakland, CA: Pacific Institute, 2000).
18. Blamey, R.K., 1996. The Nature of Ecotourism. *Occasional Paper no 21,* Bureau of Tourism Research, Canberra, Commonwealth of Australia.
19. Bhattacharya, A. K. & Saksena, Vandana, 2003. *Ecotourism auditing—A study for visitors' management In Van Vihar National Park*, Bhopal. Project Report. Madhya Pradesh Forest Department, Bhopal.
20. Jim, C.Y., 1989. Visitor's management in recreation areas. *Environmental Conservation* 16(1), 19-32 and 40.

24

Local Perception of Environmental Impacts due to Tourism Activities at Bhojpur, India

A.K. Bhattacharya, Ravi Sharma and *Kiran Sharma*

Introduction

In the last decade of the 20th Century, it was noted that tourism depends ultimately upon the environment, as it is a major tourism attraction itself, or is the context in which tourism activity takes place (Holden, 2000)[1]. As visitor's number increases in any destination area, impacts tend to become more prevalent. Single activities cause multiple impacts and each impact tends to exacerbate or compensate for other changes caused by recreation (Hammitt and Cole, 1998)[2]. The nature of the activities tourists are engaged in will greatly influence the impacts they have. McKercher (1993)[3] argued that tourism tends to overcome resources. The nature of the infrastructure that exists for tourism is significant in relation to impacts. The dynamics of the relationship between 'hosts and guests' (Smith, 1977)[4] in tourism, and the 'impacts' and consequences of tourists visits to 'host communities', are issues which preoccupied the study of tourism from its very inception (Cohen 2003)[5].

Tourism has become a major source of foreign exchange for India, and the historic places in central part of India are the major tourists centres of attractions. In the recent years scholars have attached much importance to residents' attitudes and perceptions on tourist and tourism, and have emphasized the need of including their well-found views in planning and development of resorts (Liu, Sheldon and Var 1997)[8]. The perceptions of the local people is also important as they are the real witness of the tourism scene, both as a spectator and as an actor, and eventually they are the one who are directly (or indirectly) affected by tourism (Singh 1989)[7]. So, their views should be more realistic than the observers subjective observations collected from a few visits of the area.

Historical Background

Raisen district is situated in the central part of Madhya Pradesh lying between the northern latitudes 22° 47′ and 23° 33′ and eastern longitudes 77° 21′ and 79° 49′. The tropic of cancer (23° 30′) passes through the northern part of the District. The Vindhyas which separate the northern Malwa plateau from the valley of the Narmada, almost standing like a wall, sending numerous spurs mostly in the Northern plateau. In Raisen Tehsil, one system of hills is a continuation of the northern arm of Bhojpur hill, which reaches as far north as the Bhopal-Raisen near village *Mana.*

Founded by the legendary Parmar King of Dhar, Raja Bhoj (1010-1053), and named after him, Bhojpur, 28 km from Bhopal, is renowned for the remains of its magnificent Shiva Temple and cyclopean dam. The temple, which has earned the nomenclature of the Somnath of the east, is known as the Bhojeshwar temple. In plan a simple square, with an exterior dimensions of 66 feet, it is devoid of the re-entrant angles usual in such buildings, the richly carved dome, though incomplete, has a magnificent, soaring strength of line and is supported by four pillars, the *Lingam* in the sanctum, rises to an awe-inspiring height of 7.5 feet with a circumference of 17.8 ft. Set upon a massive platform 21.5 ft. square, and composed of three superimposed limestone blocks, the architectural harmony of *Lingam* and platform creates a superb synthesis of solidity and lightness. Even wit the ravage of time, it remains one of the best examples of temple architecture of the 12th and 13th centuries.

Also, incomplete, and with a similar stone-raising ramp, is a Jain Shrine that stands close to the Bhojeshwar temple. Rectangular in plan, this temple probably belongs to the same period as the Bhojeshwar. West of the Bhojpur once lay a vast lake, but nothing remains except the ruins of the magnificent old dams by which its waters were contained.

Study Area

Bhojpur, Hindu pilgrimage tourist centre near Bhopal is renowned for its magnificent Lord Shiva temple, Jain Shrine and Cyclopean dam. In spite of these attractions a water body (Canal of river Betwa) near the temple and various other small temples of goddess near the waterbody is a major point of attraction and recreation all over the year. Tourism is in its infancy and few attractions in the region provide basic visitors facilities. Hence, the majority of visitors are engaged in self-organized activities. Whether the state of the regions natural tourism resources can be maintained over time will become a growing challenge.

Among one of these historic place, Bhojpur is one of the place of tourists' attraction. Bhojpur is endowed with scenic and aesthetic values that attract approximately much more than hundred thousand visitors annually and deemed to possess tremendous tourism potential. Though Bhojpur attracts visitors mostly for religious purpose; it also attracts attention for the following reasons:

- *Location*: Scarcely 25 km from the main centres of the capital.
- *Topography*: Interesting and aesthetic landscapes.
- *Water Resources*: Including a cyclopean dam and a canal which has arrangements of boating and also many small temples of the Goddess near the bank of the canal.

Methodology

The study was accomplished through participatory approach. Perceptions of the local people were measured through the structured surveys of local people, which were administered in the May to June of 2004. Three villages were selected for conducting the survey to know the perceptions of the locals of the area. Schedules were completed for a total of 55 households (25 from Bhojpur, 20 from Keerat Nagar and 5 from Gokulakundi village). Households were selected through the random selection and an attempt was made to include 20 per cent households in the villages. The survey included questions based on socio-economic parameters, physical environment, economy and on socio-cultural domain, soliciting residents perceptions on tourism's negative/positive impacts. Answers were ascertained and validated through the interview method from the sampled local villagers regarding their assessment of changes that have taken place or might take place in future. Most of their observations were based on subjective analysis, as it was difficult for them to measure the impacts objectively.

Responses to each questions was rated on a scale of 4 to 0. A score of 4 marks was awarded for major impact, 3 for moderate impact, 2 for marginal impact, 1 for negligible impact and 0 for no impact. Those who did not register responses or did not know how to answer the questions were ignored. Thus, most of the negative and positive effects were ranked in order (Singh 1989)[7].

It must be noted that though the questions were addressed to the heads of the groups/households, conscious efforts were made to obtain responses from all group members, to ensure a participatory approach. Also, the basic schedule was designed in English and the investigators were trained to translate and make the respondents understand the questions in their language/local dialect to elicit the correct information from them. The study also used the secondary data, but in many cases the reliability of secondary sources is not very high.

The details of the listed households (HHs) and sampled households in surveyed villages are as follows:

S. No.	*Name of Village*	*No. of Listed HHs*	*No. of Sampled HHs*
1.	Bhojpur	72	25
2.	Keerat Nagar	123	25
3.	Gokulakundi	23	05
Total		208	55 (26.44%)

Source: Census of India.

The general socio-economic facts about the selected villages are given in the following table:

S. No.	*Description*	*Bhojpur*	*Keerat Nagar*	*Gokula-kundi*
1.	No. of families	72	123	23
2.	Population	345	665	149
3.	Village area (Hectares)	556.29	610.64	271.21
4.	Total agricultural land (acres)	47.54	112.84	52.16
5.	Forest (hectares)	215.71	186.18	36.14
6.	Unirrigated (hectares)	189.89	235.14	127.86
7.	Distance from tourism spot (km)	0	1	3

Source: Summry of data collected from Census of India.

Results and Discussion

Distribution of Residents by Education

The average level of education in the sample group surveyed was of moderate level with most of the respondents belonging to the school level (44%) and high school level (34%) of education, while no one is belongs to the post graduation level of education. The distribution of respondents by the education level is shown by Fig. 24.1.

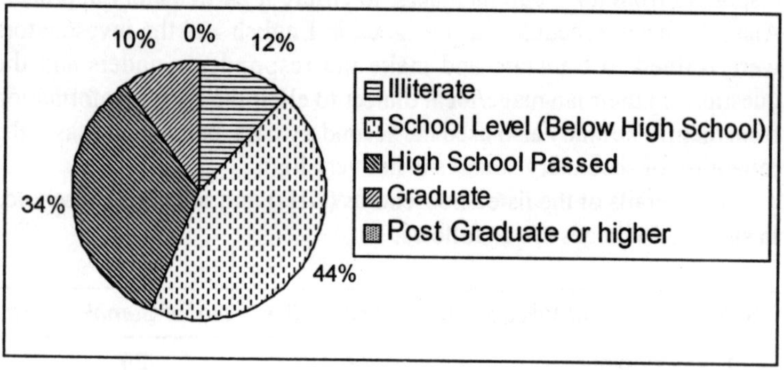

Fig. 24.1: Distribution of residents by education

Distribution of Residents by Age

In the sampled survey of locals, most of the residents were in the age group of 30-39 years followed by 40-49 years. The distribution of age groups is shown in Fig. 24.2.

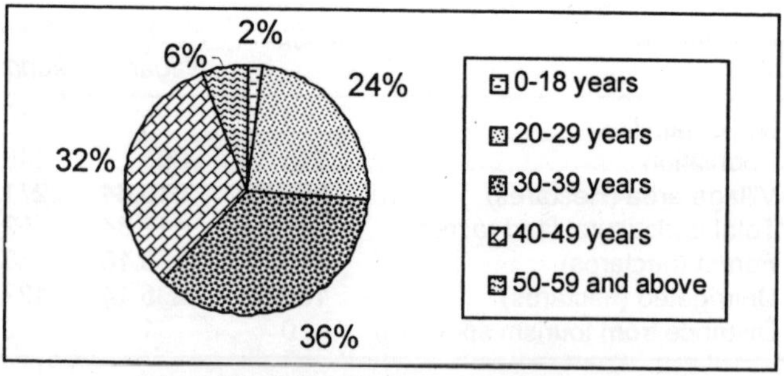

Fig. 24.2: Distribution of Residents by Age.

Distribution of Residents by Family Annual Income

Majority of the respondents among surveyed were having the annual family income of more than Rs. 25,000 and above (30%) followed by annual income of Rs. 10,000–15,000 (26%). Figure 24.3 shows the distribution of residents by the family annual income.

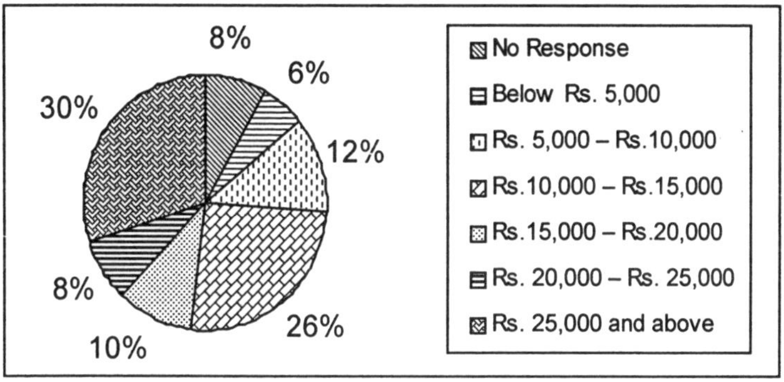

Fig. 24.3: Distribution of Residents by Family Annual Income.

Distribution of Residents by Occupation

Most of the respondents surveyed were engaged in agriculture (38%) and as labourers (38%). The others includes 18 per cent which involves shopkeepers, students etc. The distribution of residents by their occupation is shown by Fig. 24.4.

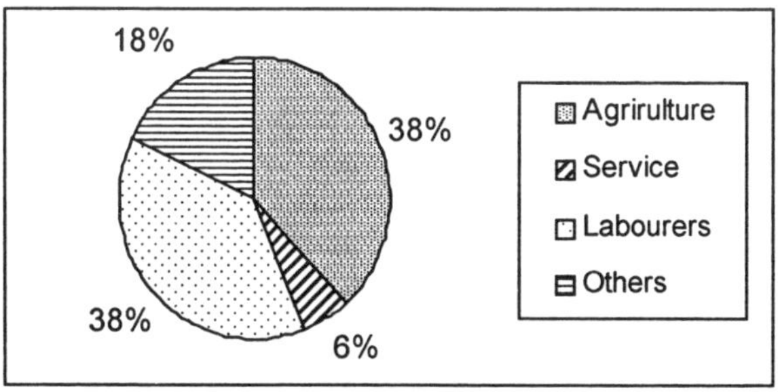

Fig. 24.4: Distribution of Residents by Occupation

Residents Engaged in Tourism Activities

From the sampled surveyed it was found that only 20 per cent residents were engaged in tourism activities like shop-keeping, parking etc., while 74 per cent of people were not engaged in tourism activities. This has also been shown in Fig. 24.5.

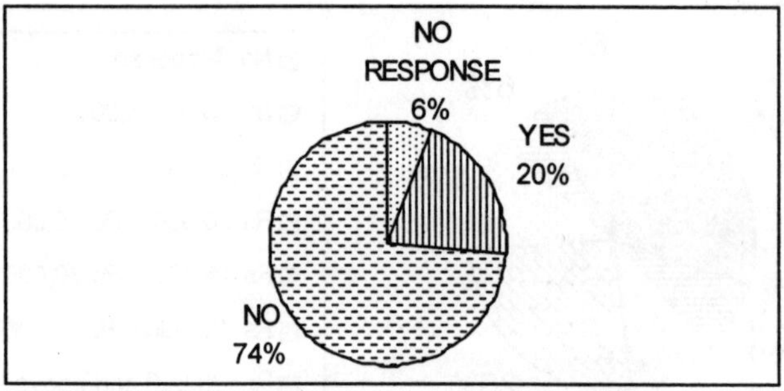

Fig. 24.5: Percentage of Residents engaged in Tourism Activities

Residents Benefiting from Tourism Activities

From the survey it has been found that 26 per cent of respondents are benefiting from the Tourism, while 68 per cent responded in negative (Fig. 24.6).

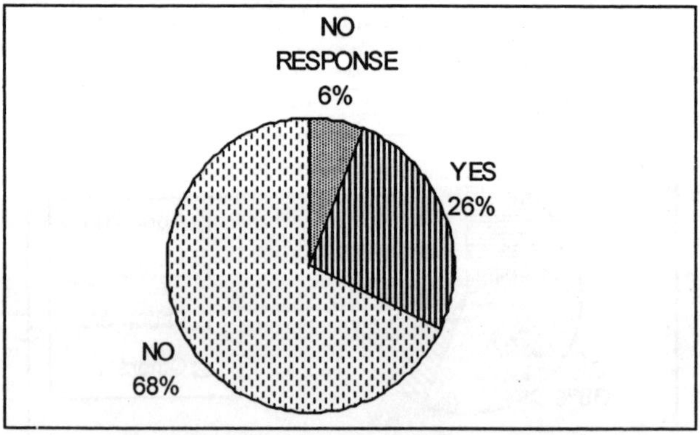

Fig. 24.6: Percentage of Residents Benefiting from the Tourism in the Area

Awareness about Tourism

Most of the residents are aware about purpose of visit to the area by the visitors. Thus, they are aware of tourism in their own perceptions.

Perception of Visitation

Through the perceptions of their interaction with the visitors, the villagers consider that people normally visit the place for religious purpose followed by outdoor picnic and due to natural beauty of the area, while the visitation for the purpose like visit for educative purpose and adventure sports were ranked very low. The perception of visitation by the residents has been depicted below in Table 24.1.

Table 24.1: Perception of Visitation Rank

Reason for Visitation	*Response*	*1*	*2*	*3*	*4*	*5*	*Score*
Pilgrimage/religious purpose	50	40	10	*	*	*	280 (37.33)
Outdoor picnic	50	9	28	13	*	*	196 (26.13)
To enjoy natural beauty	43	1	12	28	1	1	140 (18.66)
Educative purpose	29	*	*	2	21	6	54 (7.2)
For adventure sports	10	*	*	1	5	4	17 (2.26)

Note: The figures in the parentheses indicate percentage.
Scores for Ranks: 1= 5; 2= 4; 3= 3; 4= 2; 5= 1.

Perception about Impacts on the Physical Environment

The villagers have assigned higher ranks to the solid waste accumulation and littering, the noise generation and air, water pollution, site congestion thus causing loss of aesthetic value or loss of access (especially during the peak periods). The impacts like excess pressure on energy resources due to tourism activities, road degradation and soil erosion were ranked in middle order and damage to flora, fauna, vegetation and increased risk of ecological imbalance in near future were ranked very low in order. The ranking and percentage of respondents given for the impact are shown in Table 24.2 and Fig. 24.7.

Table 24.2: Residents Response to the Impacts on Physical Environment (Percentage)

Impacts on Physical Environment	*Major (4)*	*Moderate (3)*	*Marginal (2)*	*Negligible (1)*	*No Impact (0)*	*Rank*
Solid waste accumulation and littering	31 (62.0)	11 (22.0)	4 (8.0)	*	3 (6.0)	165
Noise generation and air, water pollution	8 (16.0)	14 (28%)	13 (26.0)	5 (10.0)	8 (16.0)	105
Site congestion and loss of access	8 (16.0)	12 (24.0)	16 (32.0)	3 (6.0)	10 (20.0)	103
Tourism has destroyed energy resources	6 (12.0)	5 (10.0)	13 (26.0)	12 (24.0)	13 (26.0)	77
Road degradation and soil erosion	5 (10.0)	5 (10.0)	17 (34.0)	5 (10.0)	17 (34.0)	74
Increase the risk of endangered ecological balance of nutrients	1 (2.0)	8 (16.0)	13 (26.0)	17 (34.0)	2 (4.0)	71
Damage to wildlife and vegetation 'or' encroached upon flora and fauna	2 (4.0)	6 (12.0)	8 (16.0)	18 (36.0)	15 (30.0)	60
Deforestation due to construction of roads and hotels through the forests for tourists	1 (2.0)	4 (8.0)	5 (10.0)	11 (22.0)	28 (56.0)	37

Note: The figures in the parentheses indicate percentage.

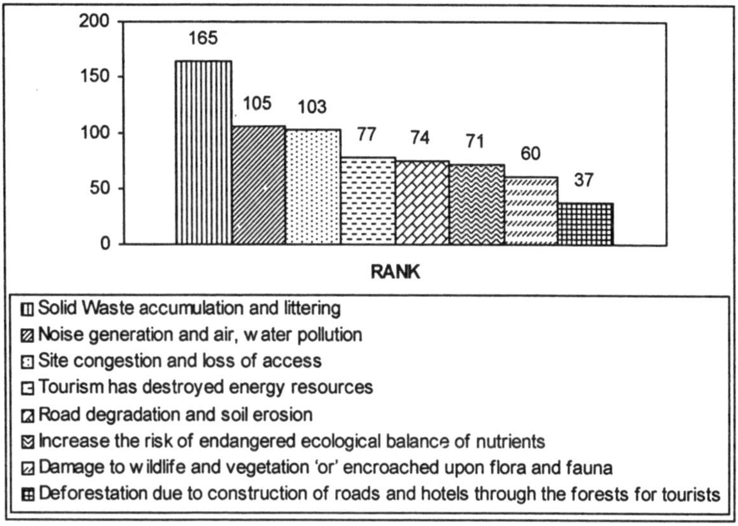

Fig 24.7: Resident's Perceived Impacts of Tourism on Physical Environment

Perception about Positive Impacts on the Physical Environment

The resident's response about the positive impacts on the physical environment has been shown below in the Table 24.3 and Fig. 24.8.

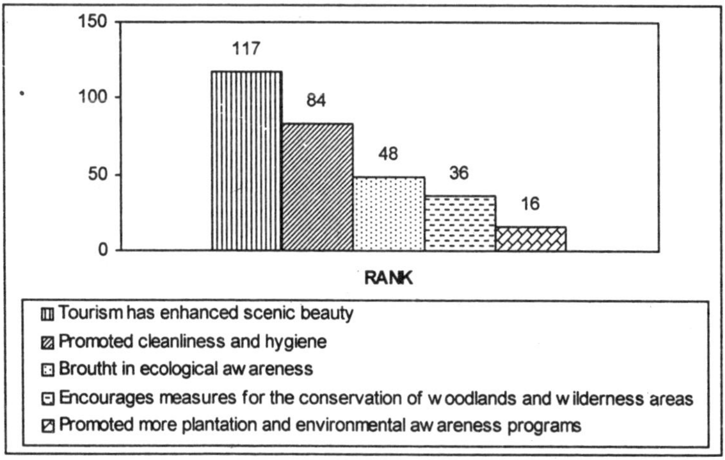

Fig 24.8: Residents response—Positive impacts on Physical Environment

Table 24.3: Residents' Response—Positive Impacts on Physical Environment

(Percentage/Rank)

Positive Impacts on physical environment	*Major (4)*	*Moderate (3)*	*Marginal (2)*	*Negligible (1)*	*No mpacts (0)*	*Rank*
Tourism has enhanced scenic beauty	14 (28.0)	17 (34.0)	3 (6.0)	4 (8.0)	10 (20.0)	117
Promoted cleanliness and hygiene	8 (16.0)	10 (20.0)	2 (4.0)	18 (36.0)	9 (18.0)	84
Brought in ecological awareness	2 (4.0)	7 (14.0)	2 (4.0)	15 (30.0)	20 (40.0)	48
Encourages measures for the conservation of woodlands and wilderness areas	1 (2.0)	3 (6.0)	4 (8.0)	15 (30.0)	20 (40.0)	36
Promoted more plantation and environmental awareness Programmes	1 (2.0)	*	2 (4.0)	8 (16.0)	33 (66.0)	16

Perception about on the Socio-Cultural Impacts

The residents' response about the Socio-cultural impacts due to the tourism has been shown below in Table 24.4 and Fig. 24.9.

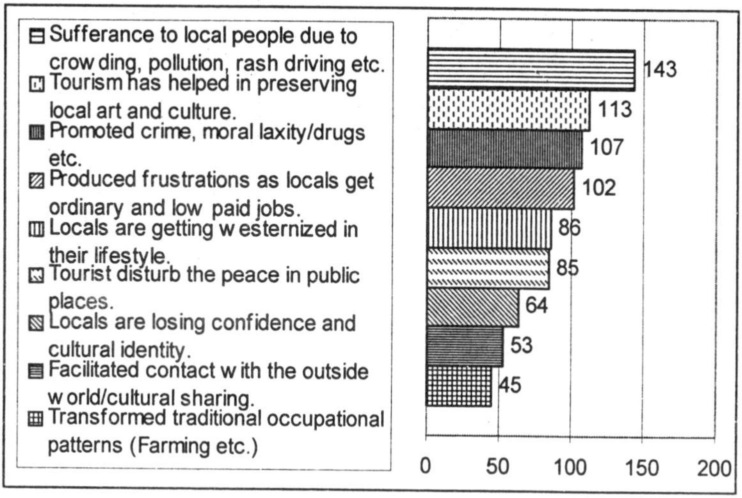

Fig. 24.9: Residents response—Socio-Cultural Impacts

Perception of Residents of Impacts on Economy

The impacts on the economy due to tourism perceived by the locals have been elaborated in Table 24.5 and Fig. 14.10.

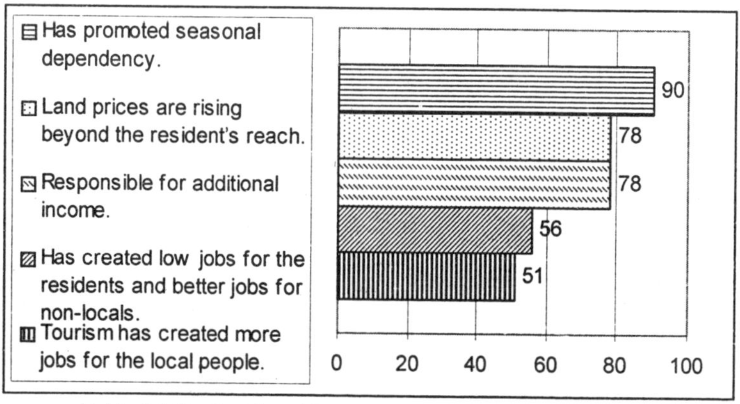

Table 24.10: Residents' Response—Socio-Cultural Impacts

Table 24.4: Residents' Response—Socio-Cultural Impacts

Socio-Cultural Impacts	*Major (4)*	*Moderate (3)*	*Marginal (2)*	*Negligible (1)*	*No Impact (0)*	*Rank*
Sufferance to local people due to crowding, pollution, rash driving etc.	22 (44.0)	13 (26.0)	5 (10.0)	6 (12.0)	2 (4.0)	143
Tourism has helped in preserving local art and culture	9 (18.0)	16 (32.0)	9 (18.0)	11 (22.0)	3 (6.0)	113
Promoted crime, moral laxity/drugs etc.	14 (28.0)	9 (18.0)	7 (14.0)	10 (20.0)	8 (16.0)	107
Produced frustrations as locals get ordinary and low paid jobs.	12 (24.0)	7 (14.0)	11 (22.0)	11 (22.0)	3 (6.0)	102
Locals are getting westernized in their lifestyle.	6 (12.0)	9 (18.0)	8 (16.0)	19 (38.0)	5 (10.0)	86
Tourist disturb the peace in public places	9 (18.0)	11 (22.0)	6 (12.0)	4 (8.0)	18 (36.0)	85
Locals are losing confidence and cultural identity	2 (4.0)	6 (12.0)	8 (16.0)	22 (44.0)	8 (16.0)	64
Facilitated contact with the outside world/ cultural sharing.	1 (2.0)	6 (12.0)	6 (12.0)	15 (30.0)	15 (30.0)	53
Transformed traditional occupational patterns (Farming etc.)	2 (4.0)	3 (6.0)	9 (18.0)	10 (20.0)	24 (48.0)	45

Note: The figures in the parentheses indicate percentage.

Table 24.5: Residents Response-Impacts on Economy (Percentage/Rank)

Impacts on Economy	*Major* (4)	*Moderate* (3)	*Marginal* (2)	*Negligible* (1)	*No Impact* (0)	*Rank*
Has promoted seasonal dependency	15 (30.0)	5 (10.0)	4 (8.0)	7 (14.0)	17 (34.0)	90
Responsible for additional income	4 (8.0)	12 (24.0)	7 (14.0)	12 (24.0)	12 (24.0)	78
Land prices are rising beyond the resident's reach	1 (2.0)	18 (36.0)	5 (10.0)	10 (20.0)	13 (26.0)	78
Has created low jobs for the residents and better jobs for non-locals	7 (14.0)	5 (10.0)	2 (4.0)	9 (18.0)	22 (44.0)	56
Tourism has created more jobs for the local people	3 (6.0)	2 (4.0)	6 (12.0)	21 (42.0)	15 (30.0)	51

Note: The figures in the parentheses indicate percentage.

Residents Interest for Protection of Forests and Natural Value of the Place

The locals are very much keen to protect the forests and the natural value of the place. Among surveyed locals, 50 per cent of the locals are very much interested, 32 per cent are somewhat interested and 8 per cent were not interested. 10 per cent of locals did not respond to the question. (Fig. 24.11).

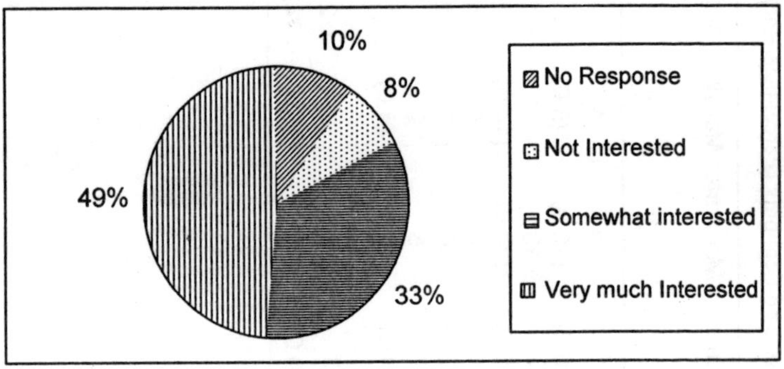

Fig. 24.11: Residents interest for protection of forests/ natural value of the place

Residents Interest for Protection of Cultural Value of the Place

The locals are very much interested (62%) for the protection of cultural values of the place. The not interested group is 8 per cent, while 6 per cent gave no response to the question (Fig. 24.12).

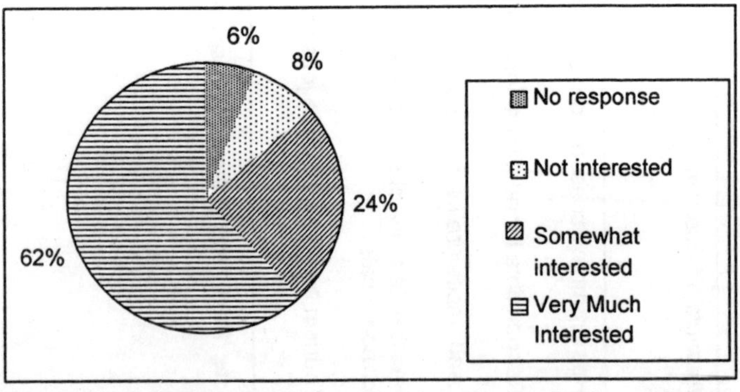

Fig. 24.12: Residents interest in protection of cultrual value of the area

Table 24.6: Steps Suggested by the Locals to Minimize the Environmental Damages in Near Future due to Development of Tourism

Suggestions	*1*	*2*	*3*	*4*	*5*	*6*	*7*	*8*	*Total Respondents*	*Score (%)**
People action can control the situation	16	10	9	1	4	6	1	2	49	295 (16.39)
Government short-term/long-term planning can help	17	10	6	4	2	3	3	4	49	289 (16.05)
To ban the use of plastics	3	8	7	18	8	1	3	2	50	255 (14.16)
Construction of buildings/facilities should be done under laws and regulations as fixed by the agencies	5	9	7	5	5	6	7	4	48	226 (12.55)
To regulate the number of tourists by the agencies/ Government	4	4	10	7	6	1	6	6	44	200 (11.11)
To discourage the mass tourism	2	1	6	8	8	8	6	5	44	172 (9.55)
To provide the alternative source of energy at hotels and for transportations	3	1	3	4	5	13	9	6	44	152 (8.44)
Use of eco-friendly substances and behaviour both by tourists and locals	1	3	2	2	9	7	9	13	46	139 (7.72)

*The figures in the parentheses indicate percentage

Scores: for Ranks: 1= 8; 2= 7; 3= 6; 4= 5; 5= 4; 6= 3; 7= 2; 8= 1.

Suggestions of the Locals to Minimize the Environmental Damage in Future Due to Development of Tourism

The ordinal ranking approach was followed for ordering the suggestions suggested by the locals to minimize the environmental damage due to tourism in the near future and the scores was assigned to each rank to rank the suggestions in order. It is significant that the sampled residents as a group valued that people's participation to control the situation (16.39%) can help followed by Government short-term/long-term planning can help (16.05%) and also to ban the use of plastics and other non-degradable wastes (14.16%) (Table 24.6).

Discussion

The results of the survey indicate that the locals are aware about the purpose of the visit to the area by the visitors and thus, they are also aware of the tourism activities going on in the area. While commenting on the perceived impacts caused by the tourism or might occur in near future on the physical environment, there were mixed opinions on issues of ecology, ecological imbalance, damage to vegetation and fauna. The major issues of concern include solid waste accumulation and littering, loss of access during peak seasons and noise generations (during peak seasons). Many of them have expressed strongly that people's participation along with the governments short and long term planning can help to control the situation in the near future. It is very significant that they realize that people's participation is important for controlling any kind of externalities in the near future. Besides, regarding the positive impacts due to tourism on the environment, they perceived that the scenic beauty and cleanliness of the area have enhanced due to the tourism. But it is also important to mention here that the locals have no or very less knowledge of any plantation or conservation awareness Programmes going on in the area, thus have given low ranking to such issues. Hence, measures for the conservation and environmental Programmes along with the people participation should be implemented in the area so as to avoid the impacts that might take place in the near future as perceived by the residents.

Regarding the socio-cultural impacts, it is well identified that tourism effects marked changes in the socio-cultural environment of the destination communities, particularly when the guests and the hosts have contrasting socio-economic background (Singh, 1989)[8]. Bhojpur's residents perceived by giving higher rank that there is strong disliking for the visitors' behaviour to spread dirt and waste, rash driving and careless drinking etc., which has also promoted the crime and alcoholism in the area. On the other positive side regarding the preservation of cultural monuments and cultural assets, they consider that tourism in the area has helped in preservation of local art and culture; this has been assigned high rank on the positive side. From the overall results on the socio-cultural impacts, it can be concluded that the major impacts like transformation of traditional occupational patterns (like farming etc.), elements of westernization in the lifestyle and loss of confidence and cultural identity were ranked very low in order. The perception of residents that tourism is not disruptive to their traditional way of life or culture, a fact that could probably be partially explained at least on the basis that the tourists stay is usually for the day only, and their time is almost completely absorbed in visits to the sites. The residents have strongly perceived that tourism in the area has promoted the seasonal dependency in the area as most of the residents in the area are labourers and agriculturists, and they are dependent on the peak seasons of visitors for the additional income from tourism (especially during the peak seasons), also the jobs created so far are semi-skilled and low paid jobs and thus very few residents (26%) are benefiting from the tourism activities in the area.

Conclusion

The study on tourism at Bhojpur attempts to capture various issues about the local perception related to the environmental impacts of tourism, which need to be addressed for the tourism-linked development of the area. The area has adequate potential for small-scale, local level tourism development and with appropriate and organized planning and implementation, conservation and economic development of the area is possible, and thus tourism can be used as a developmental tool for the area. The study also highlights the need for the implementation of conservation and environmental awareness Programmes, addressing the issues related to ecological imbalance and other impacts perceived by the residents, as otherwise these issues might assume an alarming proportion in the near future because the tourism in the area is likely to increase phenomenally in the near future. The environmental awareness programmes should be promoted especially when the level of education of majority of the residents have the is quite low (44% below high school). It is very significant that residents are able to realize that people's participation along with the government's action is important for controlling any kind of externalities in the near future. This has also been outlined by D' Amore (1983)[9]. Some of the actions suggested may be relevant for the tourism linked development of the area.

- Residents of the destination areas must be shown the socio-economic significance of the industry through promotional campaigns.
- Tourism planning should be based on overall development goals and priorities identified by the residents.
- Involve local residents in their development of tourism so that their traditions and lifestyles will be respected.
- Opportunities should be provided for broad-based community participation in tourism events and activities.
- Mitigation measures should be employed to resolve problems before any further increase occurs in tourist activity.

In addition to these guidelines, there is need for constant capacity building that will provide them opportunities for starting business at small-scale level. The youths of villages, in particular, are keen to be a part of the whole tourism venture.

However, because of the short-stay and low concentration of visitors in the core destination, the potential negative impacts from tourism in the area so far remains negligible, but the issues addressed as an outcome of the study are important for the planning sustainable tourism and sustainable development of the area.

NOTES

1. Holden, A., 2000. *Future of Tourism's Relationship with Environment.* Environment and Tourism. London, Routledge.
2. Hammit, W. & Cole, D., 1998. *Wildland recreation: ecology and management*, 2nd edition. New York, NY: John Wiley. p. 361.
3. McKercher, B., 1993. Some fundamental truths about tourism: Understanding tourism's social and environmental impacts. *Journal of Sustainable Tourism*, 1: 6-16.
4. Smith, V., 1977. *Hosts and guests: The anthropology of tourism.* Philadelphia: Univ. of Pennsylvania Press, pp. 1-14.

5. Cohen, E., 2003. Contemporary tourism and the host community in less developed areas. *Tourism Recreation Research*, Vol. 28, No. 2: 1-9.
6. Liu, J.C., Sheldon, P.J. & Var, T., 1987. Resident Perception of the Environmental Impacts of Tourism. *Annals of Tourism Research,* 14: 17-37.
7. Singh, T.V., 1989. *The Kullu Valley—Impacts of tourism development in the mountain areas.* Himalayan Books, New Delhi. Planning and development of resorts (Liu, Sheldon and Var 1987)[6].
8. D'Amore, 1983. Guidelines for planning harmony with the host community. In Murphy, P. (Ed.), *Tourism in Canada: Selected issues and options.* pp. 135-59. Victoria BC: *U. of Victoria*, Western Geographical Series 21.

25

From the Lands of Captain Forsyth

Ecotourism Bangs in Madhya Pradesh: Status Report

A.K. Bhattacharya

Prologue

India is one of the 12-mega diversity centres of the world and Madhya Pradesh (MP) with 9.38 per cent geographical area of the country accounts for 12.44 per cent of the country's magnificent forests. Within the State, 31 per cent of the geographical area amounting to 95221 sq. km is under forests. The Dry and Moist Deciduous Tropical Forests, which are abundant in the State, are extremely rich in terms of both floral and faunal biodiversity. Encompassing the rich biodiversity coupled with a range of colourful culture and heritage, the State celebrates intermittently one of the hotspots for Ecotourism as well. Arguably the State holds in its credit certain unique features in its biodiversity as well as cultural wealth, which make it a bounty of ecotourism potential territory.

MP accounts for 22 per cent tiger population of the country. Barasingha (*Cervus duvancelli branderi*) is one species which is endemic to the State. Kanha holds the only remaining population of Barasingha in peninsular India. In 1970, these were down to only 66, but now after careful management, the population numbers about 400. These are found mainly on the meadows around Kanha village, but recently they seem to have been spreading to other areas of the Park.

As far as the floral diversity is concerned, Panchmarhi, well known hill station in the North East Provinces, from the British time onwards is virtually a junction of forest representative types prevailing in the State. It is natural junction of two most important timber species, viz. Teak and Sal. The occurrence of relict population of Sal (*Shorea robusta*) in the predominant teak bearing area is a unique ecological phenomenon in the biosphere reserve. Certain endemic plants like *Psilotum* are the treasures in the State.

MP has the largest tribal population in India. The population of Scheduled Tribes alone accounts for about 20.27 per cent of the total population of the State while the area covered is about 25,652 square miles, which is, approximately, 15 per cent of the area of the State. It is said that the tribal people live in hills and forests, but the full implication of this statement is not generally understood. The coverage of forests varies from one area to the other, but it can be said that about half of the tribal area is covered with forests.

There are many important Non-Wood Forest Products (NWFP) found in the forests e.g. Tendu leaves or Bidi leaves (*Diospyros melanoxylon*), Sal Seed (*Shorea robusta*), *Chebulic Myrobolan* or Harra (*Terminalia chebula*), Gum, Chironji (*Buchanania lanzan*), flower and seeds of Mahua (*Madhuca indica*) and flowers, seeds, bark and roots of various plant species. The State is gifted with a wide variety of medicinal plants, which are sparsely distributed to the length, and breadth of the geographical area.

The History Echoes

The vibrant and versatile MP has found a prominent place even in the colonial records. Captain Forsyth of the Bengal Lancers, who happened to visit the Satpura jungles of the State, found himself attracted to the unique magnetism of the area. In his own words, which later got published in a book titled, 'The highlands of the Central India', "...Altogether, the aspect of the plateau was much more that of a fine English park than of any scene I had before come across in India." Quite impressed was he after witnessing the rich natural and archaeological splendour of the State, he could not contain himself from writing a travelogue exclusively about the State. After he went back to the plains later as the deputy commissioner, Forsyth suggested that Pachmarhi be used as a sanatorium. And the Brit exodus began, a cantonment was created, churches and bungalows were built, a road network was laid, clubs came up and with them golf, horse racing, polo, tennis, snooker..., and a hill station was born.

The colonial flavour still remains in the churches and bungalows and Pachmarhi is as unspoilt as ever. There are bridle paths leading to tranquil forest glades, innumerable water falls, glens, brooks, natural pools (wonderful for an undisturbed swim), murmur of the flowing waters, groves of wild bamboo and *Jamun*, dense *Sal* forests and delicate bamboo thickets.

Interestingly the Nobel laureate Rudyard Kipling got his clue for the world famous fiction 'The Jungle Book' from an instance ostensibly taken place at Sant Bawadi village in Seoni district of MP. Beautiful geographical landscapes of the area and the behaviour of wild animals therein received a picturesque narration in R. A. Strandell's book 'Mammalia of India and Ceylon and Denizens', and this subsequently provided a strong background for Kipling's classical work. This boy, named Mogli, brought up by a herd of wolves was found in the Sant Bawadi village by Leut. John Moor under the guidance of Col. William Sleeman in 1831. Kipling got this flamboyant description of things both from the pamphlet titled, "An account of wolves nurturing children in their dens" by Sir. William Henry Sleaman and a book on "Camp Life of Seoni" by R.A. Sterendale. The Jungle Book mentions a place where Sher Khan was killed. This place is in fact the Valley of Benganga River, near Kanhiwada village. At the present time, these places of historical significance are falling under the famous Pench National Park.

The Heart of India

MP as its name implies, lies in the heart of India. It covers an area of 3,08,000 sq. km, making it the second largest State in the country after Rajasthan, bordering seven other States—Uttar Pradesh, Bihar, Orissa, Andhra Pradesh, Maharashtra, Gujarat and Rajasthan. MP consists largely of a plateau streaked with the hill ranges of the Vindhyas and the Satpuras with the Chhattisgarh plains to the east. The hills give rise to the main river systems—the Narmada and the Tapti, running from east to west, and the Chambal, Sone, Betwa and Mahanadi, west to east. Intersected by these meandering rivers and dotted with hills and lakes, the State has a varied natural setting of great beauty. MP, as the Modern State, came into being in 1956; its cultural heritage is ancient and chequered.

The large State of MP is the geographical heartland of India. Most of the State is a high plateau and in summer it can be very hot and dry. Virtually all phases of Indian history have left their mark on MP, historically known as Malwa. There are still many pre-Aryan Gond and Bhil tribal people in the State, but MP is overwhelmingly Indo-Aryan with majority of the people speaking in Hindi and following Hinduism. Innumerable monuments, exquisitely carved temples, *Stupas,* forts and palaces on hilltops, raise in the visitor's mind visions of empires and kingdoms, of great warriors and builders, poets and musicians, saints and philosophers; of Hinduism, Buddhism, Jainism and Islam. India's immortal poet—dramatist *Kalidasa* and the great musician of the Mughal court, *Tansen,* were from these parts. One-third of the State is forested and offers a unique and exciting panorama of wildlife. In the national parks of Kanha, Bandhavgarh, Shivpuri and many others one has the rare opportunity to see the tiger; as such, MP is often referred to as the *Tiger State* of India. A wide variety of deer, antelope and bison are also the attractions of these parks.

The Cultural Heritage

Customs and beliefs in each area have added colour to the fairs and festivals. Dussehra in Bastar; Shivratri in Khajuraho, Bhojpur, Pachmarhi and Ujjain; Ramnavami in Chitrakoot and Orchha, Bhagoriya in Jhabua and the annual festival of dances at Khajuraho are events for the tourists to remember. The Malwa Festival in Indore, Mandu and Ujjain, and the Pachmarhi Festival bring alive the rich folk and tribal culture of the State in colourful celebrations. Thus, MP represents a true spirit of Northern India in its colourful life and a true place to visit and to develop as an ecotourism destination State as it has all the spices of splendour visit in it.

MP is not called the 'heart of India' just because it is located in the centre of the country, but the State has also been home to the cultural heritage of four main religions, viz. Hinduism, Buddhism, Jainism and Islam. Innumerable monuments, exquisitely carved temples, *stupas*, forts and palaces are dotted all over the State. The natural beauty of MP is equally varied. Consisting largely of a plateau, the State has—spectacular mountain ranges, meandering rivers and an abundance of dense forests. In fact, one-third of the State is forested, offering an incomparable and exhilarating panorama of wildlife in sylvan surroundings.

The Prolific Tourism Profile

The World Tourism Organisation, through the studies conducted at various levels, global, regional and national level has stated that the tourism flow globally generates a sum of Rs. 2310000 crores annually through a tourist influx of around 700 million people. The same figure for India is Rs. 13644 crores annually through the annual turn over of 2.64 million tourists. However, projections for the next 2-3 years indicate that, the income generated from this will increase to a mammoth Rs. 50000 crores nationally and the employment it will create will be more than 57 million including the direct and indirect jobs, and there will be further phenomenal increase in the figures as a result of the prolific tourism growth to 1.50 billion people by the year 2020 with expected receipt of US $ 2000 billion.

Madhya Pradesh with an annual tourism influx of around 0.11 million generates benefits equivalent to Rs. 637 crores annually. Besides, the State also has special reason to cheer up, as the projection for the next 2-3 years will bring in the State of 2.69 employment opportunities and a generated income which is quite equivalent to Rs. 2336.4 crores annually. Making an allowance for all the aforementioned facts, the State Government has already begun the work to make the most of these accruing benefits. Policy and administrative changes which have either taken place or are in the process are the cases in point.

Diverse Tourism Zones: A Right Mix of Everything

A multitude of tourism activities are reportedly taking place in the State. From the world famous heritage sites to grandeur monumental structures, the State is enriched with a range of fairs and festivals. Added to its existing tourism activities are an annual pilgrimage, which is eagerly devoured by devotees largely from within and outside the State. Taking advantage to the country's largest forested areas, the State promotes the wildlife tourism as well in a bigger way. In all, approximately 20 destinations spread across the State are offered under the 'Temptation' tourist packages offered exclusively by the State tourism development corporation. The destinations are as follows:

- *Gwalior:* A multitude of reigning dynasties, of the great Rajput clans of the Pratiharas, Kacchwahas and Tomars have left indelible etchings of their rule in this city of palaces, temples and monuments. Gwalior's tradition as a royal capital continued until the formation of present day India, with the Scindias having their dynastic seat here. The magnificent momentoes of a glorious past have been preserved with care, giving Gwalior an appeal unique and timeless.

- *Orchha:* Its grandeur has been captured in stone, frozen in time, a rich legacy to the ages.
- *Khajuraho:* Famed worldwide for the erotic sculptures on its temples. These breathtaking sculptures capture life in every form and mood.
- *Bhopal:* Bhopal, the capital of MP, combines scenic beauty, historicity and modern urban planning. Bhopal today presents a multi-faceted profile; the old city with its teeming market places, fine old mosques and palaces still bear the aristocratic imprint of its former rulers.
- *Sanchi:* The seat of Buddhist learning is famous for its stupas, monasteries, temples and pillars dating from the 3rd century BC to the 12th century AD.
- *Bhimbetka:* Over 600 rock shelters belonging to the Neolithic age were recently discovered in this place of interest. Here, in vivid panoramic detail, paintings in over 500 caves depict the life of the pre-historic cave-dwellers making the Bhimbetka group an archaeological treasure, an invaluable chronicle in the history of man.
- *Indore:* The commercial centre of MP, this bustling city has rare monuments built by the Holkar dynasty.
- *Ujjain:* Ujjain is the site of the Kumbh mela, the largest religious congregation in the world. It is also known as the temple city. Modern Ujjain is situated on the banks of the river Shipra, regarded since times immemorial as sacred.
- *Omkareshwar:* This place is a sacred island at the confluence of two rivers, Narmada and Kaveri, shaped like the holiest of all Hindu symbols, Om. At the confluence of the rivers, the devout gather to kneel before the Jyotirlinga (one of the twelve throughout India) at the temple of Shri Omkar Mandhata.
- *Narmada:* This ancient city is also mentioned in the epics Ramayana and Mahabharata. It is renowned for its bathing ghats and the famous Maheshwari sarees.
- *Mandu:* Each of Mandu's structure is an architectural gem; some are outstanding like the massive Jama Masjid and Hoshang Shah's tomb, which provided inspiration to the master builders of the Taj Mahal centuries later.
- *Amarkantak:* Amarkantak is a great pilgrim centre for the Hindus, and is the source of the rivers Narmada and Sone.

- *Bhojpur:* Founded by the legendary Parmar King of Dhar, Raja Bhoj and named after him, Bhojpur, is renowned for the remains of its magnificent Shiva temple and Cyclopean dam.
- *Chanderi:* A crafts centre famous for its sarees and boracades. Chanderi was an important military outpost, prized by rulers with power or ambition.
- *Chitrakoot:* Chitrakoot's spiritual legacy stretches back to legendary ages: it was in these deep forests that Rama and Sita spent eleven of their fourteen years of exile; here that the great sage Atri and Sati Anusuya meditated. A place with tranquil forests, rivers and streams.
- *Pachmarhi:* This city is MP's most verdant jewel, a place where nature has found exquisite expression in myriad enchanting ways. Pachmarhi is also an archeological treasure-trove. In cave shelters in the Mahadeo Hills is an astonishing richness of rock paintings.
- *Shivpuri:* This city is steeped in the royal legacy of its past, when it was the summer capital of the Scindia rulers of Gwalior.

Thus, MP provides an ideal ecotourism destination site for the visitors to the State as it has all types of tourism blended in it, archaeological (Khajuraho), wildlife (NPs), adventure etc. In addition to it, MP is culturally endowed with spectacular fairs and festivals on its land. Both the State Government through its tourism department and various organisations as well as individuals of different walk of life are promoting the annual events of cultural importance. Indeed, this has been symbolised the rich cultural heritage of the State and the tradition is progressing.

The Treasure of Festivals

Festivals are the time that bring colour and add pleasure to one's life. The State of MP can be seen enjoying in the colours of the festivities that take place throughout the year in various parts. They are in the form of the traditional festivals as well as fairs that are organised by various bodies. Some of the festive attractions of the State are as follows:

- Festival of Dances, Khajuraho—*Indian Classical Dance Performances*
- Akhil Bhartiya Kalidas Samaroh, Ujjain—*Literary Presentations along with Dramas and Indian classical programmes*

- Tansen Sangeet Samaroh, Gwalior—*Indian Classical Vocal and Musical performances*
- Dhrupad Samaroh, Bhopal—*Vocal performances of Dhrupad Style of Indian Classical Music*
- Chakradhar Samaroh, Raigarh—*Indian Classical Dance and Music*
- Ameer Khan Festival, Indore—*Indian Classical Performances*
- Rashtriya Hindi Natya Samaroh, Bhopal—*Hindi's Best Dramas' Performances*
- Pachmarhi Utsava, Pachmarhi—*Indian Folk Arts' Performances*
- Mandu Utsava, Mandu—*Indian Classical Music Performances*
- Orchha Utsava, Orchha—*Indian Music Performances*
- Keshav Jayanti Samaroh, Orchha /1 *Literary Presentation*
- Alauddin Khan Vyakhyanmala, Bhopal—*Lectures Based on Indian Classical Music*
- Bhavbhuti Samaroh, Gwalior—*Sanskrit's Literary Programmes*
- Malwa Utsava, Ujjain and Indore—*Folk Arts and Indian Classical Music*
- Makhanlal Chaturvedi Samaroh, Jabalpur—*Literary Programmes*
- Lokrang Samaroh, Bhopal—*Indian Folk Arts' Presentation and Performance*
- Pt. Kumar Gandharva Sạmaroh, Dewas—*Indian Classical Music Performances*
- Shankari Samaroh, Different Parts of MP—*Sanskrit Literary Programmes*
- Kabir Samaroh, Different Parts of MP—*Folk Arts, Music, Dances and Literary Programme*
- Rashtriya Alankaran Samaroh, Bhopal—1 *Distribution of Cultural Awards Instituted by the Government of MP*
- Krishna Rao Shankar Pandit Samaroh, Gwalior—*Indian Classical Music Performances*
- MP Sangeet Samaroh, Different Centres of MP
- Nimar Utsava, Maheshwar—*Indian Folk and Classical Performances*
- Bharat Bhawan, Bhopal—*Various Folk, Arts Exibition and Performances throughout the year*
- Film Festival, Bhopal—*Films from India and Abroad*
- Lata Mangeshkar Samaroh and Award Ceremony, Indore—*National Level Light Musical Performances*

- Bhagoria Haat, Jhabua: *This colourful festival of the Bhils and Bhilalas, particularly in the district of West Nimar and Jhabua, is actually in the nature of a mass svayamvara, a marriage market, usually held on the various market days falling before the Holi festival in March*
- Khajuraho Festival of Dances: *This festival draws the best classical dancers in the country who perform against the spectacular backdrop of the floodlit temples, every year in February / March. The finest exponents of different classical Indian styles are represented—Kathak, Bharatnatyam, Kuchipudi, Odissi, Manipuri, and many more.*
- Tansen Music Festival: *More than a monument, the Tansen Tomb is a part of Gwalior's living cultural heritage. It is the venue of the annual Indian classical festival held in November-December. Renowned classical singers of the land regale audiences through five mesmerising night-long sessions of the much-loved classical ragas.*

The Fairy Fairs

The State has been gifted with a tradition of fairs, which mark the prospects and progress achieved. Notwithstanding the religious nature of these fairs, people have attended it traditionally from different walk of life. Few of these fairs, which perhaps date back to the 18th century, still celebrated with much fanfare are, are listed below with the corresponding locations.

Fair	*Region*
Jageshwari Devi	Chanderi (Guna)
Garibnath Babu	Avantipur, Barodia (Shajapur)
Kana Babu	Sadalpur (Hoshangabad)
Dhamanu Urs	Hastan Shah Mosque (Sagar)
Nagajee	Porsa (Morena)
Hira Bhumiya	Gwalior
Ramleela	Gwalior
Pir Budhan	Shivpuri, Sanwra
Tejajee Fair	Bhamawad, Guna

(Contd.)

(*Contd.*)

Fair	*Region*
Mahamritunjya	Aahate, Rewa
Chamti Devi	Ghaghra Gao, Sidhi
Baha Sahabuddin	Neemuch, Mandsour
Kalujee Maharaj	West Nimar
Sinngajee	West Nimar
Mandhatta	East Nimar
Rajjim Fair	Rajjim, Raipur

Immense Ecotourism Potential

The Fascinating Biodiversity

India is one of the 12-mega diversity centres of the world. Quite remarkably MP with 9.38 per cent geographical area of the country accounts for 12.44 per cent of the country's forests. Within the State 31 per cent of the geographical area amounting to 9.5 million hectares is under forests. The Dry and Moist Deciduous Tropical Forests, which are abundant in the State, are extremely rich in terms of both floral and faunal biodiversity. These rich biodiversity coupled with a range of colourful culture and heritage, the State celebrates intermittently one of the hotspots for Ecotourism as well. Arguably the State holds in its credit certain unique features in its biodiversity as well as cultural wealth, which make it a bounty of ecotourism potential territory.

MP has got approximately 1026 Tigers, which account for 22 per cent tiger population of the country. Barasingha (*Cervus duvancelli branderi*) is one species, which is endemic to the State. Kanha holds the only remaining population of Barasingha in peninsular India. In 1970 these were down to only 66, but now after careful management, the population numbers about 400. These are found mainly on the meadows around Kanha village, but recently they seem to have been spreading to other areas of the Park.

As far as the floral diversity is concerned, Pachmarhi, well known hill station in the North East Provinces, from the British time onwards is virtually a junction of forest representative types prevailing in the State. It is natural junction of two most important timber species, viz. Teak and Sal. The occurrence of relict population of Sal (*Shorea robusta*) in the predominant teak bearing area is a unique ecological phenomenon in the biosphere reserve. Certain endemic plants like *Psilotum* are found in the State.

MP thus has enormous potential for Ecotourism, as it is rich in its natural diversity, cultural heritages and civilization and traditions. But so far the ecotourism potential of the State has not been explored to its point as it carries in it.

The Nowhere Else Wildlife

	National Parks		*Wildlife Sanctuaries*				
1.	Bandhavgarh	1.	Bori	10.	Chambal	19.	Son Ghariyal
2.	Fossil (Dindori)	2.	Bagdara	11.	Nauradehi	20.	Sardapur
3.	Kanha	3.	Phen	12.	Pachmari	21.	Sailana
4.	Madhav	4.	Ghatigaon	13.	Panpatha	22.	Ralamandal
5.	Panna	5.	Gandhisagar	14.	Kuno	23.	Orchha
6.	Pench	6.	Karera	15.	Pench	24.	Gangau
7.	Sanjay	7.	Ken Ghariyal	16.	Ratapani	25.	Durgawati
8.	Satpura	8.	Kheoni	17.	Sanjay Dubri		
9.	Vanvihar	9.	Narsinghgarh	18.	Singhori		

One-third of MP is forested. The State offers a unique and exciting panorama of wildlife in sylvan surroundings. Spectacular mountain ranges, meandering rivers and miles and miles of dense forests are on offer for those who visit the State.

The immediate jungle plan for the tiger country and other such attractions are part of the State's wildlife promotion drive include developing new areas in the State. The State is also serious about the future development of Kanha and Bandhavgarh National Parks, which can arguably be recognised MP's pride.

- *Bandhavgarh National Park:* One can be almost certain of seeing a tiger in this region. The density of the Tiger population at Bandhavgarh is the highest known in India. This is also White Tiger country.
- *Kanha National Park:* Kanha's Sal and Bamboo forests, rolling grasslands and meandering streams stretch over 940 sq. km in dramatic natural splendour which form the core of the Kanha Tiger Reserve created in 1974 under Project Tiger. The park is the only habitat of the rare hard ground Barasingha (*Cervus duvancelli branderi*).

More to Offer—Potential Ecotourism Sites

The various destination areas, which carry the ecotourism potential, have been listed below which have the potential to be developed as the ecotourism destination points. The following are listed some of the Sites in MP which have an enormous potential for the development of the Ecotourism sites.

Destination	*Tourism Interest*
Ratapani WLS (Obedullahgunj)	Nature, Historical
Panna NP and adjoining Forests (Panna)	Nature, Historical, Religious, Adventure
Pench WLS and Sakta Rukhad Forests (Seoni)	Nature and wildlife
Noradehi WLS (Sagar)	Nature and wildlife
Rani Durgawati WLS (Damoh)	Nature and wildlife, Historical
Satpura NP and Bori WLS (Hoshangabad)	Nature and wildlife, Religious, Adventure
Sanjay NP (Sidhi)	Nature and wildlife
Betul Forests (Betul)	Nature
Buffer zone Forests of Kanha Tiger Reserve (Mandla)	Nature
Ralamandal WLS (Indore)	Nature and wildlife
Kheoni WLS (Dewas)	Nature and wildlife

In addition to the above list, MP provides an excellent blend of nature, adventure and wildlife tourism. The major attractions of these blend can be had from the variety of identified places of interest.

The Unique Features

MP, with an infinite number of heritage sites, coupled with extensive natural flora and fauna, MP easily works out be an adventure tourism and art and culture tourism destination. With wildlife being an essential element of the State, and around one-third of the State forested, the State has a lot to offer in wildlife tourism, which can be termed as the State's USP.

- MP has got approximately 1026 tigers, which account for 19 per cent tiger population of the world.
- Barasingha (*Cervus duvancelli branderi*) is one species that is endemic to the State.

- Occurrence of relict population of Sal (*Shorea robusta*) in the predominant teak bearing area of Pachmarhi is a unique ecological phenomenon.
- *Many endemic species, like Psilotum nudum*, a very rare and primitive plant commonly known as wisk fern are found only in Pachmarhi area.

Apart from announcing its heritage policy, and hosting various road shows across the country, the Government of MP has put in place the State's eco and adventure tourism policy, declaring itself as a 'wildlife and adventure Tourism State'.

The State aims to position itself as a '*wildlife*' *destination*, especially as the '*Tiger State*' both in the national and international arena.

Disclosed an official from the State tourism department: "Today's tourists are not content with cultural or religious tourism alone—what one looks for is some thrill, fun, adventure and something other than the routine. So, in keeping with this change of attitude in tourists, the State government has decided to actively promote Ecotourism and adventure tourism. Further, to popularize and develop these forms of tourism, the government is for the first time seeking participation of private investors. Adventure tourism provides tourists a special thrill whilst participating in sporting activities like river rafting, trekking etc. MP with its rich natural flora and fauna, unexploited so far, has immense potential for such sports."

The Ambient Infrastructure

Infrastructure is an absolute necessity for any tourism activity, no matter how minimal it is. MP State also realised this aspect quite early and has taken initiatives already in this direction by involving different stakeholders. Destination specific activities and infrastructure has been promoted over the period of time. An excerpt of the destinations and activities are given below with the specific tourism category.

The Versatile Destinations and Vibrant Activities

Recently, as part of the Tourism Policy, the Government has declared special package of incentives for the tourism industry. These incentives will be available for private entreprencurs for infrastructure development activities.

The Versatility	*Places of Importance*	*Facilities*
Cultural Tourism	Gwalior, Orchha, Khajuraho, Sanchi, Mandu.	Heritage Hotels, Museums, Craft Villages.
Wildlife and Adventure Tourism	Pachmarhi, Kanha, Bandhavgarh, Panna, Satpura, Pench Valley National Parks, Tigra Lake (Gwalior), Upper Lake (Bhopal), Gandhi Sagar (Mandsaur).	Log Huts, Camping Grounds, Trekking, Water-Sports, Aero-Sports, Angling, Cruises, Caravans, Tents etc.
Leisure and Business Tourism	Pachmarhi, Khajuraho, Bhopal, Gwalior, Indore, Mainpat, Tamia, Jabalpur etc.	Convention Centres, Exhibition Grounds, Shopping, Evening Entertainment, Golf Courses, Country Clubs, Weekend Getaways etc.
Pilgrim Tourism	Ujjain, Maheshwar, Omkareshwar, Chitrakoot, Amarkantak, Rajim, Sanchi, Bhopal, Orchha etc.	Budget Accommodation, Day Shelters, Cafeterias etc.

Hotels, Motels and Restaurants

The MP Government has amplified its tourism drive by attracting bids for development from private sector in the area of hospitality investment. Inviting the hospitality industry to invest in the State is one of the most important plans of action that MP tourism has identified. The State Government has also shorted-listed locations that could attract hospitality investment.

The investment covers of the hospitality segment would be in heritage hotels, health farms, recreation centres, ropeways, golf courses, museums, amusement parks, craft villages, convention centres, and all other bona fide tourism-related activities recognised by the government.

Subsequently, hospitality giants of the country have also evinced interest in investing into the hotel sector of the State. Revealed Jamshed S. Daboo, chairman, CII tourism sub-committee, western region, and chief operating officer-leisure hotels, Indian Hotels Company Limited (IHCL): "The State has got tremendous potential in the area of tourism, so IHCL would definitely look at investing in the region in the near future. The avenues, which the State has to offer for hospitality investment, are worth exploring."

Highlighting the potentials and plans with respect to *tourism and infrastructure development,* the State has identified the hospitality industry as one of its prime movers in strengthening the tourism position of the region.

Instead of depending on its limited resources, the State Government has decided to open up this sector for private participation for optimum utilisation of these natural resources. While on one hand, with these efforts, the government hopes to attract tourists in larger numbers, on the other hand, it is expected that it would help in generating greater demand for local products, and creating new employment avenues for local communities.

In order to attract private participation, the State Government has finalised a set of simplified guidelines that are hassle free and would make it easy for entrepreneurs to conform with.

The Ecotourism Breakthroughs

The Strengths

MP, because of its central location in India, has remained a crucible of historical currents from North, South, East and West. Rich archaeological wealth has been unearthed in various parts of the State throwing light on its history. Its central location makes MP a strategic hub with so many logistical advantages.

- MP is highly benefited from its strategic location as the central part of the country. It is very well connected both by the road, rail and air to any important destinations in the country.
- MP is one among the largest producer of minerals, agricultural and forest produces and is ideal for modern biotech industries.
- MP is an ideal tourist destination from the viewpoint of hospitality industry—with two world heritage sites of Khajuraho and Sanchi and several other major monuments in addition to some major tiger reserves and spots for adventure and eco tourism. It caters a tourism package blended with various cultures, traditions, festivals, fairs and wildlife and archaeological attractions.
- MP has the advantage of promoting a Composite Tourism Package, as the State is gifted with sites worth Wildlife tourism, Cultural tourism, Archaeological tourism, Natural Tourism, Indigenous tourism, so on and so forth.

- MP dotted with 19 Industrial Growth Centres is home to a large number of companies. Well-developed industries base is available at attractive prices and a package of incentives.
- The infrastructure advantage in MP include the strongest Optical Fibre backbone at every district headquarters, a network of roads, railways and regular air services connecting Bhopal, Indore and Gwalior with rest of the country.
- MP is building a road network that can be compared to the best in the country. This network of over 10,000 kms of road taking care of 90 per cent of traffic in the State is being completed on war footing in two years time.
- MP has some of the finest educational institutions in disciplines like IT, management electronics, hospitality, law, forest management etc. New institutions in hi-tech area like biotech and microbiology are on the anvil in the State.
- MP is profited by an efficient and responsive administration.
- Striving for a growth at par with any developed States, MP offers a variety of incentives and concessions for the investors.
- On top of it all the State over the period of time developed a congenial atmosphere for the growth and development of tourism. A special policy framework that supports the tourism activity with administrative as well as legal mechanism is put in place. A separate Ecotourism policy, which will focus on the ecotourism development of the State, is on the offing.

The Gray Areas

- The poor network of road communication, both between and outside different tourism locations.
- Lack of adequate and appropriate infrastructural facilities, pulls the State back from utilising its tourism potential in its maximum.
- A concerted effort from different departments and agencies that are the stakeholders in tourism promotion is lacking.
- Involvement of the community should also be ensured, as this is one important aspect one cannot do away with.
- Comparing to other States in the country, which had achieved grand success in Tourism, MP is disadvantaged with a comprehensive promotion campaign targeting both the domestic as well as foreign tourists.

- Lack of funds to the government for the development of sustainable tourism in the State.
- Lack of efficient manpower and technical staff in the field of tourism development.
- Lack of special incentives and packages for the tour and travelling in MP.

Future Strategy

The Enabling Environment

From being the first-ever State in the country to announce its exclusive heritage policy, to that of conducting the first road show in the Capital and other metropolises in India, MP tourism has arrived with a bang. Marketing its product as 'Mystic MP' the State has firmly declared its presence on the tourism map of India. Consequently, the MP State Tourism Development Corporation (MPSTDC), the official body and tourism facilitator of the State, in association with the local CII representation, with the support of the MP State Government and other State tourism and hospitality bodies and associations have formulated concrete plans to position the State as a world class tourism destination in India.

Progressive Policies and Administrative Framework

Keeping in sync with the global trend on environmental consciousness, the North-Eastern States of Assam, Manipur, Meghalaya, Mizoram, Nagaland, Sikkim and Tripura are coming together to form a joint ecotourism policy shortly. On a similar eco-focus, like other States of Kerala, Karnataka, Maharashtra, Andhra Pradesh, Bihar, Rajasthan, Delhi and Tamil Nadu, MP is extensively working towards incorporating ecotourism as a vital approach for tourism and hospitality industries in the respective existing tourism policies.

Subsequently, the State Government in its new Ecotourism Policy has provided the much needed focus on the development of Ecotourism in the State. The State Government revealed, "In order for the State to grow as a leading domestic and international tourist destination, it needs to work in tandem to promote its greatest asset—Ecotourism."

Ecotourism Sets a Goal in the New Forest Policy

Setting things upright, the State of MP had recently come out with a new policy on forests called the MP Forest Policy, 2004. Ecotourism gets a very special mention in this policy which apparently reflects the acknowledgement of the State on its awesome potential. Reiterating the participatory approach in the forest management, the new policy finds ecotourism as a tool for better management of the forest resources with a constructive participation from the forest dependent communities. Highlighting the inherent advantages of the ecotourism explicitly, the new forest policy places the need for preparing area specific Ecotourism Management Plans at each region of the State. Evolving a permanent mechanism with appropriate participation from the stakeholders of different kind is the basic idea underlying this policy initiative. Sustaining the tourism interests of the visitors the policy proposes a composite package of experiences in the form of interpretation centres, better infrastructure so and so forth.

The Liberal Tourism Policy

In a bid to expedite the development of tourism in MP, the State Government has a proactive tourism policy in place since 1995. The new tourism policy envisages creation of an environment conducive to attracting increased private investment in the tourism sector, and a more meaningful role for the government. The policy has also conceptualised the strategy for development, wherein the focus shall be on the following areas:

- Improvement and creation of adequate basic infrastructure.
- Upgradation and augmentation of accommodation, catering and recreational facilities.
- The stepping up of transport facilities.
- Marketing of destinations to ensure optimal use of infrastructure.
- Evolving suitable policies for increasing foreign exchange earnings.
- Promotion of traditional arts and crafts of the State.

Divergent Eco and Adventure Tourism Policy

Ecotourism, which primarily connects conservation with the economic development, has got special recognition in MP. Considering the unique and abundant forest resources, the State found that the scope for Ecotourism is tremendous at par or even better than various other States in the country. This acknowledgement received administrative as well as legal support through the adoption of a new policy framework, which is called, the Eco/Adventure Tourism Policy 2001-02. Welcoming the large-scale private participation the policy enumerates the proactive role of government and its various departments in exploiting the ecotourism potential of the State. Adventure tourism, which is perhaps an ingredient in the ecotourism, got special emphasis in the policy, as the State have incredible prospects in that. Despite the fact that, it opened up a wide spectrum of opportunities and options for the private entrepreneurs, the policy evidently restricts the magnitude of those activities just to avoid all those adverse impacts.

Newer Horizons in New Ecotourism Policy

The MP Forest Department, being the custodian of the largest forest resources in the country is also coming out with an exclusive comprehensive policy on Ecotourism. The new ecotourism/nature tourism policy of the government focuses on reorienting the forest staff as well as the forest community in order to enable them in popularizing the concept of nature based tourism. The policy indeed identifies the various options and activities, which are required for the implementation of ecotourism. While listing out all the possible activities, the policy demands for equally worth infrastructural facilities as well, which suits to the requirements of the tourists of different categories? Taking a cue from the examples of various State Forest Departments, which had already become successful in the ecotourism initiative, the government through this policy wishes to bring in a range of benefits in the form of improved conservation, securing the economic development of the forest communities and accelerating the progress among them.

Strategy for Ecotourism Development

In order to promote ecotourism as a tool for sustainable development, the focus shall be on the following areas:

- Improvement and creation of adequate basic infrastructure—land, roads, water, electricity etc.
- Upgradation and augmentation of accommodation, catering and recreational facilities.
- Augmentation of transport facilities.
- Marketing of destinations to ensure optimal use of infrastructure.

- Evolving suitable policies for increasing foreign exchange earnings.
- Promotion of the arts and crafts of MP.

Role of the State Government

The State Government will confine its efforts to infrastructural development like ensuring uninterrupted electricity, water supply and provision of basic medical facilities. It will also be responsible for dissemination of information, organisation of festivals and interdepartmental co-ordination to create conditions for attracting private sector investment for the tourism sector. For this, it will provide fiscal incentives, assist in providing suitable sites and remove bottlenecks, especially those connected with infrastructural development.

Incentives

A special package of incentives for the tourism industry has been prepared. These incentives will also be available for private entrepreneurs for any of the following activities:

- Hotels, Motels and Restaurants.
- Heritage Hotels, Health Farms, Recreation Centres, Ropeways, Golf Courses, Museums, Amusement Parks, Craft Villages, Convention Centres, and all other *bona fide* tourism related activities recognised by the government.
- A/c Luxury Coaches, A/c Mini Buses, and imported Limousines purchased by registered tour operators for tourist transport.
- Activities connected with adventure tourism, such as hang gliding, para-sailing, river rafting, boating, trekking, rock climbing etc.

Special incentives are as follows:

Land

1. Government land will be offered with the approval of the Empowered Committee as equity participation on behalf of the Government at current market value for setting up Joint Ventures.
2. The companies thus formed who are given any kind of Government land will be expected to complete the project within

a period of three years from the date of transfer of land, failing which the land will revert to the government. The company will not be allowed to transfer the land to anybody else, or use it for any purpose other than that for which it is allotted.

3. For construction of approved category hotels, etc. in the urban areas, exemption under Section 20 of the Urban Land Ceiling Act will be made available on a case-to-case basis as decided by the Empowered Committee.

Luxury Tax: New projects shall be exempt from Luxury Tax for ten years from the date of commentecement of commercial operations.

Sales Tax: New projects shall be exempt from Sales Tax for ten years from the date of commencement of commercial operations.

Entertainment Tax: Entertainment Centres will be exempt from Entertainment Tax for ten years from the date of commencement of commercial operations.

Epilogue

Without having an enabling environment, the efforts for achieving development will never find the fruits. It requires a class of visionaries and energetic lot both in the governance and administration. MP, though a little rearward industrially, has found enough reasons for the apparent dull growth. Understanding the available resources, its worth and utilising it sensibly and successfully requires a quantum of confidence. And confidence emanates from an environment, which is promising and vibrant. In MP the environment has ripened enough to have a wide range of investments, especially in the sector of tourism. Ecotourism, which can be a solution for a wide variety of problems receive a boost from the State in the form of policy initiatives. These policies are demanding to create a range of facilities in the form of legal and administrative support as well as infrastructure development, which in turn can provide an atmosphere where the ecotourism can flourish. Diversity within the diversity makes MP different from others States in the country. The State is in its surge ahead for utilising these unique resources by promoting ecotourism in the big way and long way.

26

Ecological Infrastructure Designing for Sustainable Ecotourism[1]

A.K. Bhattacharya and *Muhammed Nahar J.*

Introduction

Taking into consideration the current demand for ecotourism destinations, it is crucial that the design and development of ecotourism facilities place the integrity of the given ecosystem as a top priority. It is important to note that the ecotourism industry is currently at a crossroads in its development. In the last decade, it has generated much revenue for local and regional economies worldwide, provided new incentives for governments and local communities to preserve protected areas and species, and heightened over-all local awareness of the importance of conservation. Unfortunately, it has also led to numerous problems, and placed undue pressures and threats on the natural and cultural resources that sustain it. From these often-costly lessons, we are learning that the benefits of ecotourism can only be sustained through well-planned and carefully integrated projects.

Infrastructure refers to systems in place, which allow for the efficient functioning of a business activity or concentration of people. One of the early but still often heard arguments is that this industry can be built on existing infrastructure. An important task in development planning is to determine infrastructural capacity and expected demands before proceeding with physical development. Infrastructure technology has progressed from ground-based systems to include satellite telecommunications. Infrastructural improvements are generally very expensive undertakings and often deemed the responsibility of the government.

A new approach to architecture and physical facilities planning is needed, not only in tourism, but in all human activities, if we are really going to stop the irreversible damage to the environment, further pollution, and depletion of energy sources. This new approach should be based on the concept of *eco design,* which may be defined as "any form of design that minimizes negative environmental impacts, by integrating itself into the surrounding ecosystem" (Ceballos-Lascurain, 1997)[2].

Tourism facilities should be particularly designed in an environment-friendly way, since these are frequently located in areas of great scenic beauty and ecological significance. Application of appropriate waste treatment methods and the use of alternative energy sources (especially in remote locations) are especially important items to be considered for effective infrastructure designing for ecotourism. Physical facilities should be technologically viable and adequate, and also socially acceptable and economically feasible. Joint ventures, communication and working with funding agencies can assist with addressing the expense of technologies. Physical planning and building (planning for expansion) should always be long term endeavours.

Infrastructure in Ecotourism: How far they are Significant?

The activity of creating shelter is one of the most ancient of the human races' activities. So the built form of any culture reflects its deepest roots as well as its current direction. A tourism product that can sensitively recognize this, work within it or complement it, will guarantee a unique product that is attractive to the travelling customer. Accommodation and facilities of an eco-tourist operation set the tone for the guest experience. It is essential that the facilities are well planned and designed in all respects and that they inform, and are informed by, the operations of the property and are harmonious with the culture and natural landscape of their environment.

Generally, the ecotourism structure should be a composite mixture of contemporary, vernacular and historical building forms. Most importantly it has to conceive the idea of 'low density' and 'low impact facility'. The structure which is been proposed for the tourism activity should be constructed in such a way that, the damage to the natural resources would be minimum both in the short as well as long run. The basic idea is to give emphasis on '*ECO*' in the Ecotourism.[3] The educational, cost and design factors of the built structure are also quite important. Options such as utilizing the existing structure with adequate modifications or new structures with minimal impact would be more preferable. Existing structure has the advantage of its representative value with the past and hence it should get promoted.

The structures and constructions should be reckoned with the most minimum facilities and should be in the lines of the existing architecture of the area. Invariably the cost factor will also play as a guiding force in the whole activity. Table 26.1 below encompasses few of the model types under various categories, which one can think of for ecotourism infrastructures.

Table 26.1: Ecotourism: Model Types Categories

Vernacular building types	*Historical building types*	*Contemporary structures*	*Portable and low impact structures*
Indigenous structures, Grass huts, Mud structures, Caves, Elevated halls, House boats, Reed platforms and building, Yarts, Tree platforms, Ice houses, Cliff dwellings, Stick houses etc.	Developed vernaculars, colonial architecture, residentially derived styles, commercially developed styles, Military architecture etc.	Prefabricated structures, Masonry, Glass fibre. Traditional tourism, Cottage colonies, Inns, Guest houses, Homes etc.	Rigid tents, Collapsible tents, Elevated huts, Inflatable structure etc.

(Adapted from the Weaver 2001)[3]

Sustainability: A Contested Theme

In the contemporary world, the word 'sustainability' is often seen quoted. Put in simpler terms, sustainability is providing for the best for people and the environment both now and in the indefinite future. In the terms of the 1987[4] Bruntland Report, sustainability is: "*Meeting the needs of the present generation without compromising the ability of future generations to meet their needs.*" This is very much like the "seventh generation" philosophy of the Native American Iroquois Confederacy, mandating that chiefs always consider the effects of their actions on their descendants through the seventh generation in the future.

The original term "sustainable development," was adopted by the Agenda 21 Programme of the United Nations. Some people now object to the term "sustainable development" as an umbrella term since it implies continued development, and insist that it should be reserved only for development activities. "Sustainability", then, is nowadays used as an umbrella term for all human activities. The Brundtland Committee Report had identified the following principles for the sustainable development:

- Inter-generational equity
- Intra-generational equity
- Public participation
- Environmental protection integral to economic development

Sustainable Infrastructure: Considerations Underway

Ecotourism has been hailed as one of the ways to achieve the sustainable development. Sustainable tourism, otherwise it is called Ecotourism, considers all the important principles of sustainable development in its ambit. Putting it on the other way around Ecotourism, to be successful, must promote sustainable development by establishing a durable productive base that allows local inhabitants and ecotourist service providers to enjoy rising standards of living. An ecotourism project must incorporate the social dimensions of productive organization and environmental conservation. The ecological, social and economical factors are to be given special emphasis in the sustainable tourism project.

Advocating the minimal consumption pattern of life, the concept of sustainable development demands for embracing the changes emanating from the global interdependence, environmental stewardship, social responsibility and economic viability. Global as well as local issues have to be properly represented in any development project, where sustainability matters. The underlying philosophy in the sustainable building design is to attain a balance between the human needs with the capacity of the natural and cultural environments. Minimizing the damage both in the natural as well as the cultural setting, the sustainable design encompasses the following principles:

- Utilizing the building and its architecture as a tool for educating the importance of environment and the culture.
- Reestablishing the *Human-Nature* connection for benefits ranging from spiritual, emotional and therapeutic.
- Promotion of new human values and lifestyles for achieving a harmonious relationship with the local, regional and global resources and environments.
- Awareness creation on various aspects ranging from the energy minimization to waste management and natural conservation to cultural preservation.
- Nurture living cultures to perpetuate indigenous responsiveness to, and harmony with, local environmental factors.
- Reflection on the historical as well as cultural values of the area.

Marrying the vision of the experience with the environmental sustainability and the constraints of a successful business venture is to be essentially carried out in the infrastructure development. Major question to be addressed is the complexity arising from comfort and extravaganza. There should not be any compromise on comfort, but a standard should be developed, as comfort appears different for different people.

Development strategies for Ecotourism accommodation should include following attributes:

- Maximum interface with the natural environment.
- Downsizing of central facilities offset by operational planning and efficiencies.
- Facilities and activities that minimize the emphasis on individual accommodations.
- Communal living option.
- Services and facilities provided by local development partners.
- Self-sufficiency of the occupant to minimize operational overhead and central facilities.
- Use of low impact and lower cost structures.
- Use of indigenous building types that capture local sense of place.
- Use of prefabricated building types.
- Operates to de-emphasize site requirements.
- An emphasis on landscaping.

Physical Planning

Infrastructure Design and Site Planning

Site planning and design is a process, which involves in an integrated way the issues of land use, human circulation, structures, facilities and utilities within the natural and human environment. In order to ensure harmony between tourism developments and environmental protection, it is indispensable to apply sensitive design of infrastructure, master site planning, ecologically and socially conscious site design, and landscaping.

Preserving the special characteristics of a tourism destination demands an in-depth understanding of the natural systems on the site, as well as an immersion into the time-tested cultural responses to that environment's opportunities and constraints. If we want to change the way we build traditional tourism facilities, we need a new way of thinking about site planning and design, which involves a holistic approach. Sustainable site planning and design can lead to a better integration of physical facilities for tourism and their site and surroundings and can indeed help to lessen the environmental impact of these facilities.

Site planning and design for any tourism facility must clearly indicate the process of ordering human actions and works in a specific tract of land. In addition to constituting a graphic representation (to scale) that shows location, layout, general size and shape, and orientation of the different elements of the project, site planning and design should indicate the sequence of activities that make up the project, clearly establishing a space-time interaction. Also, it should ensure that all on-site human activities should have a minimum negative impact on the natural and human environments.

The site planning and design for any tourism facility must be, first of all, an instrument that safeguards the sustainability and conservation of the surrounding natural and cultural heritage. Not only should it conserve the natural ecosystems but it must also contribute to repairing and restoring the environmental damages that may already be present in the site. The development of the site should strive to leave the site better off after development than before.

Zoning is a very important tool in the site planning and design process. It is the process of applying different management objectives and regulations to different parts or zones of a specific area.

The success of any tourism facility (including eco-lodges) often lies on the initial process of site evaluation and selection. Careful evaluation, in some instances, may reveal that the site is not appropriate for developing the facility. All considerations involved in selecting the most appropriate site will be essential in any forthcoming decisions dealing with design and construction.

Considering the increasing visitation to wilderness areas over the past decade and the resultant effects on the carrying capacities of the ecosystems, it would be prudent to select sites for developing ecotourism facilities that are situated just outside the nature preserves, although this is not always possible since some of the preserves are very large. As such, a well-conducted site evaluation can assist developers in finding alternatives to developing in protected areas. Selection of an appropriate site is critical for ensuring the sustainability and viability of an eco-lodge. The selected site should support the lodge within natural and biophysical resource limits while offering ecotourists the opportunity to experience and enjoy the environment.

Specific Examples of Best Practice from Selected Countries

1. *Bolivia*: Chalalan Eco-lodge is located in a primary rainforest area and was developed with as little ecosystem disturbance as possible during the construction process. The lodge now boasts populations of monkeys, macaws and other flagship species that are a big attraction to clients.
2. *Thailand*: The Banyan Tree Hotel in Phuket was built on a site formerly used for tin mining. In order to restore the land, the resort did extensive landscaping and planted hardy native plants such as Casuarina trees and palms. Construction materials were chosen to minimise the use of scarce local forestry resources. Instead, all villas were built of bricks, with clay roof tiles. As a result of these efforts, wildlife has been returning to the area.
3. *Nepal*: Narayani Safari Hotel and Lodge were built on the periphery of the Royal Chitwan National Park (this was preferred to building within the national park), located in land of low agricultural value. At the planning stage, permission to build tourist facilities inside the Park could have been obtained (seven tourist lodges were already located inside the park). However the company believed that the National Parks should be set aside primarily for conservation and decide to site the facilities outside of the protected area and apply for permission to use the park only for wildlife viewing trips.

Frequently, in those sites that are more appropriate for eco-lodge development, there are limited or no infrastructural elements or public services, because of typical isolation and remoteness.

It is important to analyse how much infrastructure should be provided by the local authorities, and how much by the private sector. Since the extra service demand is often only used part of the year (seasonal), and takes precedence over use by local communities, tourism providers must invest in their own infrastructure needs. Both communities and tourism sector should benefit from infrastructure development.

Best Practice Guidelines

- Ensure that the site plan is environment-friendly, minimising negative impacts on the natural landscape, local biodiversity and any existing cultural features found nearby.
- When selecting a site for any tourism facility, analyse the local availability of the following infrastructure elements and public utilities and services:
  - — Conventional systems for providing electricity, drinking water, sewage, telephone line, public lighting.
  - — Communication means: highways, roads, trails (tracks), airport, landing fields, railway, docks etc.
  - — Postal service, refuses collection and disposal, medical services, schools, commercial facilities etc.
  - — Local means of transport: land motor vehicles (bus, taxi, rent-a-car etc.); regular commercial, charter, or private flights; motor boats, cruise ships, yachts, ferries; railway (schedules of nearest railway station) etc.
- Carry out a feasibility analysis for each site option before the particular site is selected. This analysis should include biophysical features of the site (climate, land, vegetation, wildlife, etc.) as well as socio-cultural features (local communities, traditional villages, archaeological sites etc.), and business aspects (comparative analysis of the different possible sites, suitability of location, market niche, profitability, long-term economic sustainability).
- In the case of an eco-lodge, select a site, which is within, or sufficiently near, a natural area of great ecotourism potential (i.e. an area with high flora and fauna values; perhaps supplemented by important cultural values). The eco-lodge itself

should not directly affect the focal (or flagship) ecotourism attractions of the area or visually compete with these attractions. In other words, a balance must be struck between easy accessibility to outstanding natural areas and minimisation of negative biophysical and cultural impacts (including visual impacts) of the environs.

- Consider travel distances as a siting criterion and the natural and cultural features that can be accessed from the site.
- Consider proximity of the lodging facility to airports and major transportation routes in the region. However, in the case of an eco-lodge, it should not be too close to airports or major transportation routes due to the excessive noise and pollution.
- Analyse what benefits the local people will receive from the development of a tourism facility.
- Study possible environmental and cultural impacts for each site for both the construction and operation phases of the development of the eco-lodge. Consider both mid and long-term scenarios for development impacts.
- Contract appropriate professionals to develop the site plan. The site plan should prominently include an analysis of the site in regard to its major biophysical characteristics and measures to minimise impacts.
- Every site plan should carefully consider existing and future vegetation. Tree preservation and reforestation are key elements. Regarding trees, always keep in mind the quadruple principles: remove, relocate, replace, restore.
- Carry out your landscaping using native plant species. Exotic plants are always out of place in a natural environment.

Waste Management

The management of waste is a crucial conservation problem in both urban and rural areas, including natural areas (legally protected or not). The basic premise of a tourism facility operation is minimising waste generation, since this is one of the main causes of degradation of the surrounding environment. The best attitude towards trash is to avoid it as much as possible, instead of finding ways of treating it.

Try to use biodegradation practices as much as possible. In the process of biodegradation, microorganisms break down the products of other living things and incorporate them back into the ecosystem. Biodegradable or bio-convertible materials include anything that is organic. Plastics are not considered includable in this category, despite industry contention that they are.

A material doesn't become waste until it is thrown into the dustbin. If a material can be reused it is a resource, not waste.

Best Practice Guidelines

- As regards waste management, always apply the quadruple principle: recycle, reuse, refuse, and reduce.
- Reuse wastewater (both grey and black) as much as possible. Create systems in which water goes through several uses before being disposed of, utilising it for flushing toilets, and as irrigation or fertiliser for cultivations (whenever possible). Always avoid using potable water for irrigating. In case you reuse both grey and black waters, separate lines and septic systems must be installed.
- In general, use products that minimise waste and are not toxic. Whenever possible, convert biodegradable waste to compost, utilise the biomass or submit the waste to digestive anaerobic systems.
- Whenever possible, use constructed wetlands treatment systems, which are engineered systems that have been designed and constructed to utilise the natural processes involving wetland vegetation, soils, and their associated microbial assemblages to assist in treating wastewater. They are designed to take advantage of many of the same processes that occur in natural wetlands, but do so within a more controlled environment.
- Compost biodegradable wastes and make use of the resulting fertiliser or sell it.
- In order to save water, use dry toilets (and pit latrines, in cases of extreme isolation and budget restrictions), always designed to minimise negative impacts on the environment.

Eco-Design of Tourism Facilities

It is important to remember that economic benefits come from environment-friendly facilities and technologies.

Eco-lodges are often located in remote and wild areas, and therefore very few typical infrastructural elements and services found in more traditional settings are available, such as access by paved highway, public transportation services, electric and telephone lines, piped potable water, public drainage and sewage, refuse collection and disposal, nearby school and medical services, shopping areas etc.

Specific Examples of Best Practice from Selected Countries

1. *Peru*: Manu Lodge, located in the Amazonian rainforest, is a rustic facility using local building materials (wood, palm leaves, bamboos), designed in such a way that it is practically hidden in the deep forest, its building height well below the tree line. Situated in Manu Biosphere Reserve, one of the areas of major biodiversity in the world (1,100 species of birds), the lodge attracts ecotourists from around the world (especially US bird watchers), who are willing to pay US$200 per night, without having such conventional amenities as electric light (kerosene lamps are used instead), air conditioning or jacuzzi.
2. *St. John, U.S. Virgin Islands*: As an example of how apart from conventional materials (sand, gravel, wood), other materials recycled from other sources may be used, nearly all building materials at the Harmony Resort are made of innovative products from the waste stream of other industries, such as sawdust and plastic. The resort is now attempting "close the loop" on its own waste stream, by recycling bottles on-site into products such as glasses and tiles and recycling aluminium cans into table legs.
3. *Costa Rica:* As an example of "small is beautiful", currently, 65 per cent of all hotels in Costa Rica have less than 40 rooms (46% of the total offer of available rooms in the country) and 75 per cent of tourist ventures are in rural areas. This scale of facilities means providing a well distributed tourism benefits in many areas of the country, in many cases fostering family businesses. Research has shown that nature-oriented tourists prefer smaller, intimate accommodations than big, multi-storey hotels.
4. *Nepal*: In the Narayani Safari Hotel and Lodge, located just outside of the Royal Chitwan National Park, wooden beams, door and window frames from derelict houses on the land were reused for building the single-storey cottages of the hotel and lodge. Cottage roofs are made of locally made tiles. Solar panels are used for water heating, lighting is provided through kerosene lamps.

For this reason, a totally new and different approach to physical planning is required, one based on a high level of functional, energy and food self-sufficiency. Before designing and building an eco-lodge, realistically and clearly identify the specific characteristics of isolation and difficulty of access to infrastructural elements and public services and define before hand the level of self-sufficiency you wish or need to attain.

Many nature tourists do not expect, in a poor rural area, the facilities found in rich cities and beach resorts. Some enjoy roughing it for a while, and are even prepared to pay more for the privilege.

Best Practice Guidelines

- Ensure that the design allows for possible future expansion, modifications and retro-fitting.
- Minimise negative environmental impacts on site during the construction process. Clean up when you finish building each phase.
- Apply life-cycle assessments of all building materials to ensure that the best environmental and cost options are chosen.
- In the case of eco-lodges and other tourism facilities located in natural areas special attention should be paid to the conservation principles.
- Create the most appropriate access to your eco-lodge, striking the right balance between ease of approach and minimisation of negative impacts on the natural environment. Limit the number of entry points to your site (preferably only one) in order to facilitate surveillance control and management.
- Remember that paved highways are usually an invitation for all kinds of human settlements and that they stimulate branching effects. If there is no existing access paved highway and the distance from the tourism distribution points is a considerable one, sometimes there is less impact in developing a landing field for light planes than to build a paved highway.
- Keep in mind that construction of highways, roads and other motorways within a natural area or near to it causes negative impacts to the natural resources that one wishes to protect. Consequently, they should be strictly limited and justified only if there are no other viable solutions.

- Avoid building highways or motor roads with a width of more than 5m within a protected area or ecotourism destination. Wider roads become veritable barriers for wildlife mobility and also mar the natural landscape.
- Whenever possible, use waterways (fluvial, ocean or lake), ensuring the use of boats with minimum negative impact. Avoid the use of internal combustion motors as much as possible within fragile areas such as mangroves or marshes. Consider the use of electric-powered boats. For short distances and wildlife-watching excursions it is best to use a rowboat, with good stability. These waterways are often crucial to local fishermen and hence the local economy. Thus, their use is a sensitive issue and needs to be negotiated.
- Minimise impermeable surfaces when possible to reduce runoff and maximise groundwater recharge.
- Not all protected areas should have built facilities inside their boundaries. Sometimes these facilities are preferably placed in the nearby community or in the buffer zone. Always proceed according to the management plan of the protected area (if it exists).
- Interpretative centres (as part of a broader interpretative programme) are needed in most ecotourism destinations, including protected areas, but each case should be carefully analysed.
- Always endeavour to harmonise tourism facilities with the surrounding environment (both natural and cultural). Use architectural forms in harmony with the natural landscape (vegetation and land forms), designing with long-term environmental criteria. A tourism facility should always possess a sense of place.
- Ensure that your tourism facility minimises impacts on the natural and cultural environment and also acts as a tool for biodiversity conservation and to enhance natural ecosystems.
- Whenever possible, use local building materials and local hand labour.
- If local building materials are not available, bring materials from elsewhere, ensuring ease and economy of transportation and on-site assembly.

Various factors are to be taken into account before going for designing the building. The following box examines the parameters, which need to be taken into consideration for the selection of the building materials.

Environmentally-Sensitive Building Materials

The complete life cycle energy, environmental and waste implications of each building material must be examined. This cradle-to-grave analysis is the tracing of a material or product, and its byproducts, from its initial source availability and extraction through refinement, fabrication, treatment and additives, transportation, use and eventual reuse or disposal. This tracing includes the tabulation of energy consumed and the environmental impacts of each action and material.

Source of raw ingredients (renewable? sustainable? locally available? nontoxic?)

X

Raw material extraction (energy input? habitat destruction? topsoil erosion? siltation/pollution from runoff?)

X

Transportation (most local source? fuel consumption? air pollution?)

X

Processing and/or manufacturing (energy input? air/water/noise pollution? waste generation and disposal?)

X

Treatments and additives (use of petrochemicals? exposure to and disposal of hazardous materials?)

X

Use and operation (energy requirements? longevity of products used? indoor air quality? waste generation?)

X

Resource recovery/disposal (potential for recycling/reusing materials? disposal of solid/toxic wastes?)

(Adapted from Anderson, 1993)[5]

The box below depicts the Green Report Card, which can be utilized for evaluating any sites for designing the structure for ecotourism:

The Green Report Card for Evaluating Ecotourism Facilities

- Is the scale of the development appropriate for the local community and the capacity of the environment to support the facility?
- Were the members of the local community actively involved in the planning and construction of the facility?
- Are members of the local community involved in day-to-day operations of the facility?
- Is the facility to be a phased development, and if so, are the subsequent phases provided for in a manner that allows for minimal disruption of the environment and the existing facility?
- Are roads and trails placed to minimize intrusion on the environment?
- Does the facility design utilize traditional cultural building forms and materials found in the immediate area?
- Does the design of the facility encourage the visitor to look at the natural environment in a new way?
- Are there any contradictions to the ecotourism mission of conservation apparent in the facility?
- Are provisions such as a library, laboratory, discussions, guided walks or other experimental settings made to provide visitors with educational opportunities?
- Is the energy source(s) environmentally sound and sustainable?
- Are building materials free of toxic or non-biodegradable agents?
- Are appropriate technologies employed for the treatment of organic wastes and other wastes? Is recycling practiced?
- Are the building structures and paved areas properly sited to prevent erosion?
- Are the furnishings and other lodging accommodations consistent with the architectural theme and environmental parameters?
- Is accommodation made for older guests and physically-disabled individuals?
- Does the staff seem informed about ecotourism and the facility's design and operational features?

(Adapted from Anderson, 1993)[5]

Energy Conservation, Bio-Climatic Design and Alternative Energy Sources

In every type of tourism facility and operation saving energy, water and other precious resources by applying the most appropriate available technologies, should occupy the highest priority.

There are various methods for meeting the energy and vital resource needs of a tourism facility. The most common practices around the world involve recurring to conventional methods, such as public utilities (electricity, water, gas, telephone), conventional fossil fuels, and fuelwood. But there are many alternative practices that should be advanced in all types of buildings, including tourism facilities (both in urban and rural settings), since they are more environment-friendly, minimise impacts and also reduce costs. Some of these alternative practices involve design techniques that allow for natural ventilation, heating, lighting, and rain water catchments, while others use energy sources such as solar power, wind power, low-scale hydroelectricity, geothermal power, natural gas and biomass. Solar energy especially has an enormous potential (both for heating water and for generating electricity) but so far is not exploited to its full potential in the majority of countries and regions.

Tourism facilities, including eco-lodges, should make use of this widespectrum of alternative technologies. The extra energy that could be made available through the use of various conservation techniques is one of our most underused and invisible resources. Low-energy lighting should be widely preferred over conventional, incandescent bulbs. Water use in toilets and showers must also be drastically reduced and this can be achieved in many ways, including such easy techniques as placing bottles filled with sand in the flushing tank, flow-reducing shower heads, etc.

Best Practice Guidelines

- — Apply bio-climatic design criteria, which means understanding the physical setting of the facility, the local climate, including prevailing winds, solar radiation, appropriate local materials, biodegradation, surrounding vegetation, etc., and applying simple physical and biological principles.
- — Consider the sun's positioning when choosing the orientation of the tourism facilities so as to maximise use of natural light and trap incoming solar radiation to heat interior spaces, taking note that in the northern hemisphere the sun is mainly shining from the south and in the southern hemisphere the situation is reversed. To maximise the "solar window", the slope of the roof (in degrees) should be roughly equal to the latitude of the site.
- — Analyse in each case the convenience of applying solar energy for heating water and for generating electricity (photovoltaic

systems), as well as other alternative, environment-friendly energy sources, such as wind power, biomass, geothermal energy, etc.

— A good option in isolated areas with no access to a conventional power grid is the use of photovoltaic cells (based on the use of silicon) for converting solar energy in electric energy (usually 12 volts DC). Some models offer both options of 12 volts DC and 110 volts AC (using the second alternative, the battery obviously drains down faster). There is presently a choice in photovoltaic cells between four technologies: monocrystalline, polycrystalline, semicrystalline, and amorphous silicon. The four technologies are now available commercially (fundamentally Japanese, German, and U.S. producers) and all have different comparative advantages. Typical monocrystalline silicon cells have an efficiency of about 10-20 per cent, but are quite more expensive than polycrystalline cells (which have an efficiency of 4-5%). Amorphous cells are even cheaper but also less effective. In every case, electric storage is by lead-acid deep-cycle batteries, similar to those used in golf carts.

— Use techniques such as the so-called Trombe wall, which consists of storing solar energy in a sun-facing wall, made of heavy masonry material, such as brick, stone, block, or earth, with a dark-colored surface towards the sun. As the sun's rays shine on the wall, generated heat is stored and circulated passively (especially during the evening) through wall vents into the living areas.

— The architect should pay special attention to water management when planning a tourism facility, especially given the critical water situation in some nations, which at times experience severe droughts. Use water catchment methods whenever possible. At all times, maintaining the drinking quality of this scarce commodity is of paramount importance.

— In the design of the lodge or other tourism facility apply cross-ventilation, which implies placing openings in opposite and parallel walls so as to induce natural air flow from outside and cooling interior spaces. In this way you will be contributing to the elimination of air conditioning systems, which consume enormous amounts of electricity and also cause damages to the ozone layer.

Specific Examples of Best Practice from Selected Countries

1. *Australia*: J's Bay YHA Hostel in New South Wales, has designed all rooms to maximise the natural ventilation from the prevailing winds, and external screens have been fitted to shade windows from direct sunlight. Ceilings are well insulated to minimise heat transfer into the rooms. No air conditioning is required to cool the rooms. A single ceiling fan, which requires a fraction of the energy to operate, is enough to keep the rooms comfortable cool throughout the year.
2. *Hungary*: By using energy-saving light bulbs, the Budapest Hilton has reduced the energy bill by 13 per cent, or US$40,000 per year. Low flush toilets have been used in the newly restored bathrooms. Guests can choose to re-use towels or have them changed daily, which reduces water and laundry detergent use, and effluent (IHRA/UNEP, 1996).
3. Malaysia: River tours from the Sukau Rainforest Lodge in Sabah make use of quiet electric motors when viewing wildlife, since it was recognised by the owner that electric motors can significantly reduce disturbance and increase the likelihood of seeing wildlife, apart from minimising pollution effects.
4. *India*: The Hotel Guestline Days in Tirupati is designed in a Y shape to optimise the use of natural light. The ceiling on the top floor is lined with a 75mm thick layer of expanded polysterene which acts as a heat insulator, keeps the indoor temperature at a lower level and brings down air conditioning costs. Showers are installed instead of baths. Rain water is collected (especially during the monsoons) and used for many purposes, including toilet flushing.
5. *Canada*: By simply putting sand-filled bottles in toilet tanks to reduce flush volumes in 25 per cent of its rooms, the Hotel Beausejour in New Brunswick saved 100,000 litres of water per year.
6. *Australia*: At the Frogs Hollow Lodge in Darwin, up to 50 per cent of hot water requirements are met by solar water heaters on the roof.

Architectural Options: A Traditional Way

The traditional way of architectural designing could be adopted in the Ecotourism plans. The Management plans could ascribe the designs from the traditional styles of *Feng Shui* and *Vastu Shastra*. The *Feng Shui* of China and *Vastu Shatra* of India are the two architectural options which can be adopted for the eco-designing of the ecotourism infrastructures.

Feng Shui

The Chinese based the science and art of siting and orienting buildings upon the workings of earth forces, which are known as *Feng Shui*, literally "wind" and "water". Rooted deeply with ecology, geology, astronomy, and hydrology. Feng Shui aims to harmonize nature with the built environment. Feng Shui offers an ancient method of site planning that adds to balance and harmony for those who live on the land and for the earth itself.

The Chinese believe that the earth is cris-crossed with energy lines that affect and are affected by virtually all geographical and topographical phenomena. This means that humans are affected by our environments, and we in turn modify our environment by what we do within it.

Chinese believe that Chi (the Chinese word for energy) pervades every element in the cosmos and is the beginning of all life. Yin and yang are two kinds of Chi with opposite characteristics. Yin is characterized as female; negative, and passive, while yang is characterized as male: positive, and active. Only when yin and yang meet and stay in balance can life begin. Their continually complementary interaction creates the ideally harmonious site.

Vastu Shastra

In India since time immemorial, eternal principles of Vastu Shastra have been applied in the designing of villages, towns, and cities apart from temples, palaces, public buildings and residences. Vastu is derived from the verb "Vasa" meaning to "dwell" or "place of residence."

Vastu Shastra states that every form creates a concentration or dispersal of cosmic and earth energies, which are harmful or beneficial to human beings. Therefore, Vastu Shastra can bring about harmony between people, nature, and buildings. Since a property represents a fixed form, it will radiate positive as well as negative energies depending upon its shape, properties, direction and location.

Vastu Shastra recognizes with simple logic that all people live in an environment influenced by the five basic elements: *Akash* (Sky), *Prithvi* (Earth), *Pani* (Water), *Agni* (Fire), and *Vayoo* (Wind) and followers of

Vastu Shastra will respect and be in harmony with these forces when building a lodge.

Paraa Lodge, is located in Northwestern Uganda, just a few miles downstream from Murchison Falls, on the banks of the River Nile. This area has one of the largest breeding populations of the ferocious Nile crocodile. The existing, very ugly, 30-year-old lodge was in a total mess and had been ransacked by Idi Amin's army in the early 1980s. A project was undertaken to prepare an impact study for the rehabilitation and additions and suggest ways to turn the lodge into an ecotourism facility. What is interesting about this project is that the Hindu client wanted to have the entry from the North East as this is considered auspicious in Vastu Shastra.

Conclusion

The development of ecotourism as a strategy to promote sustainable resource management and tourism could positively work if brokers, locals and tourists behave to a tolerable level wherein no damage (or may be minimal and could be mitigated) to the local ecosystem is created. Moreover, the internalization of brokers, locals and tourists of appropriate behaviours corresponding to their respective statuses and roles have to be achieved by modifying the quality of their involvement in the tourism industry. The tourism system model considers that their behaviours could be modified or regulated by any of them by the proper use of power geared toward the realization of the ecotourism image desired as a departure from the tarnished image of mass tourism. In other words, power and the success and failure of tourism are not only the monopoly of tourists as traditionally perceived but by all those involved directly or not in the tourism business.

As landscape architects and planners, are we truly the stewards of the land and oceans or are we are the destroyer of these natural resources, which are essential for the existence of the mankind. Can we truly reverse the havoc on mother earth that is being caused by our fellow human beings? In this millennium, we have a lot of work ahead of us. Either we take up the challenge and create a livable, breathable world for future generations or we could take the easy road and be comfortably numb.

There is a lot to learn from the nature and the indigenous tribes of the world and their relationship to mother earth. Their landscapes were "one with the land" and we do not see why we should not be able to use these principles for the design of ecotourism facilities. A holistic approach should be adopted towards the sustainable infrastructure designing for the ecotourism facilities, so that the infrastructures conform the ecological and conservation principles and strengthen ecotourism to develop as an effective tool for sustainable development.

NOTES

1. http://www.enterprisens.org.nz/publications/other/2003/Massey_report.pdf
2. Ceballos-Lascuráin, H., 1997. *Tourism, Ecotourism, and Protected Areas*: The state of nature-based tourism around the world and guidelines.
3. Weaver, B. David, 2001. *The Encyclopedia of Ecotourism*, ed, CABI Publishing, USA.
4. Bruntland, G. (Ed), 1987. *Our Common Future*. World Commission on Environment and Development. 1987. Oxford University Press, Oxford.
5. Anderson, D.L., 1993. A window to the natural world: The design of ecotourism facilities. In *Ecotourism: A guide for planners and managers*, Volume 1, K. Lindberg and B. Hawkins (eds.), 116-133. N. Bennington, Vermont.

27

Ecotourism Marketing

With Special Reference to Madhya Pradesh, the Heart of India

A.K. Bhattacharya and *Aparupa Chatterjee*

Introduction

Ecotourism is a component in the field of sustainable tourism. However, it is important to stress that all tourism activities, be they geared to holidays, business, conferences, congresses or fairs, health, adventure or ecotourism, should aim to be sustainable. This means that the planning and development of tourism infrastructure, its subsequent operation, and its marketing should focus on environmental, social cultural and economic sustainability criteria. The strong orientation of the ecotourism field towards principles, guidelines, and certification based on sustainability standards gives it an unusual position in the tourism field. The International Ecotourism Society in 1991 produced one of the earliest definitions "Ecotourism is responsible travel to natural areas that conserves the environment and sustains the well being of local people." In the years since the concept was first defined, a general consensus has formed on the basic elements of ecotourism:

- Contributes to conservation of biodiversity.
- Sustains the well being of local people.
- Includes an interpretation/learning experience.

- Involves responsible action on the part of tourists and the tourism industry.
- Is delivered primarily to small groups by small-scale businesses
- Requires the lowest possible consumption of non-renewable resources.
- Stresses local participation, ownership and business opportunities, particularly for rural people.

Ecotourism as a Market Segment

Ecotourism is a small scale but rapidly growing industry, a niche market that is governed by market forces and regulations. It is primarily advertised as a form of nature tourism. Ecotourism is one of the fastest growing sectors in the tourism industry at present. The market for nature holidays is certainly a growing one. In 1993, the World Tourism Organization (WTO) has estimated that nature tourism generates 7 per cent of all international travel expenditure, where the total annual global earning from tourism is nearly $3 trillion and 212 million people are employed (Eagle, 1997)[1]. Year 2002 was declared as an International Year of Ecotourism (IYE) by the UN, which reflects its global importance. Some countries, companies and destinations have social and environmental policies and Programmes, while others do not. This has led to confusion worldwide about the meaning of the term ecotourism as it is applied in the marketplace. *Ecotourism has to be destination based.* India, or for that matter any country cannot be sold as a whole. A particular area has to be chosen for doing the same.

How ecotourism fits into the larger tourism marketplace? Both adventure tourism and ecotourism are components of nature tourism, while ecotourism has stronger links to rural and cultural tourism than to adventure tourism. In ecotourism the prime motivation is generally the observation and appreciation of natural features and related cultural assets, whereas in adventure tourism it is physical exercise and challenging situations in natural environments.

From a functional viewpoint, ecotourism is mostly individual or small-scale tourism (tour groups up to 25, and hotels with fewer than 100 beds) that is operated by small and medium sized companies in natural areas. It concentrates on leading and accommodating small groups in natural areas in an educational manner through interpretive materials and local specialist guides.

Ecotourism Markets are not Homogeneous

Even if a definition is established, ecotourism markets are not a homogeneous group. In the same way that there is a spectrum of products/experiences which may be termed ecotourism, there is also great variation in the activities, motivations and characteristics of markets. Indeed, individuals may be interested in a number of overlapping experiences and activities. These elements of understanding were:

- Taking vacations in natural locations
- Understanding the term ecotourism
- Attitudes towards nature and nature based-tourism
- Reasons for choosing where to take a vacation, in particular the role of nature and learning about nature
- The extent of planning for the vacation
- Nature-based activities conducted while on vacation

The Indian Tourism Market: A Brief Sketch

A study conducted in collaboration with the Confederation of Indian Industry (CII) and Federation of Indian Chambers of Commerce and Industry (FICCI) said. Indian domestic and outbound tourism will become an $80 billion industry by the year 2007. Tourism comprises only 5.2 per cent of the country's gross domestic product due to low per capita income. But things are changing as domestic travel and outbound tourism has grown by 23 and 25 per cent in the last three to five years. The trend is likely to intensify over the next five years, the study said. "Industry should see a higher growth rate in inbound travel and outbound tourism will keep pace with growth of inbound tourism."

- India averages 2.54 million International Tourists annually.
- India is one of the top 4 domestic tourism markets in the world. The domestic market has kept the tourism industry afloat with an annual average of 234 million tourists.
- Travel receipts in India amounted to $ 3.04 billion in 2001.
- Tourism activities provide employment to 10 million people directly and additional 10 million indirectly.

Table 27.1: Number of foreign tourists to India

Year	*Number of foreign tourists*
1991	17000
1992	19000
1993	18000
1994	19500
1995	21000
1996	22500
1997	24000
1998	24000
1999	25000
2000	26500

Source: CSO 2000.

Year	*Number of Tourists leaving home Worldwide*
1964	100 Million
1974	200 Million
1992	500 Million
2001	700 Million
2020	1.50 Billion expected

Table 27.2: Tourism, Ecotourism and Wildlife Tourism (2004)

	Feature	*Tourism*	*Ecotourism*	*Wildlife tourism*
WORLD	Arrivals (crores)	76.3	30.5-45.5 (40-60%)	15.0-30.0 (20-40%)
	Earnings (Rs. crore)	28,03,500	1,98,000	—
INDIA	Arrivals (lakh)	33.7	13-20	7-13
	Earnings (Rs. crores)	22,000	14500	

Table 27.3: Tourism Highlights 2004

World	*India*	*MP*
• Total arrivals: 76.5 crores • Earnings from tourism: $ 62300 crore (Rs. 30, 00,000 crore) • Growth rate: 11% • Share in export of goods and services: 6% • Share in services exports: 30%	• Total arrivals: 33.7 lakh • Earnings from tourism: $ 481.0 crore (Rs. 22,000 crore) • Domestic tourist visits: 36.7 crore • Indians going abroad: 62 lakh • Share of tourism in GDP: 4.9% • Share of India in world arrivals: 0.44%	• Madhya Pradesh with an annual tourism influx of around 0.11 million generates benefits equivalent to Rs. 637 crores annually. • The projection for the next 2-3 years will bring in the Madhya Pradesh of 2.69 million employment opportunities and a generated income, which is quite equivalent to Rs 2336.4 crores annually. • 80% of tourism in MP is ecotourism

The Madhya Pradesh Tourism Market: A Brief Sketch

Madhya Pradesh, which is considered the *heart of India*, is a major centre of attraction for tourists. There are numerous places spread all over the State, which offer rejuvenating respite to all classes who can afford. Also, it is those people who seek relaxation or a healthy refreshing change from the daily root of life or the dead routine.

Table 27.4: Wildlife Tourism in MP

Category	*2002-03*	*2003-04*	*2004-05*	*2005-06*
Indian	4,84,736	5,35,149 (+10%)	6,85,917 (+28%)	6,72,712
(-0.2%)	Foreigners	9,643	17,664 (+25%)	22,117 (+25%)
24,466 (+11%)	Revenue to Parks (Rs. Lac)	182.09	212.61 (+16%)	427.07 (+108%)
516.5 (21%)	Rev./Tourist (Rs.)	36.84	38.44 (+4.48%)	60.31 (+83%)
74.10 (23%)	Fees (Indian/ Foreign)	20/200	20/200	25/500

For them the state has much to offer. There are not less than about 450 places of tourist attraction spread all over the state. There are places of outstanding natural beauty like Pachmarhi, Bhedaghat and wildlife resorts like Kanha and Shivpuri, with their spectacular and unique wildlife as well as places of historical, archaeological or religious significance like Bhimbetka, Bagh Caves, Khajuraho, Sanchi, Bhojpur etc. Wildlife tourism has so far been the main focus in the State.

The State of Madhya Pradesh offers a variety of tourist attraction ranging from sculpture to wildlife and hence it is comparable to any top tourist region of the country. As an industry tourism has a great potentiality for earning foreign exchange and also creating employment opportunities. Based on this realisation the tourism has been marked as a priority sector for the Tenth Five Year Plan. In the state the MPTDC was set up with the objective to provide lodging catering transport facilities to nation as well as international tourists. The corporation was expected to provide information regarding tourism and organise adventure and water sports for the tourists. It was also entrusted with the responsibilities to develop new areas having tourism potential. The MPTDC has over the years tried to provide basic infrastructure and tourist facilities in such areas of tourist importance where the presence of private sector is negligible or non-existed.

Tenth Five-Year Plan and Annual Plan 2002-2003

The important provisions kept in Tenth Plan for development of tourism sector are as follows:

- It is planned to develop at least one good tourist spot in each district. An amount of Rs. 30.00 lakh has been proposed for this purpose.
- In the present age promotion and publicity has a great role to play in the marketing of tourism products. An amount of Rs. 1000 lakh has been proposed for this head.
- To develop the minimum basic essential facilities at two new tourist circuits. One at Jabalpur-Bhedaghat-Kanha-Bandhavgarh and Second at Bhopal-Bhojpur-Bhimbetka-Pachmarhi provision of Rs. 50 lakh has been made in the Tenth Plan.
- To promote youth and adventure activities an amount of Rs. 100.00 lakh has been provided.

- Provision for grants to local bodies for construction of day shelters, approach road, water supply, landscaping, construction of *Dharmshala*, development of garden etc. at important tourist centres for tourist amounting to Rs. 95.00 lakh has been made.
- During the Tenth Plan for the year 2002-03 an amount of Rs. 1400.00 lakh and for the year 2003-04 an amount of Rs. 1145.00 lakh has been provided under the 11th Finance Commission to fulfil the special needs of the tourists in the Jabalpur-Satna area.
- An outlay of Rs. 5360.00 lakh has been kept for Tenth Plan and Rs. 2073.00 lakh for Annual Plan 2002-03.

Analysis of Ecotourism Market Demand

There is a lack of systematic information on the origin and location of the demand for ecotourism. Investigations of ecotourism market demand focus on activity preferences travel motivations, demographic profiles and information sources of clients. There are five basic market characteristics for ecotourism—

- Income
- Occupation
- Education
- Age
- Gender

Ecotourism Market Profile

TIES (The International Ecotourism Society, 1994) has constituted the following ecotourists market profile:

Age: 25-54 years old, although age varied with activity and other factors such as cost.

Gender: 50 per cent female and 50 per cent male, although clear differences by activity were found.

Education: 82 per cent were college graduates, a shift in interest in ecotourism from those who have high levels of education to those with less education was also found, indicating an expansion into mainstream markets.

Household composition: No major differences were found between general tourists and experienced Ecotourists.

Party composition: A majority (60%) of experienced ecotourism respondents stated they prefer to travel as a couple, with only 15 per cent stating they preferred to travel with their families, and 13 per cent preferring to travel alone.

Trip duration: The largest group of experienced ecotourists (50%) preferred trips lasting 1-7 days.

Expenditure: Experienced Ecotourists were willing to spend more than general tourists.

Important elements of trip: Experienced Ecotourists top three responses were: (1) wilderness setting, (2) wildlife viewing, (3) hiking/trekking.

Motivations for taking next trip: Experienced Ecotourists top two responses were (1) enjoy scenery/nature, (2) new experiences/places.

Placing Strategies

MP Tourism Development Corporation can play a *major role* in the placing strategies as they are into the business for promotion of tourism in Madhya Pradesh. They have offices all over the country and have the contacts with all the major travel and tour operators of the country. They can take up new initiatives for the placing the new destination for the promotion of ecotourism in particular. For the placing, i.e. the distribution of the place to the various tourists the present *hierarchy of the tour and travel operator* should be *targeted.* The travel and tour operator at the national level should be approached and made aware about the plans and the products to be offered for the different segment of tourists. A joint effort should be made with these travel and tour operators for promotional activities to be taken up. The *corporate sector should be approached* for providing the package tours to the different destination of the state at concessive rates, as perks to their employees. The travel and tour operators of the *southern part* of the country should be approached to offer tours to the state for the people of South India, which at present form a very meagre amount of the total tourists. The strategies should also be made to involve the *tour operators* of the state like *Rajasthan, Jharkhand* and *Andhra Pradesh* so that the people of these can also be targeted from the placing point of view for the scenic spots of Pachmarhi.

Promotional Strategies

The central element in this model is the market mix of price, promotion, product and place, which provides the base for the tourism product. The price of the product is always an important decision point for consumers. The level of promotion influences consumer expectation. The product—the trip is influenced by many factors, including the philosophy of the businesses and government agencies involved. The place is very important in ecotourism because the experience is so heavily influenced by the characteristics of the site visited. Two components, the external business environment and the actual operation of the ecotourism, influence this market mix

Market evaluation model

BUSINESS ENVIRONMENT

↓

MARKET MIX

↑

ECOTOURISM OPERATION

Marketing: Mix Strategy

Product

The firm's offer to the market, which includes the product quality, design, features, branding, and packaging. The product of ecotourism will be based on three important features they are Wilderness, Adventure, and Ethnic. Typically, in and around the ecotourism areas local tribal settlements exist, this will help to promote ethnic/culture tourism.

Price

The amount of money that customer pay for the product. On the basis of the above data the price can be fixed according to the package of the product. The price can be ranging from Rs.400 per day and night stay with lodging and boarding to Rs.1000 per 3 days and 2 nights including lodging and boarding.

Place

The place is very important in ecotourism because the experience is so heavily influenced by the characteristics of the site visited. Here the infrastructure will play major role to boost and attract the tourist. Areas, which are situated near places with good infrastructure (roads, power supply etc.) and near to other places of attraction, have their own advantage to attract the eco-tourists.

Promotion

All the activities which undertake to communicate and promote its products to the target market. In this Global village it is necessary to establish its own web portal to provide the information and get registered online customer. The other means of promotion is print and electronic media as well as national and international *tourism melas.*

Designing the Product

Considering the above key elements the product will be designed on the basis of income, occupation, education, age and gender. The main target will be the domestic and middle-income groups which having the potential to boost the business. The target will be the educated people and mostly 25 to 45 years old although the age will be vary according to the situation. In gender the target will be both men and women.

Trip Characteristics

The trip characteristic will be depending upon Season, Trip length, expenditures and willingness to spend. Season is the limitation of the business. Mostly rainy and winter seasons will be targeted to attract the tourist. The trip length can range from one day to five days to attract the different tourists. One of the important factors to be considered in the product is expenditures and willingness to pay. Thus the main target is middle-income group where the rates have to be moderate, initially ranging from Rs. 400 to Rs.1000.

Activities

Typically, the most popular ecotourism activities are included in the product like admiring nature, hiking, wildlife sightseeing, camping and touring. Additionally, cultural aboriginal experiences to be included. All these features are generally available in ecotourism areas. The number of tourists coming to a place heavily depends on the kind of promotional activity of the place. There is a need to set a collaboration of different departments of the state for the promotional activity to be carried out.

For the promotion of ecotourism in the state the following activities should be taken up —

1. *The meetings and conferences* should be held with all the major national and international travel and tour operators to make them aware of the different products available in the state.
2. Promotion of the various ecotourism destination of the state in the different *travel fairs at the national and international level.*
3. The *press releases* about the different products and the initiatives taken up for development of the area and promotion of ecotourism.
4. A *brochure* should be made which tells about all the ecotourism destination of the state and the packages that can be offered. This brochure should be distributed to various travel and tour operators at the state, national and international level which can help them in planning the itineraries for the tourists. Guidelines containing the do's and don'ts for ecotourism should be distributed with the brochure. This brochure should be translated into different languages for the distribution of the various countries according to the number of tourists coming from these countries.
5. While targeting the foreign tourists special attention should be given to the *cultural activities* apart from the different other attractions of those places, which are brought in the itineraries.
6. *Articles* about the ecotourism destinations in Madhya Pradesh in various national and international travel magazines should be published.
7. A separate *website of ecotourism of Madhya Pradesh* should be launched which gives all kinds of information about the destination, charges, guidelines etc. This site can be a source of information for everyone which includes those who want to take up a package tour, those who want to come to the destination directly (walk-in-clients), travel and tour operators and the general public. The website can prove to be a source of useful and vital information for the foreigners who depend much on the Internet for any kind of information.
8. The *different stories* of all the places of *historical importance,* which are not so popular; like Kalinjar fort, Mahoba, which has got various stories of *Allha* and *Udal* who were the rulers and their sister, traditionally these have been part of folklore, Ajaigarh

fort, Birsingh palace etc. need to be popularized by means of various articles in print media, promotional material and in different travel programmes floated in the television. *Promoting such stories* about local rulers and legends would also help in reviving the culture, increase the interest of people and the responsible department will also have many things to promote. It would also help in showcasing the rich cultural heritage of the region and Madhya Pradesh could then also be promoted as Mystic '*Hazar kisso ka desh*' (Land of Thousand Tales). Mysticism unknown curious stories have always attracted peoples' attention. This would also catapult a new era for ecotourism in the state.

9. *Advertisements through television,* especially in the channels like the Discovery, National Geographic, Animal Planet etc by focusing on the different ecotourism destinations of the state and the variety of the products being offered should be highlighted in the various travel programmes. The direct target by the promotion through television will be the domestic market.
10. Promotional activities should be carried on with the focus on Bhopal as a centre for the various cultural activities with *Bharat Bhawan* and *Manav Sanghrahalaya* playing the eminent role. Information and picture gallery of other eco-destination spots should also be displayed on these venues.

Accommodation

Recent ecotourism research has found that ecotourism markets prefer more than the conventional hotel. To provide the accommodation, environment-friendly thatched, intimate and adventure-type roofed accommodation (such as bed and breakfasts, cabins, lodges etc.) could be established.

Ecotourism Certification

What is certification?

A management tool for:

- Industry

- Government
- Non-governmental organizations

What is ecotourism certification?

- The procedure by which a third party gives written assurance that a product, process, service or management system conforms to specified requirements (TIES).

Why certify?

- To keep the government out.
- To protect natural resources.
- To protect local communities.
- To get a handle on one of the largest growing business sectors.
- To create a green premium.

SWOT Analysis

After summarizing the current marketing situation, it is necessary to identify the major strengths, weaknesses, opportunities, and threats.

Opportunities/Threats: Ecotourism industry has just started in India and especially, it is a great opportunity for Madhya Pradesh to attract the tourist because of its location. Madhya Pradesh is the only state having largest forest cover, which is going to be the ultimate destination for ecotourism in the coming years. The other opportunity is that recently Bhembetika has been declared the world heritage site, which will attract the foreign tourist. The major threats are the poor infrastructure facilities but still government is providing special budget for tourism industry to develop infrastructure.

Strengths/Weaknesses: Ecotourism areas have their own strengths like large number of wildlife species and inhabitant of indigenous people, which also attracts the tourists. This also promotes cultural heritage as well as tribal development. Weakness of the product is the lack of awareness among the domestic tourists. Ecotourism is mainly based on season thus it is also a weakness of the product.

Guidelines for the Marketer

Article 1: The Nine Principles of Customer Service for the Ecotourism Industry for the Marketer:

1. *Be a Good Listener*: Most important, listen to your customers' concerns and complaints, as well as their praise. Discover what is they want. Your guests are your best source of information in assessing your services and performance. Well-written guest surveys are invaluable for getting feedback, testimonials, and referrals. Help prospects qualify themselves before signing up. Make sure they understand the rigours, responsibilities, and dangers, as well as the level of exertion, entailed in the experiences you offer.
2. *Exceed Expectations*: Under-promise and over-deliver on a consistent basis with exceptional service. This is what they'll remember and tell their friends about. It is amazing how small details can make a big impression. You work hard and spend big money to get qualified prospects to call your office or e-mail you. How professional, informative, and timely is your response? That first contact can set the tone and significantly impact a purchasing decision.
3. *Practice Safety*: Clients want to know the destination and activities are safe. They will feel more at ease during real or perceived risks if they have been educated about their new activities or environments in advance. Unfortunately, if most human beings' fears or worries are not addressed, their minds will grow weeds, not flowers. Advise on plant, animal, and natural resource cautions. Let them know that there is generally nothing to worry about if they respect the environment and heed your advice. Brief guests on your safety procedures for any excursions without staff away from home base.
4. *Provide Great Food*: Guests must eat well, never be hungry, and have adequate water and drinks. Fresh food with ample portions served at a scheduled time can make any adventure more memorable and satisfying. If meal service is going to be delayed, provide a little snack to tide guests over.
5. *Cater to Sleep Needs*: Guests need and want comfortable and peaceful sleep. Make sure everyone is comfortable with his or her bedding. Cater to individual needs by offering single accommodations. Charging an extra cost is okay. Provide easy access to bathrooms with safety procedures for nighttime use. Separate late evening social areas from sleep areas by distance and/or trees or a hill. Hand out earplugs if necessary.

6. *Think Comfort*: Provide first world standards when possible. Bathroom breaks and bathing are important on a regular basis. Give people every opportunity for comfort: hammocks, beach chairs, portable toilets, whatever you can provide. Never go too long between bathroom breaks. Some guests may have weak bladders and need more frequent breaks to feel comfortable and relaxed. Make sure everyone has adequate water, as well as appropriate clothing and footwear.
7. *Be Sensitive to Capabilities*: Never exceed physical or mental capacities of guests. Always ask groups and individuals if they are comfortable with the activity and exertion level before and during the activity. Be sensitive to the slowest as well as strongest in the group. Splitting the group into two smaller groups is a good option. Rest as needed. I believe in a trip rating or difficulty system that allows guests and outfitters to jointly participate in selecting the most appropriate trip.
8. *Be Fun*: Be friendly, helpful, courteous, and fun! Make sure your staff is there for the guests and not for themselves. Good storytellers, jokesters, and musicians can distinguish your staff and company from other companies. Be sensitive with humour and with families with kids. If you cater to kids, employ staff that loves kids. Guests sometimes need encouragement or guidance to try an activity. You are the guide and activity director combined. Discuss activities or mini-classes in advance so guests will know what is happening and when.
9. *Provide Information*: An informed guest is safer and more relaxed, and has more fun. Our guests are educated and want to learn about their new environment. Assume that guests know little or nothing about their surroundings. Look for opportunities to provide a minimum of three details about each type of surrounding element during any outing. Be prepared to talk about the plants, animals, birds, trees, rocks, geological formations, fish, marine mammals, and shells. Include information about local culture, economy, and history, too. The more we can share, the more value we are providing our guests.

Like other businesses I know and have worked with, you, too, will experience more repeat clients and referrals by applying these details of great customer service. Your business, in time, will prosper like never before!

Article 2: Options for Improvement

- Develop maps with historical, cultural, and geographic information.
- Improve marketing for cultural, historical, and recreational experiences.
- Identify suitable areas for development and areas to be protected.
- Establish a destination theme and market identity.
- Encourage communication between tourism industry and wood products industry.
- Collect reliable information concerning participation rates, user profiles, economic and environmental effects.

Article 3: Ten Commandments of Adventure Travel Trade Show Success

1. *Have beginner's minds*: Seek out expert advice and information before you exhibit. Take workshops and read all you can on trade show marketing.
2. *Establish team-created objectives*: Incorporate your most friendly, motivated, and knowledgeable staff. Ensure role-play and practiced boothmanship before show.
3. *Pre-show mailings and phone invitations to customers and prospects can triple booth attendance*: Multiple personalized letters and postcards promoting special offers redeemable only at your booth works great. Start four-five months before show.
4. *Arrive before show*: Meet with media, establish sales representation, and network with fellow adventure travel business people. When the show starts you'll be better prepared and rested then most.
5. *Show management is an asset*: Work with them on providing value to attendees through your travel presentations, workshops, sport simulator and fascinating attractions.
6. *Invest in quality*: Use exhibit with easy-to-read graphics and bold benefit-oriented copy. Use "defining statement" headline graphics. Prospect should quickly understand what types of activities you offer, where you go and level of adventure, i.e. soft, moderate, high risk.
7. *Smile:* Don't pounce on prospects. Establish rapport before qualifying prospects. Ask open-ended questions, "What do you think about foreign travel and river rafting?"

8. *Use brochures as a disengagement tool*: Use "no's or not interested" as an opportunity to move on to another prospects who's interested in your type of trip or destination.
9. *Take care of yourself*: Schedule your staff so that everyone is well rested. Drink plenty of water. Eat well-balanced meals. Avoid alcohol at all times during show. Wear comfortable shoes. Stretch your muscles while checking out other exhibits.
10. *Follow-up*: Contact key prospects immediately after show with letters, calls and brochures. Continued mailings and contacts increase chances of sales 65 per cent.

Conclusion

Keeping in view the growth rate of tourism industry Madhya Pradesh is the second largest Indian State after Kerala. It is endowed with rich and diverse forest resources. The forest area of the state is 95,221 sq. km constituting 31 per cent of the geographical area of the state. It is a reservoir of biodiversity. Thus, it has a potential to boost the Ecotourism industry by facilitating proper Ecotourism infrastructure and services under policy guidelines. This will help to conserve and maintain the biological richness of the areas as well as economic upliftment of the local people by providing employment and opportunities in the field of Ecotourism management. There are various facets of the state, which can attract tourists from the point of view of the ecotourism. One finding of the study is that the peak season for the north-eastern region is October to March which continues till April and the peak seasons for the Pachmarhi region is the summer and the rainy season which is the lean period for the other regions. This fact should be highlighted while the promotion of ecotourism and can be advertised as *"Madhya Pradesh—Destination for all Seasons and all Reasons"*.

NOTES

1. Eagles, P.J., 1997. *International Ecotourism Management: Using Australia and Africa as Case Studies, Protected Areas in the 21st Century: From Islands to Networks*, Albany, Australia.

28

Policy on Ecotourism
A Need Analysis

A.K. Bhattacharya

Background

Tourism is the faster growing industry in the world, which employs the largest number of people both directly and indirectly. But tourism has often been misunderstood as an elitist activity without realizing its economic, social and civilization significance. The purpose of tourism is varied—it can be for pilgrimage, pleasure, experience, education and business. But these are not the exclusive privilege of any class of society but a universal phenomenon. India can become the number one tourist destination in the world according to prediction of the World Travel and Tourism Council (WTTC, 2001). In this context responsible travel to nature that conserves the environment and sustains the well being of local people needs to be emphasized. The ecotourism concept provides an outlet for this concerted action. Any policy in tourism must possess three basic attributes, namely

- Environmentally, socially, culturally and economically sustainable
- Educational
- Local participatory

The tourism policy may be analyzed in terms of the above three basic attributes. The other ones being biocentrism and being symbiotic with nature.

Problem Situation

The predicted potential annual growth of tourism in India is 10.1 per cent (WTTC, 2001). The requirement is prioritizing of tourism and investment. But on the other hand the Government spending on tourism is 153rd in the world at 0.9 per cent. China spends 3.8 per cent of its budget on tourism and rates fifth in receiving the largest tourist arrivals (31.2 million) after France, U.S., Spain and Italy. India gets only 2.6 million tourists on an average per annum. The mode of tourism followed till date is largely that of mass tourism.

Mass tourism *per se* had a negative impact resulting from ignoring the sanctity of local environment, flora and fauna and indigenous people. For instance, in some Indian Wildlife Sanctuaries, tourists, often travel in vehicles exhausting thick fumes. Film-shooting camps picnicking, cooking in the open disposing of refuse along with polythene bags, and even throwing away unextinguished cigarette and painting their names on rocks or etch/carve them on tree trunks are other unethical activities indulged in. In this situation the ecotourism policies can act as panacea having economic viability as well as environmental and local community focus. Tourism policies of many countries have adopted Ecotourism approach *in lieu* of traditional tourism noteworthy of them being United Kingdom, United States and Australia.

Past Policy Performance

The ecotourism policies of states such as Switzerland based on the principle of responsible travel to nature has gained worldwide recognition and applause. India needs to follow the ecotourism policies at the earliest for holistic for holistic and fact paced development. The tourism policy of 1982 to draft ecotourism policy has made a positive lead from market focus to market focus to sustainability and conservation focus. The draft tourism policy is a laudable effort in this regard but the implementation at national, sub-national or local level is still lacking. The platform has been provided by draft policy, only the finishing touches needs to be given through public suggestions.

In November 1982 first Tourism Policy of Government of India was announced. This had marketing as its focus than have a prescription plan for development. In 1992, a National Action Plan for tourism was prepared and in 1996 the drafting of National Strategy for Promotion of Tourism was done. In 1997, a draft new tourism policy according to the ongoing economic policies of the Government along with the trends in tourism development was published for public debate and deliberations. The draft policy is currently under revision.

The tourism policy initiatives have recognized ecotourism as the core tourism policies. But, they have left much to be desired. A comprehensive ecotourism strategy has not been implemented till date. It has been found that there are conflicting policies and regulations with meagre community participation in planning. Recognition and value of ecotourism is astonishingly very low. In background of this there are few and insufficient policy initiatives. This has led to uncoordinated and arbitrary decision making. Furthermore, decision-making and role diffusion is more prevalent as a rule than an exception. For example, in the case of protected areas there is centrally controlled policy leading to process delays and red-tapism.

Problem Situation

Indian attracted 2 million foreign tourists in 2001 (WTTC, 2001). The travel and tourism operators along with other stakeholders have perceived their works form a narrow mindest of short-term commercial benefits. Excessive commercialism, taking place in the absence of effective regulatory mechanisms has led to negative consequences. Environmental degradation along with erosion of traditional social activities has been the major threats. In case of India, it has notably suffered to upland and anaesthetic growth.

Need for Analysis

India has not all geared up to reach its true potential in tourism. The issue paper is an endeavour to go for a comprehensive tourism policy with focus on ecology economics and local participation and benefit sharing. For this the questions that need to be resolved are:

- How can responsible active participation of the local people in planning and implementing tourism promotion schemes be encouraged?
- How can better municipal governance especially in and around tourism centres be promoted?
- How can vandalism be checked and promotion of responsible tourism be done?

The problem addressed in this chapter is how best to satisfy the need for a comprehensive tourism policy at national level taking into account all stakeholders in tourism along with conservation and ethical aspect involved in the activity.

The approach of analysis for solving this problem is based on the effectiveness criteria based on following factors:

- Conservation of the sites
- Development of resources and people
- Education about the activity and nature
- Local control and benefit sharing
- Economic benefits to stakeholders
- Bio-centrism
- Satisfaction to tourists

Need for Changes in the Existing Policies

The ecotourism policy should address the following issues in the existing ecotourism related policies:

1. Need for improved planning and implementation of Ecotourism activities at all levels (Local, regional and national).
2. General policies on tourism are insufficient/inadequate to deal with Community Based Ecotourism. Therefore, policy and legislation must be framed to clarify authority and responsibility of various stakeholders in Ecotourism development.
3. Tourism in wildlife areas in India is controlled and regulated by the Wildlife Tourism policy of Ministry of Environment and Forests, and also used under the stringent provisions of Wildlife (Protection) Act, 1972; the safeguards to protect ecology and the environmental the other sites are provided indirectly in various legislation's of the Central Government, viz., Forest Conservation Act, 1980; Environmental Protection Act, 1986; Air (Prevention and Control of Pollution) Act, 1981. But there is no legal mechanism to ensure protection of local culture and communities against the ill-effects of unmanaged and ill-planned tourism. This is accepted fact that tourism in Protected Areas and its surrounds should be developed and the profits earned should be recycled for the benefits of the protected areas and

the local people, which need to be done only after a detailed site specific planning.

4. The policy emphasis upon involvement of local community for the development of area. But the specific areas of community participation have to be mentioned categorically. The benefice community will receive and the mode of benefit sharing should also be worked upon.
5. Similarly, the policy talks minimizing the conflicts between resource utilization for tourism and livelihoods of local inhabitants. But what if a conflict emerges? An urgent need is of defining the priorities so that the conflicts should be avoided.
6. The specific role and benefits to most vulnerable part of society at the micro-institutional level must be defined and steps require to ensure their participation in the overall project.
7. Need to bring more elements of private and the agencies in terms of making policies little flexible to attract the more investors in the area.
8. Ecotourism in the Forest Areas (NPs and WLS), should be implemented as a recognized Programme of the Forest Department with The Private Partnerships (Andhra Pradesh APFDC model for development of Ecotourism can be taken as a model).
9. Implementation of policies which recognizes local people's right in managing areas and gives upto 50 per cent of the income generated by the NPs to the local communities.
10. There must be caution and collaboration among all the stakeholders in order to preserve the environmental heritage along with the sustainable development.
11. Plans should be developed and implemented for continuous research and training related to more effective and environmentally friendly ways of conducting Ecotourism businesses. Strategies for implementing improved practices should be developed at regional and national levels.
12. Ecotourism policy and legislation's need to be supported by research in the following areas:

    - Environmental and social impact assessment of ecotourism projects.
    - Carrying Capacity (Both Natural and Social).
    - Baselines surveys to facilitate the monitoring of change.
    - Evaluation/Research of impact on communities.

13. Proper planning of places considered to be potential Ecotourism sites and strict implementations of guidelines to preserve these areas—essential for Ecotourism to fully developed in the country.

Major Stakeholders

The major stakeholders in tourism are:

- State Functionaries (Forest Department, Archaeological Survey of India).
- The local community around the tourist sites.
- Resort owners, managers and tour operators.
- Vehicle owners and drivers.
- Persons engaged in other small income generating activities.

Setting the Policy Goals and Objectives for Ecotourism

The goal of this policy is to provide a framework to provide a tourism policy as a means of promoting social integration. Along with this conservation of heritage, natural environment and development and promotion of tourist products will also be emphasised.

The objectives of the tourism policy should be to:

- create employment opportunities and bringing about socio-economic benefits to the community.
- facilitate and expand sustainable domestic and foreign tourism in the country.
- minimise negative effects such as cultural pollution and degradation of natural environment.
- increase the tourist satisfaction and making their tourism activity pleasant and memorable.
- create a desirable tourist product supported by infrastructure.
- involve and create synergy between all agencies, public, private and government, in tourism development.
- maintain a balance between the negative and positive impacts of tourism through planning restrictions and education of the people for conservation and development.

Measures of Effectiveness

- Economic benefits and negative impact of tourism through cost benefit analysis or net economic benefits.
- Satisfaction to tourists
- Distributional benefits to the local community near tourist sites.
- Consciousness among tourist about environmental protection.
- Compliance with regulatory requirements of the site.

Potential Solutions

The potential solution is to promote tourism activity that can be responsible travel keeping ecological and social considerations in mind or just to increase the frequency and numbers so that revenue generation is augmented. Accordingly the potential alternatives have to be derived.

Policy Alternatives

There can be two broad alternatives before the policymakers, which are as follows:

- A tourism promotion policy to attract more tourists that is to develop and promote tourism within the country and to make India competitive internationally as a destination. This will have earnings as its focus or will go for mass tourism. Given the breathtaking range of tourist products, tourism could become the largest industry in India (WTTC, 2001) and the policy can be focused along these lines.
- An Ecotourism policy, which will promote sustainable tourism which will be environmentally, socially and economically sustainable as well as educational and locally participatory.

Comparison of Future Consequences

The comparative statement on these two alternatives can be viewed in terms of the consequences:

- The social/connotations described by Barbier (1987) who defines social sustainability as "the ability to maintain desired social values, traditions, institutions, cultures or other social charasteristics".

- The environmental impact of tourism.
- The costs of tourism including extensive investment in fixed assets with a low rate of return for infrastructure, transportation, accommodation, cultural institutions, exhibition centres, etc.
- Increased demands on infrastructure like land, water, health services.
- Creation of developing positive community relationships.
- Disparity and influence of the lifestyle of visitors and community.
- The possible conflicts between local community and new users of valued local resources.
- The perception of local residents about the channelisation of scarce capital resources on low priority areas like tourism.
- Cultural cost of changes in local ceremonial or traditional values.
- Loss of privacy for local communities.
- Increasing congestion and pollution.
- Maintenance of landscape and the ecological balance between man and nature.

Based on these consequences the ecotourism policy for India is proposed which will have ecology, economics as well as society at its focus. The Draft Tourism Policy 1997 states that "in the context of economic liberalisation and globalisation being pursued by the country, the development policies of no sector can remain static". The policy further states that "the emergence of tourism as an important instrument for sustainable human development including poverty alleviation, employment generation, environmental regeneration and advancement of women and other disadvantaged groups in the country" and these can only be achieved by following mantles of ecotourism.

Constraints and Political Feasibility

In the milieu that India is in there will be several constraints and feasibility questions that will crop up. These are:

- Lack of resources for infrastructural development for tourism development.
- Unsatisfactory quality of service.

- The lack of political will power for the policy makers as ecotourism policy might result conflicts with their populist measures.
- The stakeholders shift to ecotourism might be gradual rather than sudden and the time span might run in decades.
- Tourism has not been able to counteract poaching Project Tiger bears a case in point.
- Management techniques which are less user friendly or control of numbers by closing access or by multiplying the number of attractions and areas or charging higher admission fees are generally not popular with the tourist or the tour operator and are also difficult to implement because of high administrative costs.
- The private sector involvement is being promoted but the profit margin is being squeezed under the garb of ecotourism thus this bears an inherent paradox.

Implementation Strategy

The implementation strategy to be followed will be area specific in lieu of universal development plans keeping in mind the unique character of the location and its economic and social environment. This policy has to be followed both at national as well as subnational and local levels. This would help the Government to coordinate with the industry in managing visitors and their activities. NGOs working on socio-economic programmes in forest and remote areas could have a closer coordination with tourism service operators to transfer economic benefits, particularly the handicraft production and marketing sector. The linkages with nations following similar strategies has to be done taking into account geopolitical considerations also. This will catalyze a process of learning from each other's best practices.

India's comparative advantage is its mystical attraction with its ancient civilisation and culture with a history going back to 5,000 years. India had much to offer by way of spiritual and mental rejuvenation. Other attractions can be the world's highest mountains, miles of coastline with excellent beaches, tropical forests and wildlife, desert safari, lagoon backwaters, ancient monuments, forts and palaces, adventure tourism. The tropical forests in Uttaranchal, Madhya Pradesh, Karnataka, Orissa, Kerala and Rajasthan can be developed into major tourist attractions. The Indian tourism has to be promoted taking all these considerations into account.

The principles of management will be scientific planning, effective control and continuous monitoring, development of physical infrastructure, zoning and a management plan for public use of natural sites.

Provisions for Monitoring and Evaluation

The evaluation of the policy can be done through regular feedback from the stakeholders. Ministry of Tourism will act as nodal agency for this feedback both at the state and national level. The regular feedback and inviting suggestions from general public in order to plug the loopholes will act as evaluative means as well as an end in itself. The policy would be taken as a body incorporating proactive intentions will be amenable to periodic reviews.

The tourism industry is generally self-centred and not given to educational, cultural or exchange programmes on welfare basis. The general socio-economic environment, the political system and the overall policy framework influence development of a sector. Tourism policy thus will have socio-political base a statement of intent by the Government, would just form the reference point for action and criticism.

The ecotourism policy possesses a lot of implementation stage lacunae that have to be paid heed to. Moreover, the concept looks satisfying but the socio-economic milieu of India puts a great question mark over the readiness and acceptability of this. Further privatisation leads to alienation of the majority of our population and their deprivation and the employment generation generally seasonal and ill paid. In the face of the unhindered entry of international capital and successive alienation, the problems might be aggravated.

The ecotourism policy thus proposes the development of a code of conduct for the tourist, the industry and administrator. These attempts are in a very nascent stage. The kind of coordination that is required between the environmentalist and economist and societal forces is just beginning to emerge and have still to counter the myths of market forces in the field of tourism. But a beginning already exists.